Signaling and Communication in Plants

Series Editors

František Baluška
Department of Plant Cell Biology, IZMB, University of Bonn, Kirschallee 1, D-53115
Bonn, Germany

Jorge Vivanco
Center for Rhizosphere Biology, Colorado State University, 217 Shepardson Building,
Fort Collins, CO 80523-1173, USA

František Baluška • Stefano Mancuso
Editors

Signaling in Plants

Springer

Editors
František Baluška
Department of Plant Cell Biology
IZMB
University of Bonn
Kirschallee 1
D-53115 Bonn
Germany
email: baluska@uni-bonn.de

Stefano Mancuso
Universitá Firenze
Dipto. Ortoflorofrutticoltura
Polo Scientifico
Viale delle Idee, 30
50019 Sesto Fiorentino FI
Italy

ISSN 1867-9048
ISBN 978-3-642-10036-9 e-ISBN 978-3-540-89228-1
DOI: 10.1007/978-3-540-89228-1

Cover design: WMXDesign GmbH, Heidelberg, Germany

Printed on acid-free paper

9 8 7 6 5 4 3 2 1

springer.com

Stefano Mancuso dedicates this book, with the deepest admiration and appreciation, to Prof. Paolo Blasi in grateful and affectionate acknowledgement for his tireless support and friendship.

Preface

Plants are unique as their development and morphogenesis are plastic throughout their lives. They continuously monitor diverse biotic and abiotic parameters of their environment and these sensory perceptions shape their organs and bodies. Although genes are critical, the final form and architecture of above-ground organs, and especially of root systems, are determined by their sensory activities associated with motoric responses (Friml 2003; Hodge 2009). Sensory plant biology and plant electrophysiology were two lively disciplines until the late 1970s (Bünning 1959; Haupt and Feinleib 1979) but then, for somewhat obscure reasons, they showed no further development. In the last few years, however, there have been numerous advances in plant sciences. These necessitate not just a revival of plant electrophysiology and sensory biology, but also the introduction of plant neurobiology, which includes also plant sensory ecology (Baluška et al. 2006a; Brenner et al. 2006). First of all, and contrary to all "mechanistic" predictions based on the high turgor pressure of plant cells, endocytosis has been found to be an essential process of plant cells which impinges upon almost all aspects of plant life (Šamaj et al. 2005, 2006). Moreover, recent advances in plant molecular biology have identified, besides classic neurotransmitters, also several proteins typical of animal neuronal systems, such as acetylcholine esterases, glutamate receptors, GABA receptors, and endocannabinoid signaling components, as well as indicating signaling roles for ATP, NO, and reactive oxygen species (Baluška et al. 2006b). Plant action potentials have turned out to influence processes such as osmotic-force-based cell shapes, actin-cytoskeleton-based cytoplasmic streaming, organ movements, wound responses, respiration and photosynthesis, as well as flowering (Dziubinska et al. 2003; Felle and Zimmermann 2007; Davies 2004; Fromm and Lautner 2007). Last, but not least, there have been significant advances in ecological studies on plant–plant and plant–insect communications, as well as in behavioral studies on memory, planning, and learning phenomena in plants (Trewavas 2005; Gális et al. 2009; Ripoll et al. 2009). Discovery of complex plant behavior (Baluška et al. 2006a; Hodge 2009; Karban 2008; Scott 2008) implicates signal perception, processing, the integration of ambient signals (Brenner et al. 2006), and even cognition (Calvo Garzon 2007).

Recent advances in plant cell biology, molecular biology and ecology have resulted in the accumulation of a critical mass of data, which are not "digestible" within the framework of these, now classical, disciplines of plant sciences (Baluška et al. 2006a, b; Brenner et al. 2006, 2007; Baluška and Mancuso 2009; Hodge 2009). New approaches are required, characterized by systemlike analysis of information acquisition, storage, processing, and the making of decisions.

Plants retrieve properties of their environment via sensory perceptions which are critical for their survival. Especially light and gravity, two physical factors pervading the universe, are essential in this respect. Plants actively experience the environment and can both store and retrieve memories (Gális et al. 2009; Ripoll et al. 2009 to drive an active life-style (for roots see Baluška et al. 2004; Hodge 2009). Intriguingly, the translation of these physical forces into plant activities - typically differential growth responses - is based on the transcellular transport of auxin, which helps to bring about the final shape of the plant body (Friml 2003; Baluška et al. 2006b; Brenner et al. 2006).

Although the history of auxin can be traced back to the Darwin's early experiments with phototropism of coleoptiles, we still know almost nothing about its peculiar features. Let us examine the mystery of this unique molecule. Whereas auxin can probably be synthesized in each plant cell, it is tediously transported from cell to cell throughout the plant body (Friml 2003). Similarly puzzling is the well-known phenomenon that although the auxin molecule is sufficiently small to pass easily through plasmodesmatal channels, plant cells somehow manage to prevent this direct cell-to-cell means of auxin transport (Mancuso et al. 2006). Rather, plants maintain an energetically costly system based on vesicle trafficking, closely resembling neuronal and immunological cell–cell communication, to drive transcellular auxin transport (Baluška et al. 2003, 2005; Friml and Wiśniewska 2005). At least in the root apex, auxin is secreted via a synaptic neurotransmitter-like mode supported by phospholipase Dζ2 (Mancuso et al. 2007; Baluška et al. 2008). Thus, this unique information-bearing molecule is central to our understanding of sensory and communicative plants.

The next peculiarity is that when extracellular auxin hits the outside leaflet of the plasma membrane, it induces electric responses based on the ABP1 auxin-binding protein (Felle et al. 1991; Steffens et al. 2001; Baluška et al. 2004). All this suggests that auxin, besides hormone- and morphogen-like (Dubrovsky et al. 2008) properties, possesses neurotransmitter-like faculties too. Since the cell-to-cell transport of auxin translates sensory perception into adaptive motoric responses, being central for organ tropisms to light and gravity gradients, a plant neurobiology approach is needed to explain this great mystery of plants (Baluška et al. 2005, 2006b, 2008; Baluška and Mancuso 2008).

Despite having a relatively simple body organization, plants need sophisticated sets of coordinative processes. Besides their root–shoot coordination, there is also need for coordination amongst radial tissues, especially within and between the cortex and stele. Action potentials run preferentially in the axial direction and they presumably integrate activities of root and shoot apices. Plants show modular and apparently decentralized organization of their bodies. Nevertheless, there are several critical situations requiring "centralized" decisions, such as the onset of flowering and the onset or breakage of dormancy. Although these decisions are based on information retrieved via numerous distant organs, they imply some central "processor" which would reliably control the whole plant body. Importantly, any wrong decision would have detrimental consequences for the whole plant. Future studies focusing on these new aspects of plants will allow us to understand plants and their unexpected sensory, communicative, and social aspects.

Bonn, October 2008 František Baluška
Florence, October 2008 Stefano Mancuso

References

Baluška F, Mancuso S (2009) Plant neurobiology: from sensory biology, via plant communication, to social plant behaviour. Cognit Process (in press)

Baluška F, Šamaj J, Menzel D (2003) Polar transport of auxin: carrier-mediated flux across the plasma membrane or neurotransmitter-like secretion? Trends Cell Biol 13:282–285

Baluška F, Mancuso S, Volkmann D, Barlow PW (2004) Root apices as plant command centres: the unique 'brain-like' status of the root apex transition zone. Biologia 59:9–17

Baluška F, Volkmann D, Menzel D (2005) Plant synapses: actin-based domains for cell–cell communication. Trends Plant Sci 10:106–111

Baluška F, Mancuso S, Volkmann D (2006a) Communication in plants: neuronal aspects of plant life. Springer, Berlin

Baluška F, Hlavacka A, Mancuso S, Barlow PW (2006b) Neurobiological view of plants and their body plan. In: Baluška F, Mancuso S, Volkmann D (eds) Communication in plants: neuronal aspects of plant life. Springer, Berlin, pp 19–35

Baluška F, Schlicht M, Volkmann D, Mancuso S (2008) Vesicular secretion of auxin: evidences and implications. Plant Signal Behav 3:254–256

Brenner ED, Stahlberg R, Mancuso S, Vivanco J, Baluška F, Van Volkenburgh E (2006) Plant neurobiology: an integrated view of plant signaling. Trends Plant Sci 11:413–419

Bünning E (1959) Physiology of movements. Part 1. Movements due to mechanical and electrical stimuli and to radiations. In: Ruhland W (ed) Encyclopedia of plant physiology. Springer, Berlin

Calvo Garzon F (2007) The quest for cognition in plant neurobiology. Plant Signal Behav 2:208–211

Davies E (2004) New functions for electric signals in plants. New Phytol 161:607–610

Dubrovsky JG, Sauer M, Napsucialy-Mendivil S, Ivanchenko MG, Friml J, Shishkova S, Celenza J, Benková E (2008) Auxin acts as a local morphogenetic trigger to specify lateral root founder cells. Proc Natl Acad Sci U S A 105:8790–8794

Dziubinska H, Filek M, Koscielniak J, Trebacz K (2003) Variation and action potentials evoked by thermal stimuli accompany enhancement of ethylene emission in distant non-stimulated leaves of Vicia faba minor seedlings. J Plant Physiol 160:1203–1210

Felle HH, Zimmermann MR (2007) Systemic signalling in barley through action potentials. Planta 226:203–214

Felle H, Peters W, Palme K (1991) The electrical response of maize to auxins. Biochim Biophys Acta 1064:199–204

Friml J (2003) Auxin transport – shaping the plant. Curr Opin Plant Biol 6:7–12

Friml J, Wiśniewska J (2005) Auxin as intercellular signal. In: Fleming AJ (ed) Intercellular communication in plants. Blackwell, Oxford, pp 1–26

Fromm J, Lautner S (2007) Electrical signals and their physiological significance in plants. Plant Cell Environ 30:249–257

Gális I, Gaquerel E, Pandey SP, Baldwin IT (2009) Molecular mechanisms underlying plant memory in JA-mediated defence responses. Plant Cell Environ (in press)

Haupt W, Feinleib ME (1979) Physiology of movements. In: Encyclopedia of plant physiology, vol. 7. Springer, Berlin
Hodge A (2009) Root decisions. Plant Cell Environ (in press)
Karban R (2008) Plant behaviour and communication. Ecol Lett 11:727–739
Mancuso S, Barlow PW, Volkmann D, Baluška F (2006) Actin turnover-mediated gravity response in maize root apices: gravitropism of decapped roots implicates gravisensing outside of the root cap. Plant Signal Behav 1:52–58
Mancuso S, Marras AM, Mugnai S, Schlicht M, Zarsky V, Li G, Song L, Hue HW, Baluška F (2007) Phospholipase Dζ2 drives vesicular secretion of auxin for its polar cell-cell transport in the transition zone of the root apex. Plant Signal Behav 2:240–244
Ripoll C, Le Sceller L, Verdus MC, Norris V, Tafforeau M, Thellier M (2009) Memorization of abiotic stimuli in plants. A complex role for calcium. In: Baluška F (ed) Plant–environment interactions: behavioral perspective. Springer, Berlin (in press)
Šamaj J, Read ND, Volkmann D, Menzel D, Baluška F (2005) Endocytic network in plants. Trends Cell Biol 15:425–433
Šamaj J, Baluška F, Menzel D (2006) Plant endocytosis. Springer, Berlin
Scott P (2008) Physiology and behaviour of plants. John Wiley & Sons Ltd
Steffens B, Feckler C, Palme K, Christian M, Böttger M, Lüthen H (2001) The auxin signal for protoplast swelling is perceived by extracellular ABP1. Plant J 27:591–599
Trewavas A (2005) Green plants as intelligent organisms. Trends Plant Sci 10:413–419

Contents

Auxin and the Communication Between Plant Cells

Peter Nick

Abstract Multicellularity allows to assign different functions to the individual cells. Cell fate could be defined by a stereotypic sequence of cell divisions or it might arise from intercellular communication between cells. Patterning in the totipotent plant cells results mainly from coordinative signals. Auxin is central in this respect, and this chapter ventures to give a survey on the role of auxin as a coordinative signal that regulates patterning of cell differentiation, cell division and cell expansion.

Abbreviations 2,4-D: 2,4-Dichlorophenoxyacetic acid; ARF: ADP-ribosylation factor: ARP; Actin-related protein: BFA; Brefeldin A: GFP; Green fluorescent protein; IAA: Indole-3-acetic acid; NAA: 1-Naphthaleneacetic acid; NPA: Naphthylphthalamic acid; RFP: Red fluorescent protein; TIBA: 2,3,5-Triiodobenzoic acid

1 Introduction

The polar flux of auxin has been used for more than 375 million years to generate and regulate the pattern of vascular differentiation of parenchymatic cells and thus coordinates the organization of the telomes, the building block of cormophytic land plants. In addition to the patterning of vasculature, auxin mediates the coordinative signalling that controls phyllotaxis, the formation of new leaves according to an orderly, species-dependent pattern. The phyllotactic pattern is shaped by competition of young primordia for free auxin, such that the neighbourhood of an existing primordium will be depleted of auxin. Since auxin limits the formation of new primordia, this simple mechanism ensures elegantly that new structures will be laid down at a minimal distance from preexisting primordia.

P. Nick (✉)
Botanisches Institut 1, Kaiserstr. 2, 76128 Karlsruhe, Germany
e-mail: peter.nick@bio.uni-karlsruhe.de

F. Baluška and S. Mancuso (eds.), *Signaling in Plants,* Signaling and Communication in Plants,
DOI: 10.1007/978-3-540-89228-1_1,

Polar auxin transport can regulate the synchrony of cell divisions, with actin organization emerging as a central factor defining the pattern of cell division, probably by polarizing the flow of vesicles that deposit auxin-efflux carriers to the cell pole and thus determining the directionality of auxin efflux. Since the organization of actin, in turn, is regulated by auxin, a feedback loop is established that contains auxin-efflux carriers, intracellular auxin and actin filaments as central elements.

Regulated cell expansion represents the central adaptive response of the sessile plants to environmental challenges and is therefore highly responsive to stimuli, such as light or gravity. These adaptive responses involve a spatiotemporal pattern of cell expansion, which is most evident for tropistic curvature. Actually, auxin was originally identified as a signal that coordinates the pattern of cell expansion. The Cholodny–Went model explains tropism by a signal-induced redistribution of auxin fluxes across the stimulated organ. Although the Cholodny–Went model is repeatedly disputed mainly because of discrepancies between the observed response (a gradient of growth) and the amplitude of the induced gradient of auxin, it is shown that the model is still valid if the redistribution of auxin fluxes is complemented by parallel gradients of auxin responsiveness.

The chapter ends with a speculative consideration of why, during evolution, such a simple molecule as indole-3-acetic acid (IAA) has acquired such a central role for intercellular coordination. This is attributed to the molecular properties of auxin that determine its transport properties (multidirectional influx through an ion-trap mechanism, but unidirectional efflux through the localized activity of auxin-efflux carriers). On the intracellular level, this system is able to establish a clear cell polarity from even minute and noisy directional cues. On the level of tissues, this system is ideally suited to convey lateral inhibition between neighbouring cells. It was sufficient to put the localization of the efflux transporter under the control of auxin itself to reach a perfect reaction-diffusion system *in sensu* Turing (1952). Such systems are able to generate clear outputs from even minute and noisy directional cues and provide a robust mechanism to generate patterns of cell differentiation, cell division and cell expansion under the special constraints of plant development, such as signal-dependent morphogenesis and the lack of specialized and localized sensing organs.

Plant morphogenesis is not based on fixed hierarchies – there is no such a thing as a "Great Chairman" that assigns differential developmental pathways to the individual cells. Plant cells rather "negotiate" on their individual developmental fates in a fairly "democratic" manner with hierarchies being created ad hoc by mutual interactions. It seems that auxin has evolved as a central tool for this "cellular democracy" characteristic for plant development.

2 Plant Development and Cell Communication

"Why do cells exist?" – with this question Philip Lintilhac (1999) starts his thoughtful essay on the conceptual framework of cellularity. Multicellularity initially probably evolved as a strategy to increase in size and thus escape the fate of being eaten.

During growth, the volume of a cell (its "internum" *in sensu* Lintilhac) increases with the third power of the radius; its surface, though, increases only with the second power of the radius. When a cell grows, an increasing gap between consumption (by the "internum") and subsistence (through the boundary with the "externum") has to be bridged that will limit further expansion of the cell. Multicellularity allows an increase of the surface in relation to the volume – for the cell population *as an entity*. This made it possible for the cell to become bigger, again for the cell population *as an entity*. The selective advantage (not to be devoured by predators) paid off for each individual cell. However, the full potential of this achievement emerged only when the individual cells of the newborn organism began to assign different functions to individual members of the population. For the individual cell, differentiation represents a risky investment, because it implies that specific (Lintilhac coined the term "hypercellular") tasks have to be upregulated at the cost of other "hypocellular" functions that are downregulated and therefore have to be compensated by corresponding hypercellular output from neighbouring cells. This culminates in a situation where the individual cells cannot survive outside the organismal context.

Differentiation therefore requires an intensive flow of information between individual cells to maintain the subtle balance between hypercellular and hypocellular functions. Although in some systems the differentiation of individual cells seems to follow a predetermined internal programme, cell–cell communication is important at least in the initial phase, when this programme is defined and triggered. Plant cells with their principal totipotency and their comparatively diffuse differentiation have to be especially communicative. Owing to their developmental flexibility, the balance between hypercellular and hypocellular functions has to be reestablished continuously. It thus seems that cell differentiation in plants resembles more or less the ancestral situation of multicellularity. In addition, plant cells are immobile, such that temporal patterns of differentiation become manifest morphologically and are not obscured by cell migrations.

The primordial form of cell differentiation is developmental dichotomy as characteristically observed during the first formative cell division of zygotes or spores in many algae, mosses and ferns or during the first division of the angiosperm zygote. In the Volvocales, a monophyletic clade of the green algae, it is still possible to follow the evolutionary line from a cell population over cell colonies (consisting of equivalent members that are completely autonomous) to a true organism, where two cell types are coupled by hypocellular and hypercellular interactions. Genetic analysis of differentiation mutants in *Volvox carteri* has uncovered a transcription factor, *regA*, repressing nuclear encoded genes of the chloroplast in mobile, somatic cells such that growth of these cells is suppressed, leading to a delayed cell cycle (Kirk 2003). In contrast, a group of four or five *late gonidia* factors suppress the motile phase in reproductive cells and thus promote their division. The activities of *regA* and *lag* differ as early as from the first division of the mature gonidium. This primary developmental dichotomy is under the control of two or three *gonidialess* factors – mutations in those genes render the first division symmetric such that the resulting daughter organism lacks reproductive cells. In fact, the dichotomy of the first gonidial division is a cornerstone of August Weismann's concept of inheritance, where he defined the separation of the differentiating, but mortal *soma* from the non-differentiating, but immortal *germ line* as a primordial event of multicellular development (Weismann 1894).

Developmental dichotomy could be based on a gradient of developmental determinants within the progenitor cell that are then differentially partitioned to the daughter cells (formative cell division). According to this mechanism, the ultimate cause for differentiation would reside in cell lineage (Fig. 1a). Alternatively, developmental dichotomy could arise from communication between initially equipotent daughter cells and therefore would be independent of cell lineage (Fig. 1b).

As diverse as these two mechanisms might appear, it can be difficult to discriminate between them in nature since the commitment for a certain developmental pathway and the manifestation of this commitment as differentiation are not always clearly separated in time. However, the principal totipotency of plant cells is easier to reconcile with a model where differentiation is not defined a priori by a formative division (Fig. 1a), but *a posteriori* by intercellular communication (Fig. 1b).

The impact of intercellular communication on differentiation is heralded in the (prokaryotic, but plant-like) cyanobacteria during the differentiation of heterocysts. Heterocystsexpress (as hypercellular function) a nitrogenase that is able to release the constraints placed on cell division by the limited supply of bioavailable nitrogen.

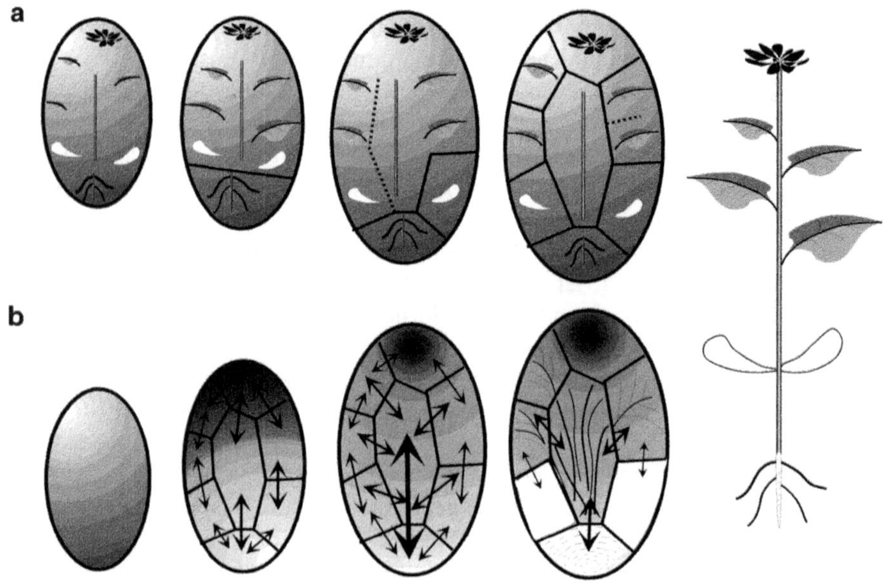

Fig. 1 Mechanisms for the establishment of developmental differentiation. (**a**) Mosaic development, where developmental fate is determined a priori and then assigned to individual daughter cells by a stereotypic sequence of formative cell divisions. (**b**) Regulative development, where developmental fate is not predetermined, but is defined a posteriori by communication between equipotent cells

This nitrogenase dates back to the earliest, anoxic phases of life on this planet and is therefore highly sensitive to oxygen; therefore, to safeguard nitrogenase activity, any photosynthetic activity has to be excluded from heterocysts. These cells are therefore hypocellular with respect to assimilation. The balance between nitrogen export and assimilate import has to be maintained although the total number of cells grows continuously. This balance is kept by iterative patterning, whereby preexisting heterocysts suppress the differentiation of new heterocysts in a range of around ten cells. When, in consequence of cell divisions, the distance between them exceeds this threshold, a new heterocyst will differentiate between them. By the analysis of patterning mutants in *Anabaena* the factor responsible for this lateral inhibition could be identified as the diffusible peptide patS (Yoon and Golden 1998). Differentiation (including the synthesis of patS) will begin in clusters of neighbouring cells; however, one of these cells will advance and then suppress further differentiation in its neighbours (Yoon and Golden 2001). This demonstrates that the differentiation of a heterocyst is not predetermined, but is progressively defined by signalling between neighbouring cells.

Developmental dichotomy in the complete absence of a predefined gradient has been impressively demonstrated for the somatic embryogenesis of embryogenic carrot cell suspensions. Here a single cell can be induced to produce an entire embryo that is indistinguishable from a sexually produced plant. Similar to zygotic development, the initial event is an asymmetric cell division giving rise to a highly vacuolated basal and a smaller apical cell endowed with a very dense cytoplasm (McCabe et al. 1997). Whereas the vacuolated cell will undergo programmed cell death, this apical cell will undergo embryogenesis. The vacuolated cell expressed a surface marker that was recognized by the monoclonal antibody JIM8 that had originally been raised as a marker of cell fate in root development. By use of ferromagnetic antibody conjugates it was possible to remove cells expressing the JIM8 marker from the suspension. These cultures lost their embryogenic potential. However, a filtrate from a culture containing JIM8-positive cells was able to restore the embryogenic potential of the JIM8-depleted culture. The JIM8 marker, a small soluble arabinogalactan protein secreted by the vacuolated cells, was therefore necessary and sufficient to confer an embryogenic fate to the densely vacuolated apical cells. Thus, the formative division is controlled by intercellular signalling.

Whereas intercellular signalling is relatively evident in these two examples, it might be more widespread. The classic experiment to dissect the role of cell lineage versus intercellular signalling in animal embryology is to transplant tissue to a different site of the embryo and to test whether the explant develops according to position (favouring a signalling mechanism) or according to its origin, as would be expected for differentiation based on cell lineage (Spemann 1936). This experiment has rarely been undertaken in plants, and so intercellular signalling might have been overlooked in many cases. For instance, the highly stereotypic cell lineage in the root meristem of *Arabidopsis thaliana* seemed to indicate that here cell fate is defined by cell lineage that could be traced back to early embryogenesis (Scheres et al. 1994). However, by very elegant laser ablation experiments (Van den Berg et al. 1995) and the analysis of mutants with aberrant definition of tissue layers (Nakajima et al. 2001) it could be shown that even in this case cell fate was defined by signals (such as the transcription factor shortroot) from adjacent cells.

Generally, the principal totipotency of plant cells is difficult to reconcile with a strong impact of cell lineage. Patterning in cells thus seems to result mainly from coordinative signals. However, as discussed in the next section, the impact of intercellular communication on development seems to reach beyond the realm of individual cells to the coordinative development of entire organs.

3 Auxin as a Pattern Generator in Cell Differentiation I: Vasculature

As consequence of their light dependency, plants increase their surface in an outward direction, which means that they have to cope with a considerable degree of mechanic load. As long as they were aquatic, this was no special challenge, because mechanical strains were counterbalanced by buoyancy, allowing for considerable size even on the basis of a fairly simple architecture. However, the transition towards terrestrial habitats increased the selection pressure towards the development of flexible and simultaneously robust mechanical lattices. Plant evolution responded to this selective pressure by generation of load-bearing elements, the so-called telomes (Zimmermann 1965). These modules are organized around a lignified vascular bundle surrounded by parenchymatic tissue and an epidermis to limit transpiration (Fig. 2a). The telomes were originally dichotomously branched, but by asymmetric branching ("overtopping") hierarchical branching systems emerged that were endowed with main and side axes. By planation and

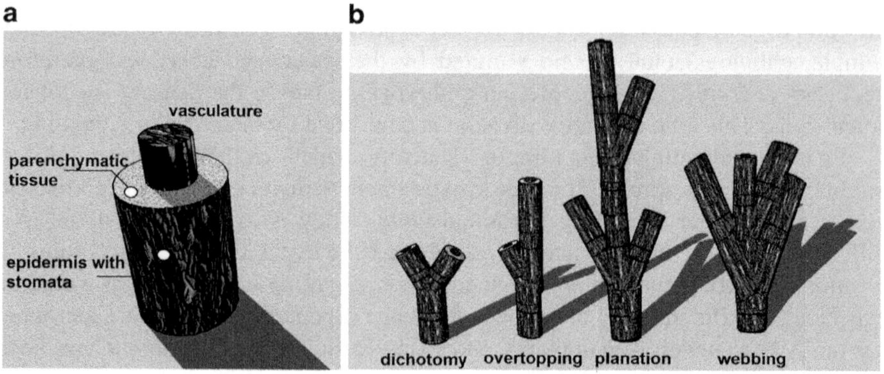

Fig. 2 Modular structure of terrestrial plants. The building block of cormophytes are the telomes (**a**), tubular elements organized around vasculature bundles that are surrounded by parenchymatic tissue and protected by an epidermis with stomata for the regulation of transpiration. By combination of the telomes in combination with simple modulations of their geometry (**b**) progressively complex hierarchical structures have been produced during the evolution of land plants

subsequent fusion of the parenchymatic tissue (so-called webbing) the telomes developed into the first leaves (Fig. 2b). This is still evident in the leaves of certain ferns and the primitive gymnosperm *Ginkgo biloba*, where, interestingly, occasionally atavistic forms are observed that uncover the original dichotomous telome structure. By spherical fusion and reduction of individual telomes globular structures arose that later evolved into sporangiophores and flower organs. Eventually, webbing of non-planar telomes generated the vasculature that has since been used throughout cormophyte evolution. In summary, the whole architecture of land plants derives from the patterned organization of these versatile modules. In other words, if one wants to understand the morphogenesis of land plants, one needs in the first place to understand the patterning of vessels as a core element of these building blocks.

Vessel patterning is central for the success of grafting in horticulture (Priestley and Swingle 1929) and therefore shifted into the focus of botanical research many years ago. As long ago as the eighteenth century regenerative events in grafting were explained by a theory where two morphogenetic factors, a heavy "root sap" and a light "shoot sap", moved towards the respective poles driven by gravity, accumulated there and triggered the formation of roots and shoots, respectively (Du Monceau 1764). In fact, the existence of such morphogenetic factors and their transport in the phloem was elegantly demonstrated by incision experiments (Hanstein 1860). By elaborate cutting and regeneration studies Goebel (1908) arrived at the conclusion that an apicobasal flux of an unknown substance defines the regeneration of new shoot and root elements. If this flux is interrupted or inverted, locally restricted inversion of shoot-root polarity becomes manifest as a gradient in the formation of vasculature and the ability to regenerate adventitious shoots or roots, respectively. The factor that was produced by developing leaves and that was able to induce the differentiation of new vasculature from parenchymatic tissue located basipetally of the leaf was later shown to be the transportable auxin IAA (Camus 1949). This finding opened up the possibility to experimentally manipulate the spatial pattern of vascular bundles, an approach that was exploited by Tsvi Sachs in a series of ingenious experiments. He could demonstrate that "differentiated vascular tissue whose source of auxin has been removed attracts newly induced vascular strands. This attraction is expressed in the joining of the new strands to the pre-existing vascular tissue. Differentiated vascular tissue which is well supplied with auxin inhibits rather than attracts the formation of new vascular strands in its vicinity" (Sachs 1968). This basic experiment and numerous experimental derivatives culminated eventually in a canalization model of auxin-dependent patterning of vasculature: if, within an initially homogeneous distribution of auxin across the parenchymatic tissue, the polar flux of auxin is increased locally (for instance by blocking other drainage paths), the increase leads to accelerated differentiation of vessels at this site. Since those developing vessels can already transport more auxin per unit time, they will deplete the neighbouring areas of auxin. A few vessels will form and mutually compete for auxin. With time, the vessels differentiate progressively and, eventually, the pattern is stabilized by lignification (Sachs 1981).

The cellular basis of this drainage model is the polarity of vascular cells that are aligned with the shoot–root axis. The vasculature of a leaf, however, does not reveal such an obvious polarity and it was therefore not clear whether the auxin canalization theory could be generalized to leaf veins as well. However, when transverse vascular strands were investigated, they were found to include adjacent vessels with opposite polarities that did not mature at the same time (Sachs 1975). Similar vessels without clear polarity could be induced experimentally when the location of the auxin source was changed repeatedly. Thus, the axis of a vessel seems to precede its polarity, and the differentiation of a network without clear directionality (as typical for dicotyledonous leaves) is thought to arise from non-synchronous auxin transport across the leaf blade. Since auxin transport has been observed to oscillate in intensity (Hertel and Flory 1968), the formation of an axial, but non-polar vessel could also originate from auxin movement in opposite directions at different times even through the same cells.

This model predicts that inhibition of polar auxin transport should impair the differentiation of leaf veins. This has been tested experimentally in *Arabidopsis* leaves treated with 1-naphthylphthalamic acid (NPA), an inhibitor of auxin transport (Mattsson et al. 1999). When the concentration of the inhibitor was raised, the vasculature was progressively confined to the leaf margin, indicating that the central regions of the leaf were depleted of auxin. The canalization model is further supported by a series of mathematical models that can explain a variety of common venation pattern (for a recent review see Roeland et al. 2007). The molecular base of canalization is generally seen by alignment of auxin-efflux transporters, such as the PIN proteins, with the flux of auxin, such that these fluxes are amplified even further. In fact, PIN1 is polarized along the putative direction of auxin flow prior to the formation of vasculature during early leaf development (Scarpella et al. 2006).

A central element of the canalization model is the feedback of auxin flux on cell polarity. This implies that the cell responds to the flow rather than to the local concentration of auxin, which poses sophisticated challenges for modelling. Alternatively, PIN proteins could localize differentially to cell membranes depending on the local auxin concentration in the cell adjacent to this membrane as proposed for phyllotaxis (Jönsson et al. 2006). When this readout of local auxin concentration is combined with an auxin-dependent expression of the channel, it is possible to model channelling patterns that are consistent with those of the classical canalization model (Roeland et al. 2007).

Although the cellular details of auxin channelling remain to be elucidated, it is clear that this pattern generator is evolutionarily very ancient. Evidence for polar auxin transport can be found in algae and mosses (for a review see Cooke et al. 2002) and polar auxin transport has been proposed to be responsible for vascular differentiation in early land plants (Stein 1993). In recent conifers or dicotyledonous plants the vasculature follows straight lines, but forms characteristic whirlpools near buds, branches or wounds when the presumed axial flow of auxin is interrupted. Identical circular patterns also occur at the same positions in the secondary wood of the Upper Devonian fossil progymnosperm *Archaeopteris*, thus providing the first clear fossil evidence of polar auxin flow (Rothwell and Lev-Yadun 2005). Thus, already 375 million years ago ancient land plants used polar auxin flux as a tool to establish and maintain a contiguous vascular pattern throughout their telomic modules.

4 Auxin as a Pattern Generator in Cell Differentiation II: Phyllotaxis

In addition to the patterning of vasculature, auxin is a central player in the coordinative signalling that controls phyllotaxis, the formation of new leaves according to an orderly, species-dependent pattern. It has been known for a long time that the position of a prospective leaf primordium in the apical meristem is defined by inhibitory fields from the older primordia proximal to the meristem (Schoute 1913). This was demonstrated by isolation of the youngest primordium by tangential incisions that shifted the position of the subsequent primordia (Snow and Snow 1931). At that time, this shift was interpreted in terms of a first available space model, where the additional space created by the incision would allow the incipient primordia to move to a position where they otherwise were excluded. However, this result is consistent with inhibitory fields emanating from the primordia. There has been a long debate on the nature of these inhibitory signals that were originally thought to be chemical agents, but were later interpreted to be of mechanical nature. Since a growing meristem is subjected to considerable tissue tension, the inhibition could be merely mechanical, because the preexisting primordia would induce stresses upon surrounding potential sites of primordium initiation. The expected stress–strain patterns can perfectly predict the position of prospective primordia (for a review see Green 1980). If the inhibition were mechanical, local release of tissue tension by beads coated with extensin should alter phyllotaxis. In fact, such beads could invert the phyllotactic pattern (Fleming et al. 1997). However, a closer look showed that the extensin-induced structures did not always develop into true leaves, but in some cases resembled mere agglomerations of tissue that did not express leaf markers such as photosynthetic proteins. True leaf development was only initiated when the extensin bead was placed in a site where according to the natural phyllotaxis a primordium would have been laid down. This meant that mere mechanical tension was not sufficient to explain phyllotaxis and this led to a rehabilitation of chemical signals as the cause of the inhibitory field emanating from preexisting primordia. Chemical inhibition was supported by studies in apices that had been freed from primordia by application of auxin-transport inhibitors (Reinhardt et al. 2000), an experimental system that allows study of the de novo generation of a pattern without the influence of preexisting structures. In this system, the coordinative signal was found to be auxin. However, against textbook knowledge, the preexisting primordia did not act as sources, but as sinks for auxin. Within the apical belt that is competent for the initiation of leaf primordia there is mutual competition for auxin as a limiting factor and this competition is biased in favour of certain sites (where, in consequence, a new primordium is initiated) by the preexisting primordia that attract auxin fluxes from the meristem (Reinhardt et al. 2003).

The phyllotactic pattern could be explained by a mechanism where PIN1 that continuously cycles between an endocellular compartment and its site of activity at the plasma membrane acts as a sensor for intercellular auxin gradients (Roeland et al. 2007). When the endocytosis of PIN1 becomes suppressed by extracellular auxin (for instance through a membrane-bound or apoplastic auxin receptor; Fig. 3a), auxin will be preferentially pumped upstream by an auxin gradient (Fig. 3b). In fact, the endocytosis of PIN1 has been shown to be suppressed by exogenous auxins (Paciorek et al. 2005) providing the positive amplification loop required for the auxin-dependent inhibitory field.

Phyllotaxis and induction of vasculature are the two classic examples for auxin-dependent pattern formation. What can be generalized from these examples? Both patterns are highly robust against stochastic fluctuations of the input, they rely

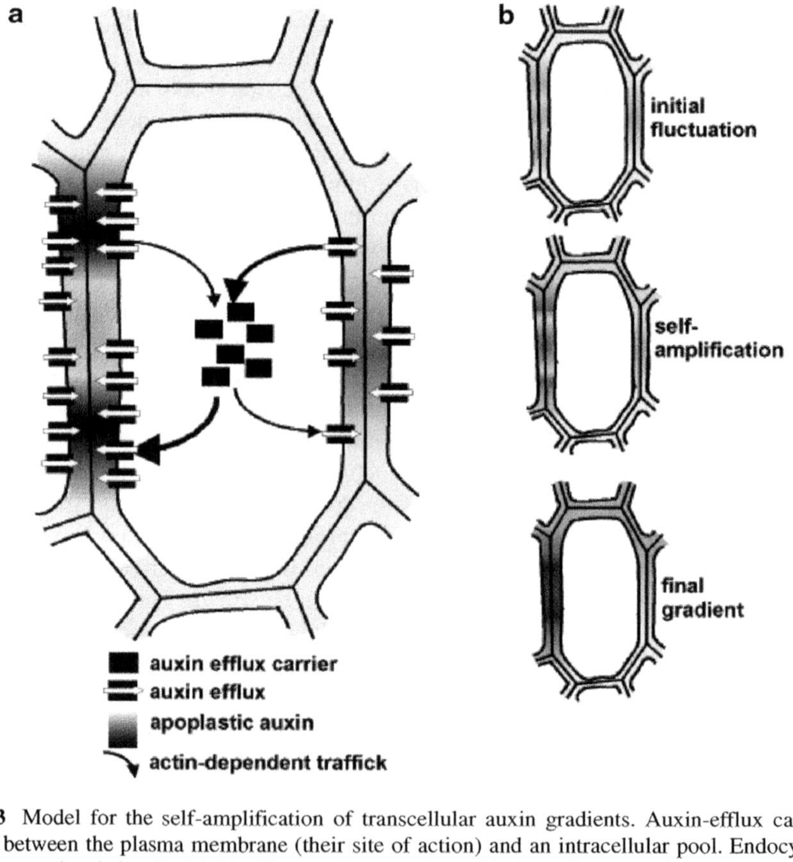

Fig. 3 Model for the self-amplification of transcellular auxin gradients. Auxin-efflux carriers cycle between the plasma membrane (their site of action) and an intracellular pool. Endocytosis of these carriers is locally inhibited by apoplastic auxin and is dependent on actin-mediated vesicle traffic (**a**). The competition between the two flanks of the cell for a limited number of the intracellular carriers in combination with local suppression of carrier endocytosis will amplify initial fluctuations of apoplastic auxin concentration progressively into clear gradients in the concentration of apoplastic auxin (**b**)

on lateral inhibition between the patterned elements, and they culminate in qualitative decisions that are probably brought about by autocatalytic feedback loops. Such mechanisms can be described by the mathematics of reaction-diffusion systems that was adapted to biology (Turing 1952), and has been quite successfully used to model various biological patterns such as foot–head patterning in *Hydra* (Gierer et al. 1972), segmentation in *Drosophila* (Meinhard 1986) and leaf venation (Meinhard 1976). In these reaction-diffusion systems, a locally constrained, self-amplifying feedback loop of an activator is linked to a far-ranging mutual inhibition (Gierer and Meinhard 1972). Auxin-dependent patterning seems to follow this model, but differs in one aspect: rather than employing an actual inhibitor as a positive entity, in auxin-dependent patterning lateral inhibition is brought about by mutual competition for the activator.

5 Auxin as a Pattern Generator in Cell Division

In addition to cell expansion, auxin can induce cell division, a fact that is widely employed for tissue culture and the generation of transgenic plants. Investigation of lateral-root formation in *Arabidopsis* suggested that auxin regulates cell division through a G-protein-dependent pathway (Ullah et al. 2003, for a review see Chen 2001). This was dissected further in tobacco suspension cells, early auxin signalling was dissected further, using the artificial auxins 1-naphthaleneacetic acid (NAA) and 2,4-dichlorophenoxyacetic acid (2,4-D). This study (Campanoni and Nick 2005) demonstrated that these two auxin species affected cell division and cell elongation differentially. NAA stimulated cell elongation at concentrations that were much lower than those required to stimulate cell division. In contrast, 2,4-D promoted cell division but not cell elongation. Pertussis toxin, a blocker of heterotrimeric G-proteins, inhibited the stimulation of cell division by 2,4-D but did not affect cell elongation. Conversely, aluminium tetrafluoride, an activator of the G-proteins, could induce cell division at NAA concentrations that were otherwise not permissive for division and even in the absence of any exogenous auxin. These data suggest that the G-protein-dependent pathway responsible for the auxin response of cell division is triggered by a different receptor than the pathway mediating auxin-induced cell expansion. The two receptors appear to differ in their affinity for different auxin species, with 2,4-D preferentially binding to the auxin receptor responsible for division and NAA preferentially binding to the auxin receptor inducing cell growth.

This bifurcation of auxin signalling (Fig. 4) appears to imply a differential interaction with the cytoskeleton as suggested by a recent detailed study on the effect of auxin on root growth in *Arabidopsis thaliana* (Rahman et al. 2007). When the contributions of cell division and cell elongation were assessed separately, the natural auxin IAAalong with NAA and the auxin-transport inhibitor 2,3,5-triiodobenzoic acid (TIBA)were observed to inhibit cell elongation while leaving filamentous actin basically unaltered. In contrast, 2,4-D and the polar transport inhibitor NP Ainhibited cell division and at the same time eliminated actin filaments.

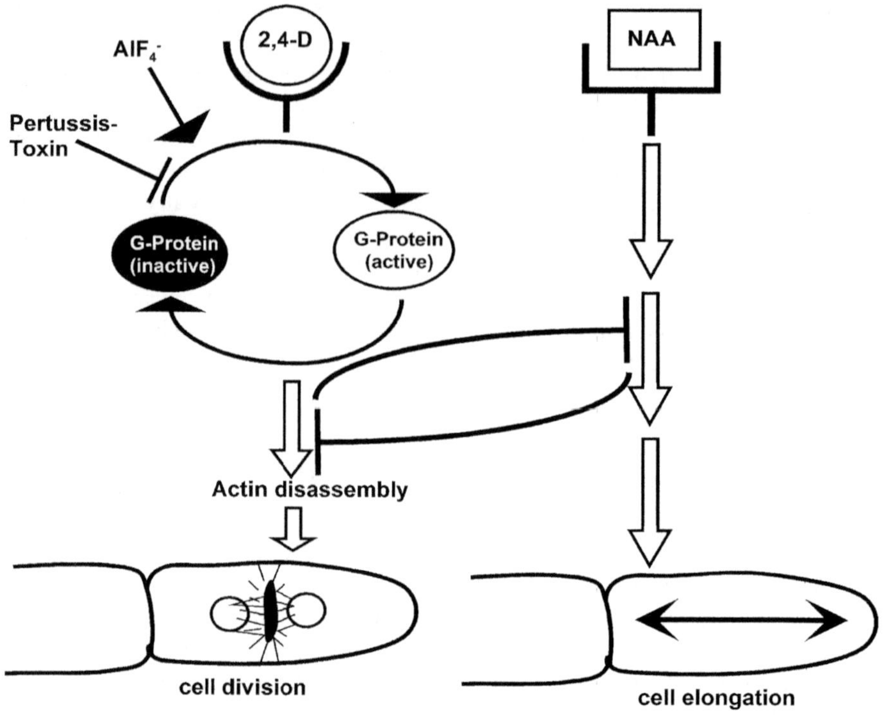

Fig. 4 Model for the bifurcation of auxin signalling in the regulation of cell division and cell elongation in tobacco cells according to Campanoni and Nick (2005) modified according to Rahman et al. (2007). The auxin receptor with high affinity for 1-naphthaleneacetic acid regulates cell elongation and is independent of G-protein activity and does not cause a disassembly of actin, whereas the auxin receptor with high affinity for 2,4-dichlorophenoxyacetic acid triggers a signal chain that involves the activity of a G-protein and triggers the disassembly of actin filaments. This signal chain is inhibited by pertussis toxin and is activated by aluminium tetrafluoride. Both pathways are mutually inhibitory

The root represents a very complex system consisting of different tissue layers that differ with respect to molecular machinery, auxin sensitivity and cytoskeletal organization. Moreover, the frequency of cycling cells, even in a rapidly growing root, is relatively modest, which makes it difficult to study the control exerted by intercellular auxin signalling on cell division on a quantitative level. However, a clear pattern of cell divisions is evident, with the cells of the quiescent centre acting as stem cells for the generation of proliferative tissues. As pointed out already, in the primary root of *Arabidopsis*, where this phenomenon has been dissected most intensively, this pattern can be traced back to early embryogenesis, whereas it seems to be more flexible in meristems of the Graminea. Nevertheless, the pattern of cell division is already established when the root meristem becomes accessible to cell-biological inspection and it is very difficult, if not impossible, to manipulate these patterns in a fundamental manner. Thus, root meristems represent a beautiful system to study pattern perpetuation, but for the analysis of pattern induction simpler systems that are less determined might be more appropriate. Suspension lines of tobacco are such models to study the primordial stages of division patterning and, in general, cellular aspects of cell division. These lines usually proceed from unicellular stages through a series of axial cell divisions towards cell files that are endowed with a clear axis and, in most cases, with a clear polarity. As will be explored in more detail below, these cell files are not a mere aggregation of autonomous, independent, cells, but display holistic properties such as an overall directionality and a pattern of cell division. In other words, these files are nothing other than a very reduced, but entire version of a multicellular "organism". Owing to this extreme reduction in the level of complexity, it may be easier to study the intercellular negotiations of hypercellular and hypocellular functions rather than in a highly complex and differentiated meristem. Two tobacco cell lines have been studied in more detail with respect to cell–cell communication:

The cell line VBI-0 (Opatrný and Opatrná 1976; Petrášek et al. 1998) derives from stem pith parenchyma, i.e. the cells that can differentiate into vascular tissue in response to auxin flow. These cells have preserved the ability to generate the structured cell-wall thickenings characteristic for xylogenesis (Nick et al. 2000). In the same way as its parenchymatic ancestor cells, this cell line grows in files where fundamental characteristics of patterning, such as clear axis and polarity of cell division and growth, are manifest. The progression into the culture cycle, the duration of the lag phase, the rate of cell division and the length of the exponential phase (Campanoni et al. 2003), but also cell polarity and axiality (Petrášek et al. 2002), can be controlled by auxin. The cell files are formed from singular cells, such that positional information inherited from the mother tissue probably does not play a role. If there are patterns of competence within a cell file, they must originate de novo during the culture cycle.

The widely used cell line BY-2 (Nagata et al. 1992) has generated a wealth of data on the role of phytohormones during the plant cell cycle. Compared with VBI-0, the temporal separation between cell division and cell expansion phases is less pronounced (probably as a consequence of the extremely high mitotic activity and short culture cycle). Moreover, the subsequent differentiation of these cells cannot be observed because they very rapidly lose viability if they are not subcultured directly after the logarithmic phase. However, BY-2 is transformed much more easily than VBI-0, such that a broad panel of different transgenic lines expressing fluorescently tagged marker proteins has become available. Moreover, although not as clearly manifest as in VBI-0, the basic features of patterning as well as file axis and polarity can be observed as well in this line.

During the work with these two cell lines, the cell divisions within the file were found to be partially synchronized, leading to a much higher frequency of cell files with even cell numbers than cell files with uneven cell numbers (Campanoni et al. 2003; Maisch and Nick 2007). The experimental data could be simulated using a mathematical model derived from non-linear dynamics, where elementary oscillators (cycling cells) were weakly coupled, and where the number of these oscillators was not conserved, but increased over time. The model predicted several non-intuitive

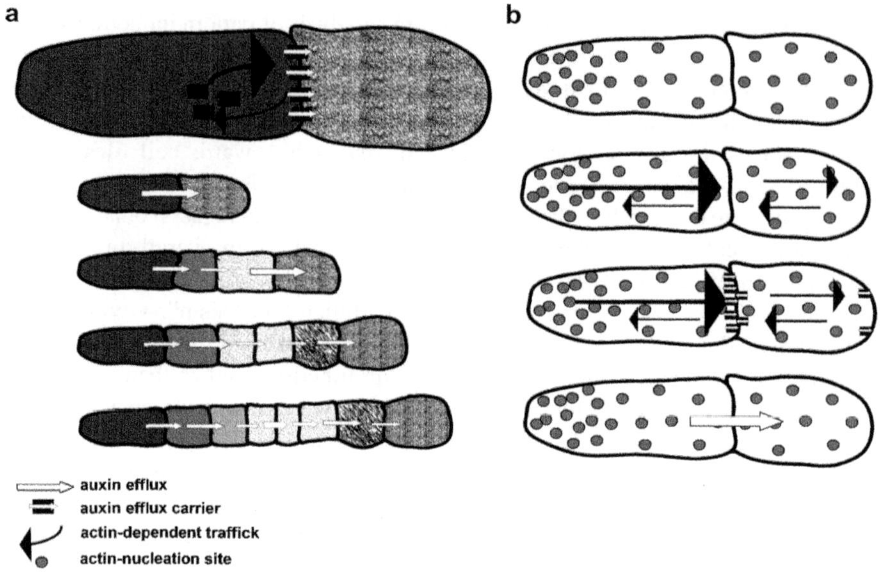

Fig. 5 Model for the regulation of cell division patterns in tobacco cell cultures by polar auxin transport. **a** Actin-dependent cycling of auxin-efflux carriers results in a polar distribution of the carrier and a polar flow of auxin through the cell file. Divisions of neighbouring cells are synchronized by this flow such that even cell numbers become more frequent than uneven cell numbers. **b** Actin-related protein 3 as marker for actin-nucleation sites is distributed in a gradient in the polarized tip cells, but not in the other cells of the file. The gradient of actin nucleation should result in a gradient of actin-dependent traffic that in turn will generate a graded distribution of auxin-efflux carriers such that auxin flow is polarized along the file axis

properties of the experimental system, among them that this coupling is unidirectional, i.e. that the coordinating signal was transported in a polar fashion. The coupling corresponds to a phase shift in the cell cycle, i.e. a dividing cell will cause its downstream neighbour to accelerate its cell cycle such that it will also initiate mitosis. The synchrony of cell divisions could be inhibited by low concentrations of the auxin-efflux inhibitor NPA. Although it has been known for a while that auxin is necessary for the progress of the cell cycle, and thus can be used to synchronize the cell cycle in plant cell cultures (for a review see Stals and Inzé 2001), this was the first time that auxin was shown to coordinate the divisions of adjacent cells. The modelling and the time courses of cell division showed that the noise in this system was considerable, with high variation in the cycling period over the cell population. Nevertheless, the division of adjacent cells was synchronized to such a degree that files with uneven cell numbers were rare compared with files with even numbers (Fig. 5a). Frequency distributions over the cell number per file thus exhibited oscillatory behaviour with characteristic peaks at the even numbers.

Since auxin efflux carriers cycle between the plasma membrane and an endocytotic compartment, auxin signalling has been linked to the organization of actin (for a review see Xu and Scheres 2005). However, this presumed link has recently been questioned by experiments, where PIN1 and PIN2 maintained their polar localization, although actin filaments had been eliminated by 2,4-D or NPA (Rahman et al. 2007). For the phytotropins TIBAand 2-(1-pyrenoyl) benzoic acid, it was shown very recently that they induce actin bundling not only in plants, but also in mammalian and yeast cells, i.e. in cells that are not to be expected to utilize auxin as a signalling compound (Dhonukshe et al. 2008). This has been interpreted as supportive evidence for a role of actin filaments in polar auxin transport. However, it was mentioned in the same work that NPA failed to cause actin bundling in non-plant cells, suggesting that its mode of action must be different.

Irrespective of the suggested direct effect of TIBA and 2-(1-pyrenoyl) benzoic acid on microfilaments, actin organization has been found to be highly responsive to changes in the cellular content of auxins (which would explain the NPA effect, for instance). This finding is actually quite old. During the classical period of auxin research, Sweeney and Thimann (1937) proposed that auxin might induce coleoptile growth by stimulating cytoplasmic streaming that is indeed very prominent in the coleoptile epidermis. In a series of publications, the late Kenneth Thimann returned to this idea and showed that elimination of actin very efficiently blocked auxin-dependent growth and argued that microfilaments are necessary for cell growth (Thimann et al. 1992; Thimann and Biradivolu 1994). These findings contrasted with laser tweezer measurements, where the rigour of actin limiting cell expansion was shown to be released by auxin (Grabski and Schindler 1996). In the framework of this actin-rigour model, the elimination of actin would be expected to stimulate rather than inhibit auxin-dependent growth. On the other hand, at that time there was no alternative model that could explain how actin filaments would support cell growth.

To get insight into the role of actin in the control of cell growth, the phytochrome-triggered cell elongation of maize coleoptiles was studied in more detail (Waller and Nick 1997), leading to a physiological definition of two microfilament populations that were functionally different. In cells that underwent rapid elongation, actin was organized into fine strands that became bundled in response to conditions that inhibited growth. This transition was rapid and preceded the changes in growth rate. Moreover, this response was confined to the epidermis, i.e. to the target tissue for the signal control of growth (Kutschera et al. 1987). Later, these two actin populations could be separated biochemically owing to differences in sedimentability (Waller et al. 2002). The fine actin filaments correlated with a cytosolic fraction of actin, whereas actin became trapped on the endomembrane system and was partitioned into the microsomal fraction in conditions that induced bundling. The transition between the two states of actin could be induced, in a dose-dependent manner, by light (perceived by phytochrome), by fluctuations of auxin content, or by the secretion inhibitor brefeldin A (BFA). The bundling of actin was accompanied by a shift of the dose-response of auxin-dependent cell elongation towards higher concentrations and thus to a reduced auxin sensitivity *in sensu strictu*. This led to a model whereby auxin signalling caused a dissociation of actin bundles into finer filaments that were more efficient with respect to the polar transport of auxin-signalling/transport components. Thus, any modulation of cellular auxin content (such as that induced by phytotropins) is expected to repartition the ratio between bundled and detached actin filaments.

This short excursion makes clear that although the organization of actin seems to play a role in the polarity of auxin fluxes, there is also a clear effect of auxin on the organization of actin filaments. This bidirectionality in the relation between actin and auxin has to be considered to avoid flaws in the interpretation of inhibitor effects. The feedback circuit between auxin and actin was addressed using patterning of cell division as a sensitive trait to monitor changes of polar auxin fluxes (Maisch and Nick 2007). If actin were part of an auxin-driven feedback loop, it should be possible to manipulate auxin-dependent patterning through manipulation of actin (Fig. 5a). This prediction was tested using a transgenic BY-2 cell line stably expressing a fusion between the yellow fluorescent protein and the actin-binding domain of mouse talin (Ketelaar et al. 2004). In this cell line, the microfilaments were constitutively bundled, and the synchrony of cell division was impaired in such a way that the characteristic oscillations described above disappeared. When transportable auxin was added (auxin *per se* was not sufficient), both a normal organization of actin and the synchrony of cell division could be restored. This demonstrated that actin is not only responsive to changes in the cellular content of auxin, but that it also actively participates in the establishment of the polarity that drives auxin transport.

When actin organization is relevant for the synchrony of cell division (mediated by a polar transport of auxin), the factors that regulate the organization and polarity of actin filaments are highly relevant for patterning. A central player might be the actin-related protein (ARP) 2/3 complex, a modulator of the actin cytoskeleton shown by immunofluorescence to mark sites of actin nucleation in tobacco BY-2 cells (Fišerová et al. 2006). ARP2/3 caps the pointed end such that the actin filament grows in the direction of the barbed end. Tobacco Arp3 was cloned and fused to red fluorescent protein (RFP) as a marker for bona fide sites of actin nucleation (Maisch J, Fišerová J, Fischer L, Nick P, submitted). By biolistic transient transformation of tobacco cells it was possible, for the first time, to visualize ARP3 in living plant cells. With use of dual-fluorescence visualization of actin [by a green fluorescent protein (GFP) fusion of the actin-binding site of fimbrin] the RFP–ARP3 could be shown to decorate actin filaments in vivo. When actin filaments were transiently eliminated (either by treatment with cytochalasin D or by cold treatment) and then allowed to recover, RFP–ARP3 marked the sites from which the new filaments emanated.

With use of this marker, the behaviour of actin-nucleation sites could be followed through patterned cell division in comparison with AtPIN1::GFP-PIN1 as a marker for cell polarity. This uncovered a qualitative difference between the terminal (polarized) cells of a file and the (isodiametric) cells in the centre of a file (Fig. 5b). The density of ARP3 was increased in the apex of terminal cells in a gradient opposed to the polarity monitored by PIN1 (which was concentrated at the opposite, proximal cross wall). Upon disintegration of the file into single cells, the graded distribution of ARP3 persisted, whereas PIN1 was redistributed uniformly over the plasma membrane of these cells. In contrast, the isodiametric cells in the file centre did not exhibit a graded distribution of the ARP3 signal, and the accumulation of PIN1 at the cross wall was much fainter than at the terminal cells, indicating that they are caused by residual amounts of PIN1 laid down by the (polar) progenitors of these cells.

The relationship between actin, vesicle flow and polar auxin transport appears to be interwoven by a bifurcated signal chain: vesicle trafficking mediated by ADP-ribosylation factors (ARFs) is required for the polar localization of Rho-related GTPases in plants which control regulators of the ARP2/3 complex (Frank et al. 2004). On the other hand, ARF-mediated vesicle trafficking also controls the localization of PIN proteins which is known to rely on the activity of the serine-threonine kinase PINOID (Friml et al. 2004) and on the function of P-glycoproteins/multiple drug resistance proteins (Noh et al. 2001). When the function of these ARFs is impaired, in consequence of either treatment with the fungal toxin BFAor a mutation in one of the guanine nucleotide exchange factors that activate the ARFs, PIN1 becomes mislocalized and is trapped in intracellular compartments (Geldner et al. 2001). This cellular effect accounts for the phenotype of the corresponding *Arabidopsis* mutant, *gnom*, that suffers from a drastic loss of cell and organ polarity and, in consequence, is not able to establish an organized *Bauplan*. Thus, ARF-dependent vesicle flow controls actin nucleation (through the activity of the ARP2/3 complex) and, in parallel, the localization of PIN proteins. However, the initial cue that controls the spatial pattern of ARF activity remains unknown. ARP3 maintained an intracellular gradient in the polar terminal cells of BY-2, whereas PIN1 was redistributed (Maisch J, Fišerová J, Fischer L, Nick P, submitted), indicating that actin nucleation might be upstream of the events that culminate in a polar distribution of PIN1. However, owing to the split signalling of the ARFs on the Rho-related GTPases and on the ARP2/3 complex, ARP3 and PIN1 might as well be parallel downstream targets of unknown factors that are expressed in response to cell polarity.

Irrespective of these uncertainties in the molecular details, actin filaments have emerged as central players for the directional vesicle flow by which the polar localization of auxin-efflux carriers is established and perpetuated. The cycling of PIN1 is suppressed by exogenous auxin such that PIN1 remains longer in the plasma membrane (Paciorek et al. 2005) and is therefore able to pump auxin more efficiently into the apoplast. On the other hand, the localization of PIN1 depends on the activity of actomyosin and the organization of the actin tracks is in turn under the control of auxin. These interactions will therefore establish a feedback loop with auxin-efflux carriers, intracellular auxin and actin filaments as central elements (Fig. 3a). This feedback loop is nothing other than a reaction-diffusion system *in sensu* Turing and might represent the cellular pacemaker of auxin-mediated pattern formation.

6 Auxin as a Pattern Generator in Cell Expansion

Once a plant cell has been born by cell division, it undergoes rapid expansion by uptake of water. This expansion is impressive: plant cells can increase in size by up to 4 orders of magnitude (Cosgrove 1987). Regulated cell expansion represents the central adaptive response of the sessile plants to environmental challenges and is therefore highly responsive to stimuli, such as light or gravity, and internal factors, including developmental signals and plant hormones. Whereas the mechanisms driving and regulating cellular expansion have been investigated in great detail over several decades, relatively little attention has been paid to the coordinative aspects of cell expansion. However, historically it was exactly this coordination of cell expansion that led to the discovery of auxin. In their famous *The Power of Movement in Plants*, the Darwins demonstrated for the phototropism of graminean seedlings that the direction of light is perceived in the very tip of the coleoptile, whereas the growth response to this directional stimulus occurs at the coleoptile base (Drawin and Drawin 1880). The signal transported from the tip to the base of the coleoptile must transmit not only information about the fact that the coleoptile tip has perceived light, but also information about the direction of the light stimulus.

Simultaneously, but independently, Cholodny (1927), for gravitropism, and Went (1926), for phototropism, discovered that this transmitted signal must be a hormone. By means of the famous *Avena* biotest this hormone was later found to be IAA (Kögl et al. 1934; Thimann 1935). The Cholodny–Went model explains tropistic curvature by an alignment of auxin transport with the stimulation vector. The resulting gradient of auxin between the two flanks of the stimulated organ will then induce a growth differential that drives bending in the direction of the inductive stimulus.

Since its beginnings, the Cholodny–Wentmodel has been challenged by attempts to explain tropism independently of cell communication by mere summation of cell-autonomous responses (Fig. 6a). For instance, when light causes an inhibition of growth, a gradient of light should produce a gradient of growth that would not require the exchange of intercellular signals (Blaauw 1915). Alternatively, each cell could perceive the direction of the stimulus and produce a directionality on its own – without interaction with the other cells – and the individual cell polarities would then add up to the polarity of the entire organ (Heilbronn 1917). This debate stimulated an ingenious experiment by Johannes Buder, where the gradient of light across the tissue and the direction of light were opposite (Fig. 6b). He irradiated the coleoptile from inside-out using a prototype of a light-pipe (Buder 1920). Under these conditions, the coleoptiles bent towards the lighted flank, i.e. according to the gradient of light and opposite to its direction. The outcome of this experiment demonstrated clearly that the direction of light is sensed in the coleoptile tip owing to extensive communication between the perceptive cells and strongly argues against cell-autonomous models of tropistic perception (Nick and Furuya 1996).

The transverse polarity built up in response to phototropic or gravitropic stimulation in the perceptive tissue subsequently redistributes the basipetal flow of auxin and thus transmits the directional information into the responsive tissue at the coleoptile base. This gradient of auxin flow is well established, starting from bioassays (for instance Dolk 1936) and ending up with tracer experiments using radioactively labelled auxin (Goldsmith and Wilkins 1964; Parker and Briggs 1990; Iino 1991; Godbolé et al. 2000) or direct measurements of free auxin across tropistically stimulated coleoptiles (Philippar et al. 1999; Gutjahr et al. 2005).

The Cholodny–Went model has been under continuous debate (see also Trewavas 1992), mainly because there is a discrepancy in amplitude between the gradient of the growth rate and the gradient of auxin concentration. The difference in auxin content between the two flanks of a tropistically stimulated coleoptile is in the

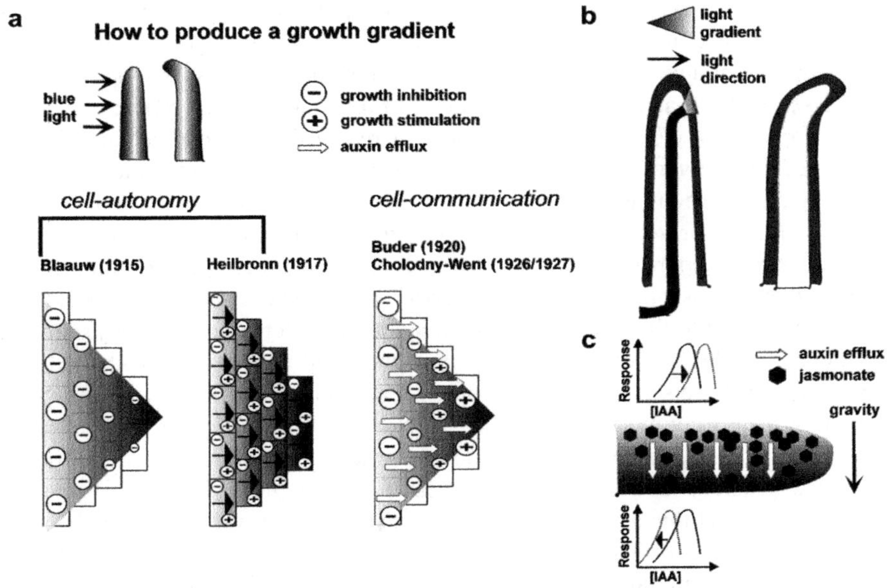

Fig. 6 Patterns of cell expansion during tropistic curvature of graminean coleoptiles. **a** Models for the formation of a growth gradient. According to Blaauw (1915), phototropic curvature emerges from a summation of growth inhibitions in response to the local intensity of light without interaction between cells. According to Heilbronn (1917), individual cells perceive the direction of light and respond by an intracellular gradient of growth. In contrast to these models that are based on complete cell autonomy, Buder (1920) explains curvature by interactions between individual cells across the coleoptile, and Cholodny (1927) and Went (1926) imply a lateral transport of a growth substance ("auxin"). **b** The experiment of Buder (1920), where the light direction and the light gradient across the organ are opposed. The bending is determined by the gradient, not by the direction of the light, contradicting the model postulated by Heilbronn (1917). **c** Extended Cholodny-Went model of gravitropism (according to Gutjahr et al. 2005). The lateral transport of auxin across the stimulated coleoptile is accompanied by a counterdirected gradient of jasmonate abundance and a gradient of auxin responsiveness across the tissue, with elevated responsiveness in the lower flank and reduced responsiveness in the upper flank

range of about 1:2 (Goldsmith and Wilkins 1964; Parker and Briggs 1990; Gutjahr et al. 2005), whereas growth is completely shifted from one flank to the other, i.e. the decrease in growth rate in one flank corresponds to the increase in growth rate in the other flank (Digby and Firn 1976; Iino and Briggs 1984; Himmelspach and Nick 2001). Elongation growth in coleoptiles increases more or less proportionally to the logarithm of auxin concentration (Wang and Nick 1998), such that the observed doubling of auxin concentration in the one flank would not succeed in causing the observed changes in growth rate. Moreover, when gravitropically stimulated hypocotyls (Rorabaugh and Salisbury 1989) or coleoptiles (Edelmann 2001; Gutjahr et al. 2005) were submersed in high concentrations of auxin, they showed positive gravitropism, i.e. they behaved as if they were roots. This is difficult to reconcile with a gradient of auxin as a unique cause for tropistic curvature.

With use of a classic biotest for auxin, the split-pea assay, in gravitropically stimulated rice coleoptiles, it could be demonstrated that, in parallel to the redistribution of auxin itself, a gradient of auxin responsiveness develops (Gutjahr et al. 2005) with elevated responsiveness at the lower flank and reduced responsiveness at the upper flank (Fig. 6c). This gradient of responsiveness can account for the strong redistribution of growth even for relatively modest gradients of auxin. It can even explain the peculiar sign reversal of bending for incubation with high concentrations of auxin beyond the optimum – for such superoptimal concentrations, the elevated auxin responsiveness at the lower flank should result in an inhibition of growth that is less pronounced in the upper flank, where the responsiveness is lower. In parallel to the gradient of auxin, a counterdirected gradient of jasmonate developed with higher concentrations at the upper flank as compared with the lower flank. Jasmonate acts as a negative regulator for auxin responsiveness, because both signal pathways compete for signalling factors such as AXR1 (Schwechheimer et al. 2001). Thus, the observed jasmonate gradient might well account for the observed gradient in auxin responsiveness across a gravitropically stimulated coleoptile. To test this assumption, the jasmonate gradient was either equalized by flooding the coleoptiles with exogenous methyl jasmonate or eliminated in consequence of a mutation that blocks jasmonate synthesis (Gutjahr et al. 2005). In both cases, the gravitropic response was delayed by about 1 h, but was eventually initiated and proceeded normally. This indicates that the jasmonate gradient, although not necessary for gravitropism, acts as a positive modulator. When auxin transport was inhibited by NPA, the jasmonate gradient nevertheless developed, suggesting that it is induced by gravitropic stimulation in parallel to and not in consequence of lateral auxin transport.

In summary, the Cholodny–Went model has to be extended by signal-triggered, modulative gradients of auxin responsiveness, but remains valid in its central statements. This means that tropistic responses, representing nothing other than a patterned distribution of cell expansion over the cross-section of the stimulated organ, can be explained in terms of auxin-dependent cell communication.

The analysis of auxin-dependent cell communication in cell division has identified a feedback loop between actin and auxin. This loop represents also a central element of patterned cell expansion. Actually, it was cell elongation in coleoptiles where the regulation of actin organization by auxin was discovered first (Sweeney and Thimann 1937; Thimann et al. 1992; Thimann and Biradivolu 1994; Waller and Nick 1997; Wang and Nick 1998; Holweg et al. 2004). Treatment with BFA, a fungal inhibitor of vesicle budding, caused, despite the presence of auxin, a rapid bundling of microfilaments and shifted actin from the cytosolic fraction into the microsomal fraction (Waller et al. 2002) depending on the dose of auxin and of BFA. In parallel, BFA shifted the dose-response curve of auxin-dependent growth to higher concentrations. In other words, BFA decreased auxin sensitivity *in sensu strictu*, consistent with an actin-dependent transport of auxin-signalling components such as auxin-efflux carriers. Again, a self-amplification loop emerges, consisting of auxin-dependent organization of actin filaments and actin-dependent transport of auxin-signalling components.

A prediction from this model for the actin-auxin feedback loop would be that a bundling of actin should be followed by a reduction in the activity of polar auxin transport. This prediction is supported by the suppression of division synchrony in tobacco cell lines that overexpress mouse talin (Maisch and Nick 2007). However, to measure auxin transport directly, it would be necessary to administer radioactively labelled auxin to one pole of the cell file and to quantify the radioactivity recovered in the opposite pole of the file. This is not possible in a cell culture that has to be cultivated as suspension in a liquid medium. This approach would be feasible, however, in the classical graminean coleoptile system, where auxin transport can be easily measured by following the distribution of radioactively labelled IAA fed to the coleoptile apex. Transgenic rice lines were generated that expressed various levels of the actin-binding protein talin (Nick P, Han MJ, An G, submitted). In those lines, as a consequence of talin overexpression, actin filaments were bundled to variable extent, and this bundling of actin filaments was accompanied by a corresponding reduction in the polar transport of auxin and gravitropic curvature (as a physiological marker that relies on the activity of auxin transport). When a normal configuration of actin was restored by addition of exogenous auxin, this restored auxin transport as well. This rescue was mediated by transportable auxin species, but not by 2,4-D, which lacks polar transport. With use of this approach, the causal relationship between actin configuration and polar auxin transport could be shown directly.

A further prediction of the actin–auxin feedback model is oscillations in transport activity, because auxin will, through the reorganization of actin, stimulate its own efflux such that the intracellular level of auxin will drop, which in turn will result in a bundled configuration of actin, such that auxin-efflux carriers will be sequestered, culminating in a reduced efflux such that auxin received from the adjacent cells will accumulate and trigger a new cycle. The frequency of these oscillations should depend on the dynamics of actin reorganization (around 20 min; Nick P, Han MJ, An G, submitted), and the speed of PIN cycling (in the range of 5–10 min; Paciorek et al. 2005) and is therefore expected to be in the range of 25–30 min. In fact, classic experiments on basipetal auxin transport in coleoptiles report such oscillations with a period of 25 min (Hertel and Flory 1968).

We therefore arrive at a model of a self-referring regulatory circuit between polar auxin transport and actin organization, where auxin promotes its own transport by shaping actin filaments. Thus, similar to the patterning of cell division, the actin-auxin oscillator seems also to be the pacemaker for the patterning of cell expansion.

7 Why Auxin - or Order Without a "Great Chairman"

Already in multicellular algae, a polar transport of auxin can be detected (Dibb-Fuller and Morris 1992; Cooke et al. 2002) and seems to play a role in the establishment of polarity (Basu et al. 2002), indicating that the central role of auxin in cell communication is evolutionarily quite ancient and had already been developed prior to the colonization of terrestrial habitats. Why has evolution selected such a simple molecule for such a central role in intercellular coordination?

Although we are far from providing a full answer to this question, it seems that the answer is related to plant-specific features in the organization of signalling and development: In animal development, cell differentiation is typically controlled by precise and defined regulatory networks that are structured by predetermined hierarchies. In contrast, plant cells are endowed with a pronounced developmental flexibility that is maintained basically throughout the entire life span of a plant cell. Moreover, there are hardly any fixed hierarchies – plant development does not know of such a thing as a "Great Chairman" that assigns differential developmental pathways to the individual cells. Plant cells rather "negotiate" their individual developmental fates in a fairly "democratic" manner with hierarchies being created ad hoc by mutual interactions. It seems that auxin is a central tool in these "negotiations", because it represents a versatile tool to establish ad hoc hierarchies on the background of the high degree of "cellular anarchy" and noise that is characteristic for plant development.

Why is plant development so "noisy"? The strong developmental noise seems to be the tribute paid to indeterminate morphogenesis. The manifestation of the *Bauplan* in an individual plant depends strongly on the environmental conditions encountered during development. This developmental flexibility includes a rapid response of cell expansion, complemented by a somewhat slower addition of morphogenetic elements, such as cells, pluricellular structures and organs. This patterning process can integrate signals from the environment, and must therefore be both highly flexible and robust. More specifically, the patterning process has to cope with signals that can vary over several orders of magnitude for the strength of the control signal, and new elements have to be added such that the pattern formed by the preexisting elements is perpetuated and/or complemented.

Plant sensing occurs in a rather diffuse manner – there are no such things as eyes, ears or tongues; there are, instead, populations of relatively undifferentiated cells that sense environmental cues and signals. Nevertheless, plant sensing is surprisingly sensitive. When this high sensitivity of signalling is reached without specialized sensory organs, the individual cells must already be endowed with very efficient mechanisms for signal amplification that are active already during the first steps of the transduction chain. This strong signal amplification will inevitably result in all-or-none outputs of individual cells. The efficient amplification of weak stimuli on the one hand, with the simultaneous necessity to discriminate between very strong stimuli of different amplitude, poses special challenges for plant signalling. If all cells of a given organ were absolutely identical and homogeneous, even an extremely weak stimulation would yield a maximal response of the whole organ. It is clear that such a system would not have survived natural selection – the amplitude of the output must vary according to variable amplitudes of the input signal, because the plant has to respond appropriately to stimuli that vary in intensity, even if these stimuli are strong. One way to reconcile the requirement for high sensitivity with the requirement for a graded, variable output would be to assign the two antagonistic tasks to different levels of organization: the high sensitivity to the individual cells that perceive the signal; the graded, variable output to the population of cells (i.e. the tissue or organ) by an integration over the individual cell responses. But this works only when the sensory thresholds of individual cells differ over the population; in other words, when the individual cells are highly heterogeneous with respect to signal sensitivity and thresholds.

This heterogeneity was actually observed when photomorphogenesis was investigated on a cellular level for phytochrome-induced anthocyanin patterns in mustard cotyledons, a classic system of light-dependent plant patterning (Mohr 1972; Nick et al. 1993) or for microtubule reorientation in coleoptiles triggered by blue light or auxin depletion (Nick 1992). Even adjacent cells exhibited almost qualitative differences although they had received the same dose of the signal. However, when the frequency of responsive cells in a given situation was scored and plotted against the strength of the stimulus, a highly ordered function emerged. Thus, the realm of individual cells was reigned over by chaos; order emerged only on the level of the whole organ. This highly stochastic, all-or-none type response of individual cells becomes especially manifest for an early response to a saturating stimulus or as the final result of weak induction (Nick et al. 1992). It thus appears that early signalling events are highly stochastic, when assayed at the level of individual cells. These responses are not merely "noisy" because the flexible nature of plant cells can tolerate this. These "noisy" responses rather represent an innate system property of plant signalling. However, this "noisy" inputs poses especial challenges for any ordering principle.

It might be these challenges that have rendered IAA a central integrator and synchronizer of plant development. A molecule that is easily transported through the acidic environment of the apoplast, but that is readily trapped in the cytoplasm and then has to be actively exported is ideally suited to convey lateral inhibition between neighbouring cells. It was sufficient to put the localization of the efflux transporter (whatever its molecular nature may be) under the control of auxin itself to reach a perfect reaction-diffusion system *in sensu* Turing (1952). On the intracellular level, this system is able to establish a clear cell polarity from even minute and noisy directional cues. On the level of tissues, this cell polarity will generate patterns in a manner that meets the special constraints of plant development, i.e. noisy inputs as a consequence diffuse sensing and progressive addition of new elements to the pattern. Since the natural auxin IAA can enter the cell from any direction (because it can enter the cell even independently of import carriers such as AUX1), but will exit in a defined direction defined by the localized activity of the efflux carrier, it can collect the input from several neighbours and focus this input into a clear directional output. It is this property that makes auxin a versatile and robust integrator for cell–cell communication.

References

Basu S, Sun H, Brian L, Quatrano RL, Muday GK (2002) Early embryo development in *Fucus distichus* is auxin sensitive. Plant Physiol 130:292–302

Blaauw AH (1915) Licht und Wachstum II. Bot Zentralbl 7:465–532

Buder J (1920) Neue phototropische Fundamentalversuche. Ber Dtsch Bot Ges 28:10–19

Campanoni P, Nick P (2005) Auxin-dependent cell division and cell elongation: NAA and 2,4-D activate different pathways. Plant Physiol 137:939–948
Campanoni P, Blasius B, Nick P (2003) Auxin transport synchronizes the pattern of cell division in a tobacco cell line. Plant Physiol 133:1251–1260
Camus G (1949) Recherche sur le role des bourgeons dans les phenomenes de morphogenèse. Rev Cytol Biol Veg 11:1–199
Chen JG (2001) Dual auxin signalling pathways control cell elongation and division. J Plant Growth Regul 20:255–264
Cholodny N (1927) Wuchshormone und Tropismen bei den Pflanzen. Biol Zentralbl 47:604–626
Cooke TJ, Poli DB, Sztein AE, Cohen JD (2002) Evolutionary patterns in auxin action. Plant Mol Biol 49:319–338
Cosgrove D (1987) Assembly and enlargement of the primary cell wall in plants. Annu Rev Cell Dev Biol 13:171–201
Drawin C, Drawin F (1880) Sensitiveness of plants to light: it's transmitted effect. In: The power of movement in plants. Murray, London, pp 574–592
Dibb-Fuller JB, Morris DA (1992) Studies on the evolution of auxin carriers and phytotropin receptors: transmembrane auxin transport in unicellular and multicellular Chlorophyta. Planta 186:219–226
Digby J, Firn RD (1976) A critical assessment of the Cholodny–Went theory of shoot gravitropism. Curr Adv Plant Sci 8:953–960
Dhonukshe P, Grigoriev I, Fischer R, Tominaga M, Robinson DG, Hašek J, Paciorek T, Petrašek J, Seifertová D, Tejos R, Meisel LA, Zažímalová E, Gadella TWJ, Stierhof YD, Ueda T, Oiwa K, Akhmanova A, Brocke R, Spang A, Friml J (2008) Auxin transport inhibitors impair vesicle motility and actin cytoskeleton dynamics in diverse eukaryotes. Proc Natl Acad Sci U S A 105:4489–4494
Dolk HE (1936) Geotropism and the growth substance. Recl Trav Bot Neerl 33:509–585
Du Monceau D (1764) La physique des arbres. Winterschmidt, Nuremberg, pp 87–93
Edelmann H (2001) Lateral redistribution of auxin is not the means for gravitropic differential growth of coleoptiles: a new model. Physiol Plant 112:119–126
Fišerová J, Schwarzerová K, Petrášek J, Opartný S (2006) ARP2 and ARP3 are localized to sites of actin filament nucleation in tobacco BY-2 cells. Protoplasma 227:119–128
Fleming AJ, McQueen-Mason S, Mandel T, Kuhlemeier C (1997) Induction of leaf primordia by the cell wall protein expansin. Science 276:1415–1418
Frank M, Egile C, Dyachok J, Djakovic S, Nolasco M, Li R, Smith LG (2004) Activation of Arp2/3 complex-dependent actin polymerization by plant proteins distantly related to Scar/WAVE. Proc Natl Acad Sci U S A 101:16379–16384
Friml J, Yang X, Michniewicz M, Weijers D, Quint A, Tietz O, Benjamins R, Ouwerk PB, Ljung K, Sandberg G, Hooykaas PJ, Palme K, Offringa R (2004) A PINOID-dependent binary switch in apical-basal PIN polar targeting directs auxin efflux. Science 306:862–865
Geldner N, Friml J, Stierhof YD, Jürgens G, Palme K (2001) Auxin transport inhibitors block PIN1 cycling and vesicle trafficking. Nature 413:425–428
Gierer A, Meinhard H (1972) A theory of biological pattern formation. Kybernetik 12:30–39
Gierer A, Berking S, Bode H, David CN, Flick K, Hansmann G, Schaller H, Trenkner E (1972) Regeneration of hydra from reaggregated cells. Nature 239:98–101
Godbolé R, Michalke W, Nick P, Hertel R (2000) Cytoskeletal drugs and gravity-induced lateral auxin transport in rice coleoptiles. Plant Biol 2:176–181
Goebel K (1908) Einleitung in die experimentelle Morphologie der Pflanzen. Teubner, Leipzig, pp 218–251
Goldsmith MHM, Wilkins MB (1964) Movement of auxin in coleoptiles of *Zea mays* L. during geotropic stimulation. Plant Physiol 39:151–162
Grabski S, Schindler M (1996) Auxins and cytokinins as antipodal modulators of elasticity within the actin network of plant cells. Plant Physiol 110:965–970
Green PB (1980) Organogenesis – a biophysical view. Annu Rev Plant Physiol 31:51–82
Gutjahr C, Riemann M, Müller A, Weiler EW, Nick P (2005) Cholodny–Went revisited – a role for jasmonate in gravitropism of rice coleoptiles. Planta 222:575–585

Hanstein J (1860) Versuche über die Leitung des Saftes durch die Rinde. Jahrb Wiss Bot 2:392
Heilbronn A (1917) Lichtabfall oder Lichtrichtung als Ursache der heliotropischen Reizung? Ber Dtsch Bot Ges 35:641–642
Hertel R, Flory R (1968) Auxin movement in corn coleoptiles. Planta 82:123–144
Himmelspach R, Nick P (2001) Gravitropic microtubule orientation can be uncoupled from growth. Planta 212:184–189
Holweg C, Süßlin C, Nick P (2004) Capturing *in-vivo* dynamics of the actin cytoskeleton. Plant Cell Physiol 45:855–863
Iino M (1991) Mediation of tropisms by lateral translocation of endogenous indole-3-acetic acid in maize coleoptiles. Plant Cell Environ 14:279–286
Iino M, Briggs WR (1984) Growth distribution during first positive phototropic curvature of maize coleoptiles. Plant Cell Environ 7:97–104
Jönsson H, Heisler MG, Shapiro BE, Meyerowitz EM, Mjolsness E (2006) An auxin-driven polarized transport model for phyllotaxis. Proc Natl Acad Sci U S A 103:1633–1638
Ketelaar T, Anthony RG, Hussey PJ (2004) Green fluorescent protein-mTalin causes defects in actin organization and cell expansion in *Arabidopsis* and inhibits actin depolymerizing factor's actin depolymerizing activity in vitro. Plant Physiol 136:3990–3998
Kirk DL (2003) Seeking the ultimate and proximate causes of *Volvox* multicellularity and cellular differentiation. Integr Comp Biol 43:247–253
Kögl F, Hagen-Smit J, Erxleben H (1934) Über ein neues Auxin ("Hetero-Auxin") aus Harn. Hoppe-Seylers Z Physiol Chem 228:104–112
Kutschera U, Bergfeld R, Schopfer P (1987) Cooperation of epidermal and inner tissues in auxin-mediated growth of maize coleoptiles. Planta 170:168–180
Lintilhac PM (1999) Towards a theory of cellularity – speculations on the nature of the living cell. Bioscience 49:60–68
Maisch, J., Fišerová, J., Fisher, L., Nick, P. Actin-related protein 3 labels actin-cucleating sites in tobacco BY-2-cells. J. Exp. Botany, in press
Maisch J, Nick P (2007) Actin is involved in auxin-dependent patterning. Plant Physiol 143:1695–1704
Mattsson J, Sung ZR, Berleth T (1999) Responses of plant vascular systems to auxin transport inhibition. Development 126:2979–2991
McCabe PF, Valentine TA, Forsberg S, Pennella R (1997) Soluble signals from cells identified at the cell wall establish a developmental pathway in carrot. Plant Cell 9:2225–2241
Meinhard H (1976) Morphogenesis of lines and nets. Differentiation 6:117–123
Meinhard H (1986) The threefold subdivision of segments and the initiation of legs and wings in insects. Trends Genet 3:36–41
Mohr H (1972) Lectures on photomorphogenesis. Springer, Heidelberg
Nagata T, Nemoto Y, Hasezava S (1992) Tobacco BY-2 cell line as the "Hela" cell in the cell biology of higher plants. Int Rev Cytol 132:1–30
Nakajima K, Sena G, Nawy T, Benfey PN (2001) Intercellular movement of the putative transcription factor SHR in root patterning. Nature 413:307–311
Nick P, Furuya M (1996) Buder revisited – cell and organ polarity during phototropism. Plant Cell Environ 19:1179–1187
Nick P, Schäfer E, Furuya M (1992) Auxin redistribution during first positive phototropism in corn coleoptiles: microtubule reorientation and the Cholodny–Went theory. Plant Physiol 99:1302–1308
Nick P, Ehmann B, Schäfer E (1993) Cell communication, stochastic cell responses, and anthocyanin pattern in mustard cotyledons. Plant Cell 5:541–552
Nick P, Han M, An G. Auxin stimulates its own transport by shaping actin filaments. Plant Physiology, under revision
Nick P, Heuing A, Ehmann B (2000) Plant chaperonins: a role in microtubule-dependent wall-formation? Protoplasma 211:234–244
Noh B, Murphy AS, Spalding EP (2001) Multidrug resistance-like genes of *Arabidopsis* required for auxin transport and auxin-mediated development. Plant Cell 13:2441–2454

Opatrný Z, Opatrná J (1976) The specificity of the effect of 2,4-D and NAA on the growth, micromorphology, and occurrence of starch in long-term *Nicotiana tabacum* L. cell strains. Biol Plant 18:359–365

Paciorek T, Zažimalová E, Ruthardt N, Petrášek J, Stierhof YD, Kleine-Vehn J, Morris DA, Emans N, Jürgens G, Geldner N, Friml J (2005) Auxin inhibits endocytosis and promotes its own efflux from cells. Nature 435:1251–1256

Parker KE, Briggs WR (1990) Transport of indole-3-acetic acid during gravitopism of intact maize coleoptiles. Plant Physiol 94:1763–1768

Petrášek J, Freudenreich A, Heuing A, Opatrný Z, Nick P (1998) Heat shock protein 90 is associated with microtubules in tobacco cells. Protoplasma 202:161–174

Petrášek J, Elčkner M, Morris DA, Zažímalová E (2002) Auxin efflux carrier activity and auxin accumulation regulate cell division and polarity in tobacco cells. Planta 216:302–308

Philippar K, Fuchs I, Lüthen H, Hoth S, Bauer CS, Haga K, Thiel G, Ljung K, Sandberg G, Böttger M, Becker D, Hedrich R (1999) Auxin-induced K^+ channel expression respresents an essential step in coleoptile growth and gravitropism. Proc Natl Acad Sci U S A 96:12186–12191

Priestley JH, Swingle CF (1929) Vegetative propagation from the standpoint of plant anatomy. USDA Tech Bull 151:1–98

Rahman A, Bannigan A, Sulaman W, Pechter P, Blancaflor EB, Baskin TI (2007) Auxin, actin and growth of the *Arabidopsis thaliana* primary root. Plant J 50:514–528

Reinhardt D, Mandel T, Kuhlemeier C (2000) Auxin regulates the initiation and radial position of plant lateral organs. Plant Cell 12:507–518

Reinhardt D, Pesce ER, Stieger P, Mandel T, Baltensperger K, Bennett M, Traas J, Friml J, Kuhlemeier C (2003) Regulation of phyllotaxis by polar auxin transport. Nature 426:255–260

Roeland MH, Merks RMH, Van de Peer Y, Inze D, Beemster GTS (2007) Canalization without flux sensors: a traveling-wave hypothesis. Trends Plant Sci 12:384–390

Rorabaugh PA, Salisbury FB (1989) Gravitropism in higher plant shoots VI. Changing sensitivity to auxin in gravistimulated soybean hypocotyls. Plant Physiol 91:1329–1338

Rothwell GW, Lev-Yadun S (2005) Evidence of polar auxin flow in 375 million-year-old fossil wood. Am J Bot 92:903–906

Sachs T (1968) On the determination of the pattern of vascular tissue in peas. Ann Bot 33:781–790

Sachs T (1975) The control of the differentiation of vascular networks. Ann Bot 39:197–204

Sachs T (1981) The control of the patterned differentiation of vacular tissues. Adv Bot Res 9:152–262

Scarpella E, Marcos D, Friml D, Berleth T (2006) Control of leaf vascular patterning by polar auxin transport. Genes Dev 20:1015–1027

Scheres B, Wolkenfelt H, Willemsen V, Terlouw M, Lawson E, Dean C, Weisbeek P (1994) Embryonic origin of the *Arabidopsis* root and root meristem initials. Development 120:2475–2487

Schoute JC (1913) Beiträge zur Blattstellungslehre. I. Die Theorie. Rec Trav Bot Neerl 10:153–325

Schwechheimer C, Serino G, Callis J, Crosby WL, Lyapina S, Deshaies RJ, Gray WM, Estelle M, Deng XW (2001) Interactions of the COP9 signalosome with the E3 ubiquitin ligase SCF^{TIR1} in mediating auxin response. Science 292:1379–1382

Snow M, Snow R (1931) Experiments on phyllotaxis. I. The effect of isolating a primordium. Philos Trans R Soc Lond Ser B 221:1–43

Spemann H (1936) Experimentelle Beiträge zu einer Theorie der Entwicklung. Springer, Berlin

Stals H, Inzé D (2001) When plant cells decide to divide. Trends Plant Sci 6:359–364

Stein W (1993) Modeling the evolution of stelar architecture in vascular plants. Int J Plant Sci 154:229–263

Sweeney BM, Thimann KV (1937) The effect of auxin on cytoplasmic streaming. J Gen Physiol 21:123–135

Thimann KV (1935) On the plant growth hormone produced by *Rhizopus suinus*. J Biol Chem 109:279–291

Thimann KV, Biradivolu R (1994) Actin and the elongation of plant cells II: the role of divalent ions. Protoplasma 183:5–9

Thimann KV, Reese K, Nachmikas VT (1992) Actin and the elongation of plant cells. Protoplasma 171:151–166

Trewavas, AJ (1992) Forum: What remains of the Cholodny-Went theory? Plant Cell Environ 15: 759–794

Turing AM (1952) The chemical basis of morphogenesis. Philos Trans R Soc Lond Ser B Biol 237:37–72

Ullah H, Chen JG, Temple B, Boyes DC, Alonso JM, Keith RD, Ecker JR, Jones AM (2003) The beta-subunit of the *Arabidopsis* G protein negatively regulates auxin-induced cell division and affects multiple developmental processes. Plant Cell 15:393–409

Van den Berg C, Willensen V, Hage W, Weisbeek P, Scheres B (1995) Cell fate in the *Arabidopsis* root meristem is determined by directional signalling. Nature 378:62–65

Waller F, Nick P (1997) Response of actin microfilaments during phytochrome-controlled growth of maize seedlings. Protoplasma 200:154–162

Waller F, Riemann M, Nick P (2002) A role for actin-driven secretion in auxin-induced growth. Protoplasma 219:72–81

Wang QY, Nick P (1998) The auxin response of actin is altered in the rice mutant *Yin-Yang*. Protoplasma 204:22–33

Weismann A (1894) Aufsätze über Vererbung und verwandte Fragen. Fischer, Jena

Went F (1926) On growth accelerating substances in the coleoptile of *Avena sativa*. Proc Kon Ned Akad Weten 30:10–19

Xu J, Scheres B (2005) Cell polarity: ROPing the ends together. Curr Opin Plant Biol 8:613–618

Yoon HS, Golden JW (1998) Heterocyst pattern formation controlled by a diffusible peptide. Science 282:935–938

Yoon HS, Golden JW (2001) PatS and products of nitrogen fixation control heterocyst pattern. J Bacteriol 183:2605–2613

Zimmermann W (1965) Die Telomtheorie. Fischer, Stuttgart

Integrated Calcium Signaling in Plants

Narendra Tuteja

Abstract Calcium ion (Ca^{2+}) is the most important universal signal carrier used by living organisms, including plants, to convey information to many different cellular processes. The cytosolic free Ca^{2+} concentration ($[Ca^{2+}]_{cyt}$) has been found to increase in response to many physiological stimuli, including stress. The Ca^{2+} spikes normally result from two opposing reactions, Ca^{2+} influx through channels or Ca^{2+} efflux through pumps. The removal of increased Ca^{2+} from the cytosol to either the apoplast or intracellular organelles requires energized "active" transport. Ca^{2+}-ATPases and Ca^{2+}/H^+ antiporters are the key proteins catalyzing this movement. The increased level of Ca^{2+} is recognized by some Ca^{2+} sensors or calcium-binding proteins, which can activate many calcium-dependent protein kinases. The regulation of gene expression by cellular Ca^{2+} is also crucial for plant defense against various stresses. In this chapter several aspects of calcium signaling, such as Ca^{2+} requirement, Ca^{2+} transporters/pumps (Ca^{2+}-ATPases, Ca^{2+}/H^+ antiporter), Ca^{2+} signature, Ca^{2+} memory, and various Ca^{2+}-binding proteins, are presented.

1 Introduction

The maintenance of intracellular ionic homeostasis is fundamental to the physiological function of a living cell. The Ca^{2+} is a versatile signal ion regulating many physiological processes, such as ionic balance, gene expression, metabolism of lipids, proteins, and carbohydrates, cell growth, cell division, cell development, and secretion (Poovaiah and Reddy 1993; Pardo et al. 1998; Mahajan et al. 2006a; Tuteja and Mahajan 2007; Tuteja 2007). Many external stimuli, including light and various stress factors, can bring about changes in cellular Ca^{2+} level, which can affect plant growth and development (Sanders et al. 2002; Rudd and Franklin-Tong 2001). Ca^{2+} serves as a second messenger and its concentration is delicately balanced by the presence of "Ca^{2+} stores" such as vacuoles, endoplasmic reticulum (ER), mitochondria,

N. Tuteja (✉)
Plant Molecular Biology Group, International Centre for Genetic Engineering and Biotechnology (ICGEB), Aruna Asaf Ali Marg, New Delhi, 110 067, India
e-mail: narendra@icgeb.res.in

F. Baluška and S. Mancuso (eds.), *Signaling in Plants,* Signaling and Communication in Plants, 29
DOI: 10.1007/978-3-540-89228-1_2, © Springer-Verlag Berlin Heidelberg 2009

and cell wall. Recently, Xiong et al. (2006) demonstrated that the organelles surrounded by a double membrane (e.g., mitochondria, chloroplasts, and nuclei) are also equipped to generate calcium signal on their own, which is delimited by a double membrane. Ca^{2+} ion represents an important signaling species and a convergence point of many disparate signaling pathways. A generic signal transduction pathway has the following steps: (1) perception of the signal by the membrane receptors, (2) generation of second messengers, (3) a cascade of protein phosphorylation/dephosphorylation events that may target transcription factors controlling a specific set of stress-regulated genes, and (4) stress tolerance, plant adaptation, and other phenotypic responses (Tuteja and Mahajan 2007).

In plants, Ca^{2+} plays many essential roles. For example, it is an essential plant nutrient required for growth and development, especially of the root and the shoot tip, which are meristematic, where cell division occurs by mitosis. Ca^{2+} helps in the formation of microtubules and microtubules in turn are essential for the anaphasic movement of chromosomes. Ca^{2+} is an important divalent cation and is required for structural roles in the cell wall and membranes, where it exists as calcium pectate. Ca^{2+} accumulates as calcium pectate in the cell wall and binds the cells together. It is also required as a countercation for inorganic and organic anions in the vacuole and as an intracellular messenger in the cytosol (Mahajan and Tuteja 2005). Ca^{2+} is required for pollen tube growth and elongation (Sanders et al. 2002). Several genes are reported to be upregulated in response to calcium in plants. An analysis of transcriptome changes revealed 230 calcium-responsive genes, of which 162 were upregulated and 68 were downregulated (Kaplan et al. 2006). These include known early stress-responsive genes as well as genes of unknown function. Recently, a blue light receptor phototropin, which regulates growth and development of plants, has also been shown to be involved in calcium signaling in higher plants (Harada and Shimazaki 2007). The Ca^{2+} signaling pathway can also regulates a K^+ channel for low-K response in *Arabidopsis*. Calcium is also reported to be an essential component of the sucrose-signaling pathway that leads to the induction of fructan synthesis (Martinez-Noel et al. 2006). Calcium signaling is also involved in the regulation of cell cycle progression in response to abiotic stress. In appreciation of the immense significance of Ca^{2+} ions, this chapter is solely dedicated to the salient features associated with integrated calcium signaling. Various aspects regarding the Ca^{2+} requirements of plant and signal-induced changes, Ca^{2+} transporters/efflux pumps, Ca^{2+}/H^+ antiporters, Ca^{2+} signatures, Ca^{2+} memory, and Ca^{2+} sensor and transducer proteins are briefly covered.

2 Ca^{2+} Concentrations in Plant Cells and Signal-Induced Changes

Ca^{2+} is also an essential plant macronutrient element which is taken up by roots from the soil and is delivered to the shoot via xylem. In general, the cytosolic Ca^{2+}, typically kept at submicromolar levels in plant cells, is a well-established link for various signal transduction networks (Poovaiah and Reddy 1993; Bush 1995). The cytosolic Ca^{2+} ($[Ca^{2+}]_{cyt}$) concentration in plant cells is usually maintained at 200 nM (Bush 1995) or in the range of 0.1–1 μM (Poovaiah and Reddy 1993). However, the $[Ca^{2+}]_{cyt}$ is far higher than this because of the high affinity for Ca^{2+} of a range of Ca^{2+}-binding proteins. With use of various techniques it has been shown that Ca^{2+} concentrations in apoplasts range between 0.1–1.0 and 1–3 mM in ER lumen and these levels can go up to 50 mM (Sarwat and Tuteja 2007). Plastids and mitochondria contain millimolar levels of Ca^{2+} but much less than ER. Ca^{2+} concentrations within the vacuole range from 0.1 to 10 mM. In sieve tubes, the Ca^{2+} level is reported to be 20–100-fold higher than it is in typical plant cells (Knoblauch et al. 2001). Plants vary in their Ca^{2+} requirements and the ability to extract Ca^{2+} from complex soil environments. In particular, monocots require less Ca^{2+} for optimal growth than do dicots (see Tuteja and Mahajan 2007).

The environmental and hormonal signals are known to induce changes in cytosolic Ca^{2+} (Poovaiah and Reddy 1993). Regulation of Ca^{2+} level is important for the survival of the cell. The plant cell contains a number of vesicular compartments, which store Ca^{2+} that can be released into the cytoplasm when required. Specific channels/pumps regulate the movement of Ca^{2+} in and out of cells and organelles (Mahajan et al. 2006a; Tuteja 2007). While the source of Ca^{2+} resulting in its increase in the cytosol after a signal has not been extensively studied, in general we know that Ca^{2+} release can be primarily from an extracellular source (apoplastic space) as addition of ethylene glycol bis(2-aminoethyl ether) *N*,*N*,*N'*,*N'*-tetraacetic acid or 1,2-bis(*O*-aminophenoxy)ethane-*N*,*N*,*N'*,*N'*-tetraacetic acid was shown in many cases to block calcium effects (Fig. 1). Ca^{2+} release may also result from activation of PLC, leading to hydrolysis of phosphatidylinositol bisphosphate to inositol triphosphate (IP_3) and subsequent release of Ca^{2+} from intracellular Ca^{2+} stores (Fig. 1). In plants, resting intracellular Ca^{2+} concentration ($[Ca^{2+}]_i$) oscillates diurnally. Tang et al. (2007) reported that in *Arabidopsis thaliana*, $[Ca^{2+}]_i$ oscillations are synchronized to extracellular Ca^{2+} concentration oscillations largely through the Ca^{2+}-sensing receptor CAS. CAS regulates concentrations of IP_3, which in turn directs release of Ca^{2+} from internal stores. Furthermore, calcium-binding proteins (calcium sensors) can provide an additional level of regulation in the calcium signaling. These sensor proteins recognize and decode the information provided in the calcium signatures and relay the information downstream to initiate a phosphorylation cascade, leading to the activation of the signal-induced genes, products of which can directly or indirectly provide the signal response (e.g., stress tolerance). The signal response could also be growth inhibition or cell death, which will depend upon what kinds of genes are upregulated or downregulated (Mahajan et al. 2006a; Tuteja 2007) (Fig. 1). Overall, the calcium-induced response could be a coordinated action of many genes.

Mechanical stimuli (wind and touch) and fungal elicitors have also been reported to induce rapid and transient increases in cytosolic Ca^{2+} (Poovaiah and Reddy 1993). The source for the increase in cytosolic Ca^{2+} by different signals may be different, for example, the source for cold signal is extracellular and that for wind signal is intracellular. In response to abscisic acid (ABA) an increase (37–80%) in cytosolic Ca^{2+} in the opened stomatal guard cells has been observed. Light (phototropism) and gravity (geotropism) in maize coleoptiles have been shown to be responsible for increasing the cytosolic Ca^{2+}. Gibberellic acid increases the cytosolic Ca^{2+} of

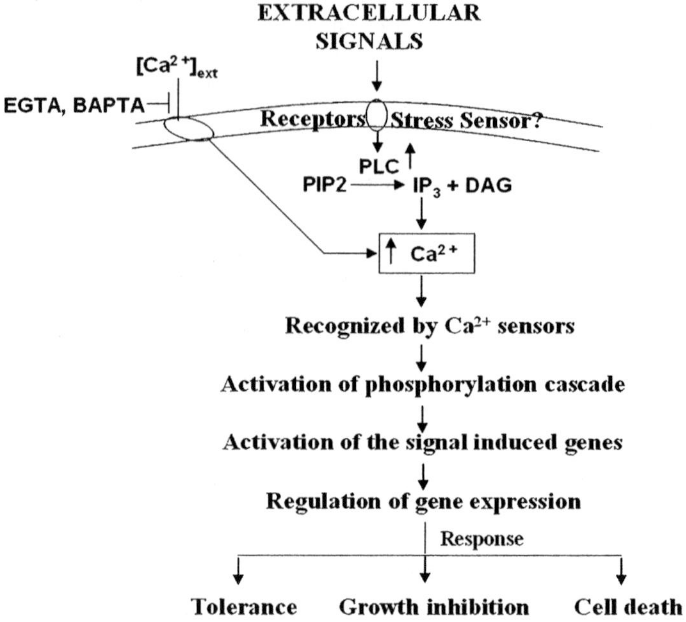

Fig. 1 Generic pathway for calcium signaling. The extracellular stress signal is first perceived by the membrane receptors and then activates a large and complex signaling cascade intracellularly including the generation of secondary signal molecules. High-salinity (Na^+) stress initiates the calcium signaling network. The signal first activates phospholipase C, which hydrolyzes phosphatidylinositol bisphosphate to generate inositol triphosphate and diacylglycerol, resulting in an increase in the level of Ca^{2+} ions. This increase in Ca^{2+} ion concentration is sensed by calcium sensors, which interact with and activate some protein kinases. The signal cascade results in the expression of multiple stress-responsive genes, the products of which can provide the phenotypic response of stress tolerance directly or indirectly. The stress response could also be a growth inhibition or cell death, which will depend upon how many and what kinds of genes are upregulated or downregulated. Overall, the stress response could be the coordinated action of many genes which may talk with each other

barley aleurone protoplast threefold (50–150 mM), while ABA decreases the cytosolic Ca^{2+} of barley aleurone protoplast (for all the above, see Poovaiah and Reddy 1993). Blue light is also known to increase cytosolic Ca^{2+}. Recently, Chena et al. (2008) suggested that inositol polyphosphate 5-phosphatase-13, a key enzyme in the phosphatidylinositol metabolic pathway, antagonizes PHOT1-mediated effects on calcium signaling under blue light.

3 Calcium Transporters/Pumps

Active transport of Ca^{2+} from the cytosol is a prerequisite for the restoration of low levels cytosolic calcium after the signaling event. Ca^{2+}-ATPases and Ca^{2+}/H^+ antiporters are the key proteins catalyzing this movement. By the removal of Ca^{2+} from the cytosol several important functions are performed by these proteins (Hirschi 2001). Some of the functions are as follows:

1. These proteins maintain a low cytosolic free Ca^{2+} concentration ($[Ca^{2+}]_{cyt}$) in the resting or the unstimulated cell which is appropriate for the cytoplasmic metabolism.
2. These proteins restore $[Ca^{2+}]_{cyt}$ levels to the resting normal levels following a $[Ca^{2+}]_{cyt}$ perturbation, thereby influencing the kinetics of the magnitude and subcellular location of $[Ca^{2+}]_{cyt}$ signals.
3. These proteins replenish intracellular and extracellular Ca^{2+} stores for subsequent $[Ca^{2+}]_{cyt}$ signals and permit the generation of localized $[Ca^{2+}]_{cyt}$ oscillations through their interplay with Ca^{2+} channels (Harper 2001).
4. These proteins provide Ca^{2+} in the ER for the secretory system to function.
5. These proteins remove some divalent cations, such as Ni^{2+}, Zn^{2+}, Mg^{2+}, and Mn^{2+}, from the cytosol to prevent mineral toxicity (Hirschi 2001).

Hirschi (2001) suggested that the Ca^{2+}-ATPases, which have high affinity (K_m=1–10 μM) but low capacity for Ca^{2+} transport, are responsible for maintaining $[Ca^{2+}]_{cyt}$ homeostasis in the resting cells, whereas the Ca^{2+}/H^+ antiporters, which have lower affinities (K_m=10–15 μM) but high capacities for Ca^{2+} transport, are likely to remove Ca^{2+} from the cytosol during $[Ca^{2+}]_{cyt}$ signals and thereby modulate $[Ca^{2+}]_{cyt}$ perturbations. This hypothesis is supported by the fact that the Ca^{2+}/H^+ antiporter, but not the vacuolar Ca^{2+}-ATPase, resets $[Ca^{2+}]_{cyt}$ in yeast following hypertonic shock (Denis and Cyert 2002).

3.1 Ca^{2+}-ATPases

Ca^{2+}-ATPases are estimated to represent less than 0.1% of the membrane protein and are thus 30–100-fold less abundant than H^+-ATPases in the plasma membrane (3%) and the endomembranes (5–10%). ATPase pumps are condemned to push Ca^{2+} uphill for eternity into the ER or out of the cell. Plant Ca^{2+}-ATPases belong to two major families: (1) the P-type ATPase IIA family and (2) the P-type ATPase IIB family (Axelsen and Palmgren 2001).

3.1.1 P-Type ATPase IIA Family

The nucleotide specificity of these pumps is broad (30–60% activity achieved with GTP and ITP). The pumps are inhibited by erythrosine B (IC_{50} ≤1μM) and Ca^{2+} affinity is estimated to be in the range of 0.4–12μM. The first family. The P-type ATPase II A family lacks an N-terminal autoregulatory domain. Four members of this family have been identified in the *Arabidopsis* genome (termed "AtECAs 1–4" by Axelsen and Palmgren (2001). They are likely to be present in the plasma membrane, tonoplast ER, and the Golgi apparatus.

3.1.2 P-Type ATPase IIB Family

This second family is characterized by an autoinhibitory N-terminal domain that contains a binding site for Ca-calmodulin (CaM) and in addition a serine-residue phosphorylation site. Their catalytic activity can be modulated by $[Ca^{2+}]_{cyt}$ either through activation upon binding CaM or by inhibition following phosphorylation by Ca^{2+}-dependent protein kinases (CDPKs) (Hwang et al. 2000). Since CaM binding sites are generally quite diverse, each type IIB Ca^{2+}-ATPase may have different affinity for CaM or may bind a different CaM isoform. Ten members of the type IIB Ca^{2+}-ATPase family have been identified in the *Arabidopsis* genome and are termed "ACASs 1, 2, and 4" and "ALACAs 7–13" (Axelsen and Palmgren (2001). These Ca^{2+}-ATPases reside on various cellular membranes, including the plasma membrane (AtACA8), the tonoplast (AtACA4), and the plastid inner membrane (AtACA1). The relative molecular mass of type IIB Ca^{2+}-ATPase pumps has been estimated to be between 115,000 and 135,000Da (Rasi-Caldogno et al. 1995).

The abundance of Ca^{2+}-ATPase isoforms suggests that individual isoforms are functionally distinct and may respond differentially to distinct cellular processes involving specific Ca^{2+} signals. It also implys a requirement for CaM-independent and CaM-dependent regulation of Ca^{2+}-ATPase activities in the modulation of $[Ca^{2t}]_{cyt}$ perturbations during cell signaling. The expression of many Ca^{2+}-ATPases is increased upon exposure to high salinity or high $[Ca^{2+}]_{cyt}$ and some Ca^{2+}-ATPase genes are expressed only under stress conditions (Garciadeblas et al. 2001). This may reflect a role in maintaining $[Ca^{2+}]_{cyt}$ homeostasis or in reducing Na^{+} influx to the cytosol in saline environments.

3.2 Ca^{2+}/H^{+} Antiporter

Calcium regulation of ion homeostasis by various ion pumps is depicted in Fig. 2. Ca^{2+}/H^{+} antiporters are efflux transporter and are different from Ca^{2+}-ATPases in that they do not require ATP and they are not sensitive to vanadate. *CAX1* (calcium exchanger 1) was the first plant Ca^{2+}/H^{+} antiporter which was cloned and functionally expressed (Tuteja and Mahajan 2007). The gene was identified by its ability to restore growth on a high Ca^{2+} medium to a yeast mutant defective in vacuolar Ca^{2+} transport. The Ca^{2+}/H^{+} antiporters present in the plasma membrane and tonoplast have been characterized biochemically (Sanders et al. 2002). These have a lower affinity for Ca^{2+} than Ca^{2+}-ATPases and may also transport Mg^{2+}. Eleven genes encoding putative Ca^{2+}/H^{+} antiporters (*AtCAX*) have been identified in the genome of *Arabidopsis thaliana* (Hirschi 2001). The transporters AtCAX1, AtCAX2, and AtCAX4 are located at the tonoplast. The AtCAX1 antiporter exhibits both high affinity and high specificity for Ca^{2+}, whereas the AtCAX2 transporter is a high-affinity, high-capacity H^{+}/heavy metal cation antiporter. The *AtCAX* genes have homologues in other plant species and their physiological roles have been investigated using transgenic plants (Hirschi 2001). Transgenic tobacco overexpressing *AtCAX1* exhibits Ca^{2+}-deficiency disorders, which include tip burn, metal hypersensitivity, and susceptibility to chilling, that can be reversed by increasing Ca^{2+} supply. Increase of the Ca^{2+} supply resulted in the expression of *AtCAX1* and *AtCAX3* (but not *AtCAX2* or *AtCAX4*) genes being increased (Hirschi 2001). Ca^{2+}/H^{+} antiporters utilize the H^{+} gradient generated by the tonoplast V-type H^{+} pump and by a

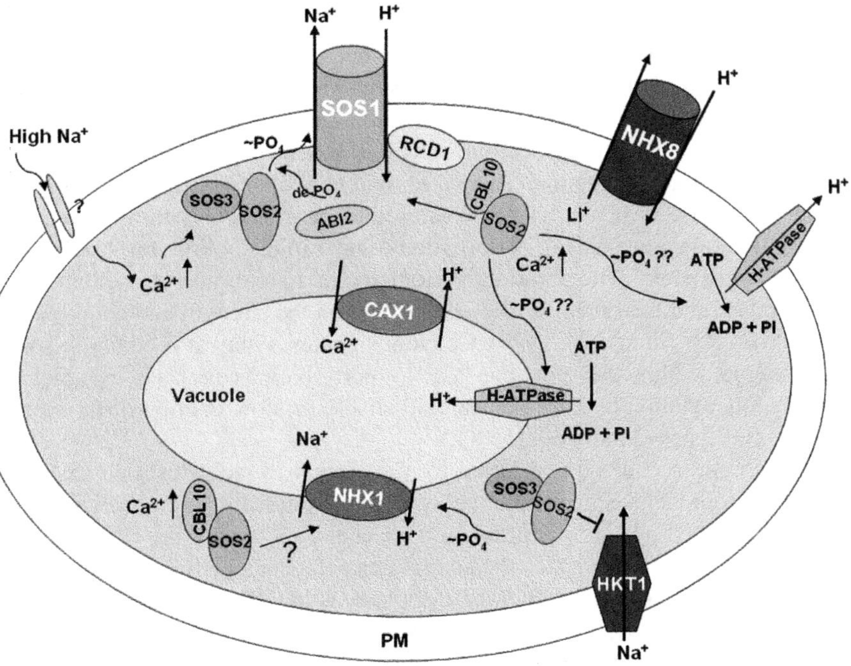

Fig. 2 Calcium and SOS pathway in regulation of ion homeostasis by various ion pumps. Under saline condition, the increase in the cytosolic Na^+ level causes damage to several cellular processes. One of the major salt detoxifying mechanisms in the cell is the calcium-activated SOS3-SOS2 protein complex, which activates SOS1, a Na^+/H^+ antiporter on the plasma membrane responsible for extrusion of Na^+ from the cell. At the same time the SOS3-SOS2 complex is also involved in inhibiting HKT1, low-affinity potassium transporter, which transports Na^+ ion under high salt condition. Recently, another member of the SOS3 family, CBL10, has also been found to form a complex with SOS2. This complex is speculated to regulate both extrusion of Na^+ ion (by regulating SOS1) and sequestration/compartmentalization of Na^+ ion into the vacuole (activating the NHX-type transporter which pumps Na^+ ion into the vacuole). SOS1 also interacts with RCD1 and imparts protection against reactive oxygen species and oxidative stress. SOS2 also works to activate H^+-ATPases under salt stress and helps in reinstating ionic homeostasis. For details, refer to the text and Mahajan et al. (2008)

proton-pumping pyrophosphatase to sequester Ca^{2+} in the vacuole (Tuteja and Mahajan 2007). In many plant cells the vacuole occupies more than 50% of the cell volume, and it is evident that *trans* tonoplast Ca^{2+} transport makes a very significant contribution to the regulation of cytosolic Ca^{2+} concentrations.

4 Ca^{2+} Signature

It is now clear that transient increase in $[Ca^{2+}]_{cyt}$ is a universal response to stress and also critical for the production of a physiological response. Perturbation in cytosolic calcium levels (termed "$[Ca^{2+}]_{cyt}$ signature") is unique and results in an appropriate physiological response to a particular stimulus. The uniqueness is manifested in the subcellular location and/or the kinetics of the magnitude of the $[Ca^{2+}]_{cyt}$ perturbation (Rudd and Frankin-Tong 2001). The transient increase in $[Ca^{2+}]_{cyt}$ can be single (spike), double (biphasic), or multiple (oscillations). The perturbations generated may differ in their cellular location, role, and extent of propagation and in their amplitude during propagation. Several abiotic challenges result in an immediate, transient increase in $[Ca^{2+}]_{cyt}$ that is restored to basal levels within minutes. Such changes include mechanical perturbations and rapid cooling for brief periods, termed "cold shock." The duration, periodicity, and amplitude of oscillations vary considerably, and their form is often dependent on the strength and combination of specific stimuli (Allen et al. 1999). Calcium signatures may also be tissue-specific. For example, within the root, the $[Ca^{2+}]_{cyt}$ perturbations induced by mechanical perturbation, salinity, osmotic stress, cold shock, or slow cooling differ markedly between cell types (Moore et al. 2002).

An increase in $[Ca^{2+}]_{cyt}$ is effected by Ca^{2+} influx to the cytosol either from the apoplast, across the plasma membrane, or from the intracellular organelles. The Ca^{2+} influx is mediated by Ca^{2+} permeable ion channels, and their type, cellular localization, and abundance influences the spatial characteristics of $[Ca^{2+}]_{cyt}$ perturbations. Since the diffusion of Ca^{2+} within the cytoplasm is low, and the buffering of Ca^{2+} in the cytoplasm is high (0.1–1 mM) (Trewavas 1999), the opening of the Ca^{2+} channel produces a local increase in $[Ca^{2+}]_{cyt}$ that dissipates rapidly after the channel has been closed. The subcellular localization of the Ca^{2+} channel is therefore critical for the targeting of different cellular processes. Cytosolic calcium "waves" are produced within the cytoplasm by the successive recruitment of particular Ca^{2+} channels to coordinate cellular responses. It was suggested that a local elevation of $[Ca^{2+}]_{cyt}$ might generate soluble second messengers, such as IP_3 or cADPR (cyclic ADP Ribose), that diffuse through the cytoplasm to activate a relay of spatially separated Ca^{2+} channels (Trewavas 1999). This theory was supported by plant cells responding to salt stress (Tuteja and Mahajan 2007). In addition to these subcellular waves, "waves" of cells with high $[Ca^{2+}]_{cyt}$ may also propagate through the plant tissue. This can be induced in root tissues by mechanical stimulation or saline shock, in cotyledons by cold shock, and in leaves by chilling plant roots briefly. Electrical action potentials, osmotic perturbations, or chemical signals may trigger these waves. Although elevated $[Ca^{2+}]_{cyt}$ is necessary for signal transduction, a prolonged increase in $[Ca^{2+}]_{cyt}$ is lethal. Sustained high $[Ca^{2+}]_{cyt}$ is implicated in apoptosis, both during the normal development and in hypersensitive responses to pathogens (Tuteja and Mahajan 2007).

5 Ca^{2+} Memory

There is considerable evidence that $[Ca^{2+}]_{cyt}$ signatures are modified by previous experience. A diminished $[Ca^{2+}]_{cyt}$ elevation upon repetitive stimulation by the same environmental challenge or a developmental cue is a common observation. All these suggested the existence of Ca^{2+} memory in plant cells. The term "memory" was put forward first by Knight et al. (1996). The attenuated response of $[Ca^{2+}]_{cyt}$ after repeated stimulation by various elicitors forms a part of cellular memory and the cells are able to retain the previous information. This memory is significant and helps the cells respond better to a particular stress without disturbing the delicate balance of Ca^{2+} levels. Some examples to support this are as follows:

1. The magnitude of the $[Ca^{2+}]_{cyt}$ perturbation elicited by the wind-induced motion becomes progressively smaller upon repeated stimulation and a refractory period of several minutes is required before a full response is observed again. A second exposure to an elicitor does not influence $[Ca^{2+}]_{cyt}$ for several hours after its initial application (Blume et al. 2000).
2. Plant cells challenged with H_2O_2 fail to respond to H_2O_2 for several hours (Price et al. 1994).
3. There is also evidence that the $[Ca^{2+}]_{cyt}$ signatures elicited by one environmental challenge can be modified by prior exposure to a contrasting one. For example, the magnitude of the $[Ca^{2+}]_{cyt}$ perturbations in response to oxidative stress was reduced by prior exposure to hyperosmotic stress and the opposite was also found to be true. These observations also imply cross talk between the signaling cascades.

6 Calcium-Binding Proteins

In general, Ca^{2+} binding triggers changes in protein shape and charge. Several reports indicate that Ca^{2+} regulates the transcription of target genes by affecting changes in the phosphorylation status of specific transcription factors. ABA-induced expression of two genes in aurea mutants of tomato, *rd-29A*, a desiccation-induced gene, and *kin2*, a cold-responsive gene, is also reported to be Ca^{2+}-regulated. Analysis of transcriptome changes in response to increase in cytosolic Ca^{2+} revealed that ABA response element related sequences also act as calcium-responsive *cis* elements (Kaplan et al. 2006). It seems therefore that Ca^{2+} can control many cellular processes in plant cells by interacting with various proteins and kinases, which results in the regulation of expression of various genes. Recently a novel family of transcription factors that bind CaM, named "CaM-binding transcription activators," have been identified and indicated to be involved in stress-induced calcium-mediated gene expression in *Arabidopsis* (Bouché et al. 2002; Yang and Poovaiah 2002).

An increase level of the cytoplasmic Ca^{2+} in response to signals is sensed by an array of Ca^{2+} sensors, which are small Ca^{2+}-binding proteins that change their conformation in a Ca^{2+}-dependent manner. Specificity in the signaling pathway is provided by the uniqueness in calcium signatures and also by plethora of Ca^{2+} sensors, which can decode the Ca^{2+} perturbations quite precisely. Once Ca^{2+} sensors have decoded the elevated $[Ca^{2+}]_{cyt}$, Ca^{2+} efflux into the cell exterior and/or the sequestration into cellular organelles such as vacuoles, ER, and mitochondria to restores its level to that of the resting state. The EF-hand is the most frequent motif found in Ca^{2+}-binding proteins. There are at least two types of Ca^{2+}-binding proteins, one which contains EF-hand and other which does not contain EF-hand.

6.1 EF-Hand Containing Ca^{2+}-Binding Proteins (Ca^{2+} Sensors)

Most of the Ca^{2+} sensors bind Ca^{2+} using a helix-loop-helix motif termed the "EF hand" (named after the E and F regions of parvalbumin), which binds a single Ca^{2+} ion with high affinity (Tuteja and Mahajan 2007). The Ca^{2+} sensors utilize the side-chain oxygen atoms of the EF-hand motif for Ca^{2+} coordination. In 1973 Kretsinger and Nockolds (1973) first discovered the EF-hand structural motif in the crystal structure of parvalbumin. The properties of EF-hands are described below:

1. The EF-hand motifs are mostly found back-to-back in antipararallel pairs with β-sheet-like hydrogen-bonding occurring between the loops of the coupled sites. These motifs help in the stabilization of the protein structure.
2. The EF-hand is a highly conserved 29 amino acid motif consisting of an α helix E (residues 1–10), a loop (residue 10–21), which binds the Ca^{2+} ion, and a second α-helix F (residues 19–29).
3. The Ca^{2+} ion is coordinated by an oxygen atom or by a bridging water molecule of the side chains (Tuteja and Mahajan 2007).
4. In most of the functional EF-hand motifs, the first amino acid is aspartate and the 12th is glutamate. Glutamate contributes both its side-chain oxygen atoms to the metal ion coordination.
5. Most of the EF-hand proteins are characterized by the relatively high percentage of acidic residues.
6. Several isoforms of an EF-hand protein may exist in a single organism.

The Ca^{2+} binding affinities of the EF-hand protein vary substantially ($K_d = 10^{-4}$–10^{-9}M) and depend on the amino acid sequence of the protein, especially with regard to the 12-residue consensus loop that provides all the acids that directly ligate to Ca^{2+} ions. There are many EF-hand-containing calcium sensors in plants. The major families of Ca^{2+} sensors include CaM, CaM-like proteins, CDPKs, and calcineurin B-like proteins (CBL). These are briefly described next.

6.1.1 Calmodulin

CaM (17 kDa) is a prototypical Ca^{2+}-sensor protein that can control many important biological functions by binding to hundreds of target proteins. It is a highly conserved acidic protein with two globular domains, each containing two EF-hands, connected by a flexible α-helical linker (Luan et al. 2002). CaM is found in the apoplast, the cytosol, the ER and the nucleus of plant cells. The specific biological functions of plant CaM are not well known. Yoo et al. (2005)

isolated a complementary DNA (cDNA) encoding a CaM-binding transcription factor, MYB2, that regulates the expression of salt- and dehydration-responsive genes in *Arabidopsis*. CaM works in a calcium-dependent or calcium-independent manner by binding to and regulating the activity of target proteins called "CaM-binding proteins." Protein phosphatase PP7 is the first protein serine/threonine phosphatase to be found to interact with CaM in plants. PP7 is reported to be involved in thermotolerance in *Arabidopsis* (Liu et al. 2007). Katou et al. (2007) have shown that a CaM-binding mitogen-activated protein kinase phosphatase is induced by wounding and regulates the activities of stress-related mitogen-activated protein kinases in rice. Within the cytosol, the estimated CaM concentration is 5–40 μM (Zielinski 1998; Rudd and Frankin-Tong 2001). The role of CaM has been implicated in many physiological processes, such as those affected by light, gravity, mechanical stress, phytohormones, pathogens, osmotic stress, heat shock, and chilling (Zielinski 1998; Rudd and Frankin-Tong 2001). The structure of CaM revealed that all four EF-hands are saturated by Ca^{2+} ions. CaM appears to be regulatory protein and induces large changes in interhelical angles as Ca^{2+} is bound. The affinity of CaM for Ca^{2+} is influenced by the presence of particular target proteins (Zielinski 1998). CaM can also regulate gene expression by binding to specific transcription factors (Bouché et al. 2002).

6.1.2 CaM-Like Proteins

Plants also possess CaM-like proteins, which differ from CaM in containing more than 148 amino acid residues and have between one and six EF-hand motifs. They possess limited homology to CaM (75% identity) with canonical CaM isoforms (Luan et al. 2002). In *Arabidopsis*, they include CaBP-22, TCH2, TCH3, AtCP1, NADPH oxidases, and Ca^{2+} binding protein phosphatases such as ABI 1 and ABI 2. These proteins have been implicated in cellular responses to diverse environmental, developmental, and pathological challenges.

6.1.3 Ca^{2+}-Dependent Protein Kinases

Five different types of Ca^{2+} regulated protein kinases have been reported in plants. These include (1) CDPKs independent of CaM, (2) CDPK-related protein kinases (CRKs), (3) CaM-dependent protein kinases (CaMKs), (4) Ca^{2+}/CaM-dependent protein kinases (CCaMK), and (5) SOS3/CBL interacting protein kinases (SIPKs/CIPKs).

1. CDPKs independent of CaM. There are at least 34 genes encoding CDPKs in the *Arabidopsis* genome (Cheng et al. 2002) and similar numbers in other plant species. These CDPKs are ubiquitous in plants and generally have four EF-hands at their C-terminus that bind Ca^{2+} and activate the serine/threonine

kinase activity of the enzyme. These kinases require micromolar concentrations of Ca^{2+} for their activity and have no requirement for CaM or lipids. They have a unique structure as the N-terminal protein kinase domain is fused with the C-terminal autoregulatory domain and a CaM-like domain, which has a Ca^{2+}-binding EF-hand or helix-loop-helix motif. The autoinhibitory domain of CDPKs is a 30 amino acid sequence, which acts as a pseudo-substrate (Harper et al. 1994). The N-terminal domain of CDPKs is variable and provides specificity to different CDPK isoforms. These enzymes show several-fold stimulation with Ca^{2+} and show autophosphorylation. The binding of Ca^{2+} to some of CDPKs is modulated by lipids or phosphorylation (Cheng et al. 2002). Ca^{2+} binding to CDPK effects conformation of the kinase and relieves the inhibition caused by the autoinhibitory region. CDPKs are implicated in pollen development, control of cell cycle, phytohormone signaling, light-regulated gene expression, gravitropism, thigmotropism, cold acclimation, salinity tolerance, drought tolerance, and responses to pathogens (Xiong et al. 2002).

2. *CRKs*. CRKs are similar to CDPKs expect that the CaM-like region is poorly conserved with degenerate or truncated EF-hands that may not be able to bind Ca^{2+}. There are at least seven CRKs in the *Arabidopsis* genome, and orthologues of these are present in many plant species. However, the regulation and function of these kinases are not known (Harmon et al. 2001).
3. *CaMKs*. Several CaMKs have been cloned from *Arabidopsis* and other plants. Kinase activity of CaMKs is stimulated by CaM-dependent autophosphorylation and their catalytic activity is also modulated by CaM. They are highly expressed in rapidly growing cells and tissues of the root and flower (Zhang and Lu 2003). Recently, Liu et al. (2008) showed that CaM-binding protein kinase 3 is part of heat shock signal transduction in *Arabidopsis thaliana*.
4. *CCaMKs*. These are a group of Ca^{2+}-dependent kinases which in addition to Ca^{2+} also require CaM for their activity. Thus, CaM besides acting directly could also exert its effect by binding to protein kinases and modulating their activity. A CCaMK was characterized from lily and other plant species (Lu et al. 1996). Sequence analysis revealed the presence of an N-terminal catalytic domain, a centrally located CaM-binding domain, and a C-terminal visinin-like domain containing only three EF-hands. Biochemical studies of CCaMK established that Ca^{2+} and CaM stimulate CCaMK activity. In the absence of CaM, Ca^{2+} promotes autophosphorylation of CCaMK. The phosphorylated form of CCaMK possesses more kinase activity than the nonphosphorylated form. Recently, Jeong et al. (2007) reported a novel *Arabidopsis* CCaMK (AtCK) which is presumably involved in CaM-mediated signaling.
5. *SIPKs/CIPKs*. CBL were found to interact specifically with a class of serine/threonine protein kinases known as CIPKs (Mahajan and Tuteja 2005; Mahajan et al. 2006a, 2008). Recently, a novel CIPK from pea was reported and was found to interact and phosphorylate the pea CBL (Mahajan et al. 2006b). Genetic analysis confirmed that SOS1-SOS3 function in a common pathway of salt tolerance. This pathway also emphasizes the significance of Ca^{2+} signal in reinstating cellular ion homeostasis. The SOS pathway is depicted in Fig. 2.

The increased level of cytosolic Ca^{2+} caused by salt stress is sensed by a calcium sensor such as SOS3, which interacts with SOS2 protein kinase. This SOS3-SOS2 protein kinase complex phosphorylates SOS1, a plasma membrane Na^+/H^+ antiporter, resulting in efflux of excess Na^+ ions (Mahajan et al. 2008). This complex interacts with and influences other salt-mediated pathways, resulting in ionic homeostasis. This complex inhibits HKT1 activity (a low-affinity Na^+ transporter), thus restricting Na^+ entry into the cytosol. SOS2 also interacts and activates NHX (vacuolar Na^+/H^+ exchanger), resulting in sequestration of excess Na^+ ions, further contributing to Na^+ ion homeostasis. In a cell, phosphorylation events are often coupled with dephosphorylation events for the maintenance of homeostasis. SOS2 also interacts with ABI2. It is possible that ABI2 may dephosphorylate the proteins that are phosphorylated by SOS2 in order to restore homeostasis after a stress condition. CAX1 (Ca^{2+}/H^+ antiporter) has been identified as an additional target of SOS2 activity reinstating cytosolic Ca^{2+} homeostasis under salt stress (Fig. 2). Research has led to further elucidation of this classic SOS pathway. SOS1 also has a role in detoxification of ROS as it interacts with RCD1 under salt and oxidative stress (Katiyar-Agarwal et al. 2006). RCD1 is an important transcriptional regulator of oxidative stress responsive genes and has been shown to interact with the C-terminal tail of SOS1, thus clearly emphasizing the involvement of SOS1 in detoxification of reactive oxygen species and prevention of oxidative stress injury (Fig. 2) (Mahajan et al. 2008).

In silico analysis as well as yeast two-hybrid interactions revealed the presence of 25 CIPK genes in the *Arabidopsis* genome and 30 CIPK genes in rice genome (Kolukisaoglu et al. 2004). In contrast to CBL, which are mainly concentrated on chromosomes 4 and 5, the 25 CIPK genes are dispersed among all five *Arabidopsis* chromosomes. In *Arabidopsis* AtCIPK24 is homologous to SOS2. CIPKs do not harbor any decipherable localization signal or any target motif (Kolukisaoglu et al. 2004); therefore, the localization of CIPKs could exclusively be dependent on their respective interaction partner, which would thus serve a dual role as a calcium sensor and as an anchoring protein, regulating the localization and activity of the CIPK at different locations within the cell.

6.1.4 Calcineurin B-Like Proteins

These are a relatively new class of calcium sensors discovered in *Arabidopsis* originally, in the search for the genes imparting salt tolerance and maintaining cellular ion homeostasis (Kudla et al. 1999; Mahajan et al. 2008). Molecular analysis of the salt overlay sensitive mutants opened a new chapter in relation to salt stress signaling that led to the discovery of a pathway that transduces a salt-stress-induced Ca^{2+} signal to reinstate cellular ion homeostasis. The *SOS3* gene, which was identified by Jiang Zhu and colleagues, shares significant sequence homology with *CBL4* from *Arabidopsis* (Mahajan et al. 2006a, 2008). Currently ten CBL and 25 CBLK genes have been reported from *Arabidopsis* (Mahajan et al. 2006a, 2008). The CBL-CIPK network is also widely distributed among higher plants, but except for *Arabidopsis*, the complexity and characterization of this pathway remains largely unrevealed. As different plants vary in their genome complexity, phenotype, species-specific function or functional diversification can be expected. The essential role imparted by CBL-CIPK genes in stress tolerance necessitates their detailed characterization from higher plants. In fact, there has been no report on experimental characterization of CBL from any higher plant except *Arabidopsis*. *AtCBL3*, in particular has been largely overlooked even in *Arabidopsis*. Recently, we have reported the cloning and characterization of a novel CIPK and its interacting partner CBL from pea. Pea CIPK showed autophosphorylation and could phosphorylate pea CBL. Both pea CBL and pea CIPK were found to be coordinately upregulated in response to various stresses such as cold and salinity but were not coordinately upregulated in response to dehydration stress (Mahajan et al. 2006b).

The localization of a Ca^{2+}-sensor protein to a specific compartment of the plant cell plays an important role in decoding the spatially distinct Ca^{2+} signatures. In silico analysis indicates that some structural features specify subcellular localization for these proteins. The CBL harboring the myristoylation sites, i.e., AtCBL1, AtCBL4, AtCBL5, and AtCBL9 have been localized predominantly at the plasma membrane (Liu and Zhu 1998; Kim et al. 2000). Moreover, other CBL lacking this myristoylation motif may be primarily cytosolic. This pattern of localization of CBL and their interacting kinases allows specific decoding of Ca^{2+} signatures, which are differentiated spatially within a given cell (Mahajan et al. 2008).

6.2 Without EF-Hand Ca^{2+}-Binding Proteins

There are several proteins that bind Ca^{2+} but do not contain EF-hand motifs. These include the phospholipase D (PLD), annexins, pistil-expressed Ca^{2+}-binding protein (PCP), calreticulin (CRT), calnaxin, and forisomes.

6.2.1 Phospholipase D

The activity of PLD, which cleaves membrane phospholipids into a soluble head group and phosphatidic acid, is regulated by $[Ca^{2+}]_{cyt}$ through a Ca^{2+}/phospholipids binding site termed the "C2 domain" (Wang 2001). PLD activity is implicated in cellular responses to ethylene and ABA, α amylase synthesis in aleurone cells, stomatal closure, pathogen responses, leaf senescence, and drought tolerance (Wang 2001). Plants possess several PLD isoforms that differ in their affinity for Ca^{2+} and their modulation by phosphoinositides, free fatty acids, and lysolipids (Wang 2001). These biochemical modulators of PLD activity are the substrates or products of phospholipase C, which generates IP_3, diacylglycerol, phospholipase A_2, and diacylglycerol kinase, both of which (substrates or products) are regulated by CaM. It is suggested that $[Ca^{2+}]_{cyt}$ signaling cascades might coordinate the activities of these diverse enzymes to effect specific responses to the environmental stimuli (Ritchie et al. 2002).

6.2.2 Annexins

Annexins are a family of proteins in plants and animals that bind phospholipids in a Ca^{2+}-dependent manner and contain four to eight repeats of approximately 70 amino acids (Clark and Roux 1995). Annexins are encoded by 12 genes in vertebrates and by eight in higher plants. These proteins are involved in organization and function of biological membranes. Although the exact function of plant annexin is not known, annexins are implicated in secretory processes and some have ATPase peroxidase activities. On the basis of existing literature and experimental evidence, Górecka et al. (2007a) have proposed that plant annexins may have a role in stress response. In fact, Górecka et al. (2007b) have shown that annexin At1 of *Arabidopsis thaliana* (AnnAt1), which is one of eight proteins of this family in *Arabidopsis thaliana*, plays important role in pH-mediated cellular response to environmental stimuli. The same group also suggested that the pH-sensitive ion channel activity of AnnAt1 might play a role in intracellular ion homeostasis.

6.2.3 Pistil-Expressed Ca^{2+}-Binding Protein

A 19-kDa novel PCP expressed in anthers and pistil was isolated (Furuyama and Dzelzkalns 1999). PCP is a high-capacity (binds 20mol of Ca^{2+} per mole of PCP), low-affinity Ca^{2+}-binding protein. PCP has been implicated in pollen-pistil interactions and pollen development.

6.2.4 Calreticulin

CRT is an abundant Ca^{2+}-binding protein (Ca^{2+} sensor) and was first detected in the ER of rabbit skeletal muscle (Ostwald and MacLennan 1974). CRT plays a crucial role in many cellular processes, including Ca^{2+} storage and release, protein synthesis, and molecular chaperone activity. Besides its main location in the ER (Opas et al. 1996), CRT has been found to reside in the nuclear envelope (Napier et al. 1995), the spindle apparatus of the dividing cells (Denecke et al. 1995), the plasmodesmata in the root apex (Baluska et al. 1999), and on the cell surface (Gardai et al. 2007), indicating that CRT is essential for normal cell function. In plants it was first detected and isolated from spinach leaves (Menegazzi et al. 1993). Later, cDNA clones of CRT were isolated from *Arabidopsis* (Huang et al. 1993), barley (Chen et al. 1994), maize (Napier et al. 1995), tobacco (Denecke et al. 1995), pea (Hassan et al. 1995), *Brassica rapa* (Lim et al. 1996), *Ricinus communis* L. (Coughlan et al. 1997), and rice (Li and Komatsu 2000). Plant CRT shares the same structural domain features and basic functions identified for animal CRTs; therefore, the plant CRT might also be involved in regulation of Ca^{2+} homeostasis and Ca^{2+}-dependent signal pathways (Wyatt et al. 2002). Plant CRT is highly expressed during mitosis in tobacco (Denecke et al. 1995), embryogenesis of barley (Chen et al. 1994), *Nicotiana plumbaginifolia* (Borisjuk et al. 1998), and maize

(Dresselhaus et al. 1996), and in flower tissues, including pollen tubes as well as anthers (Nardi et al. 2006). Increasing evidence also indicates that this protein is involved in the plant response to a variety of stress-mediated stimuli. Recently, wheat CRT (TaCRT) was shown to be involved in the plant response to drought stress, indicating a potential in the transgenic improvements of plant water-stress (Jia et al. 2008).

6.2.5 Calnexin

Calnexin (CNX) is one of the important ubiquitous calcium-binding proteins, and was initially identified as an ER type I integral membrane protein (Sarwat and Tuteja 2007). It acts as a chaperone that share several functions, including Ca^{2+} binding, lectin-like activity, and recognition of misfolded proteins. CNX binds to monoglucosylated carbohydrate on newly synthesized glycoproteins. It consists of a large (50-kDa) N-terminal calcium-binding lumenal domain, a single transmembrane helix, and a short (90-residue), acidic cytoplasmic tail. CNX together with its teammates, such as ERp57 (a protein disulfide isomerase like protein resident in the ER), and CRT comprise the so-called CNX/CRT cycle which is responsible for the correct folding of newly synthesized proteins and glycoproteins before their translocation to the secretory pathway (Sarwat and Tuteja 2007). Huang et al. (1993) isolated a cDNA encoding a CNX homologue CNX1p from *Arabidopsis thaliana*, a transmembrane protein of type I topology, and it showed 48% identity with dog CNX. Sequences encoding CNX homologues have also been cloned from maize (Kwiatkowski et al. 1995), soybean (Goode et al. 1995), and *Pisum sativum* (Ehtesham et al. 1999). Li et al. (1998) demonstrated the possible role of CNX in folding and assembly of vacuolar H^+-ATPase from oat seedlings along with another ER chaperone, BiP. Pea CNX was shown to be a constitutively expressed 72.5-kDa phosphoprotein. The acidic domain of CNX at the N-terminus present in the ER lumen may be involved in Ca^{2+} binding. *Arabidopsis thaliana* CNX lacks both of the acidic domains; hence, it is unlikely to be involved in low-affinity Ca^{2+} binding. Similarly, pea CNX also lacks the C-terminal acidic domain, but possesses a potential Ca^{2+}-binding domain at the N-terminus (Ehtesham et al. 1999).

6.2.6 Forisomes

Forisomes are giant contractile motor protein bodies specific for the sieve tubes of Fabaceae plants (legume) and function as a gatekeeper in phloem sieve tubes. They are unique protein bodies which act as cellular stopcocks, by undergoing a Ca^{2+}-dependent and ATP-independent conformational switch in which they plug the sieve element (Eckardt 2001; Knoblauch et al. 2001). In living cells, this reaction is probably controlled by Ca^{2+} transporters in the cell membrane. Knoblauch et al. (2003) showed that an influx of calcium into legume sieve elements stimulates the rapid and reversible dispersal of crystalloid P-protein aggregates to occlude sieve plate pores, which may be produced by wounding or pathogen-mediated mechanical injury. Overall, this helps the plant to stop hemorrhaging nutrients and to prevent entry of pathogens; because of this property the name "forisome" has been given to the proteins (Latin *foris* meaning "the wing of a gate," Greek *soma* meaning "body") (Knoblauch et al. 2003). Forisomes also control flux rates in the phloem of faboid legumes by reversibly plugging the sieve tubes. These proteins undergo an anisotropic shape transition (longitudinally expanded to contracted) in response to ion concentration changes (Ca^{2+}, H^{+}, etc.). Forisomes were mostly located close to sieve plates, and occasionally were observed drifting unrestrainedly along the sieve element, suggesting that they might be utilized as internal markers of flow direction. Recently, forisome protein aggregates received broader attention owing to their ability to convert chemical into mechanical energy. The *for1* gene is highly conserved among Fabaceae species and appears to be unique to this phylogenetic lineage since no orthologous genes have been found in other plants, including *Arabidopsis* and rice (Noll et al. 2007). It is hypothesized that forisome plugs are removed once the cytosolic calcium level has returned to the initial level in those sieve tubes (Furch et al. 2007). Peters et al. (2007) isolated tailed forisomes of the sword bean *Canavalia gladiata* and reported that Ca^{2+} induced a sixfold volume increase within about 10–15 s; the reverse reaction following Ca^{2+} depletion proceeded in a fraction of that time. They suggested that *Canavalia gladiata* provides a superior experimental system, which will prove indispensable in physiological, biophysical, ultrastructural, and molecular studies on the unique ATP-independent contractility of forisomes.

7 Conclusions

Evolution has adopted positively charged calcium as the primary signaling element of cells. Ca^{2+} is one of the principal second messengers for functioning as a central node in the overall "signaling web" and plays an important role in providing stress tolerance to plants. Analysis of $[Ca^{2+}]_i$ dynamics has demonstrated its signaling role in plant cells in response to a wide array of environmental cues. In general, the stress leads to increased cytosolic Ca^{2+}, which initiates the stress signal transduction pathways for the stress tolerance. Ca^{2+} is the most tightly regulated ion within all membrane-bound organisms and binds to several proteins to effect changes in localization, association, and function. It is now clear that Ca^{2+} signaling affects almost every aspect of the cellular metabolism of living organisms. In fact, an indirect increase in $[Ca^{2+}]_{cyt}$ in plant cells is mediated by the increase in intracellular concentrations of other second messengers such as cyclic GMP, cyclic AMP, and IP_3, in response to stimuli. The elevation of Ca^{2+} concentration in cytoplasm is found to be a key event in the plant cell for transduction of various signals to a biological effect. This suggests that Ca^{2+} is a communication point in the cytoplasm for the cross talk between different signal transduction pathways and finally helps in the interchanging of the information. The Ca^{2+} signaling involves many sensor proteins that decode temporal and spatial changes in cellular Ca^{2+} concentration. Cells usually invest much of their energy to effect changes in Ca^{2+} concentration. Recently, in yeast it has been reported that there is cross talk between Ca^{2+}-calcineurin signaling, nutrient sensing, and regulation of the cell cycle (Zhang and Rao 2008). Overall, the involvement of Ca^{2+} in various metabolic processes in plant might have general implications.

Acknowledgements I thank Renu Tuteja for critical reading of and corrections to the article. This work was partially supported by the grant from the Department of Biotechnology (DBT), Government of India. I apologize to the many scientists whose work I was not able to credit owing to space restrictions. In most cases, reviews have been cited at the expense of the original work.

References

Allen GJ, Kwak JM, Chu SP, Llopis J, Tsien RY, Harper JF, Schroeder JI (1999) Cameleon calcium indicator reports cytoplasmic calcium dynamics in *Arabidopsis* guard cells. Plant J 19:735–747

Axelsen KB, Palmgren MG (2001) Inventory of the superfamily of P-type ion pumps in *Arabidopsis*. Plant Physiol 126:696–706

Baluska FS, Amaj J, Napier R, Volkmann D (1999) Maize calreticulin localizes preferentially to plasmodesmata in root apex. Plant J 19:481–488

Blume B, Nürnberger T, Nass N, Scheel D (2000) Receptor-mediated increase in cytoplasmic free calcium required for activation of pathogen defense in parsley. Plant Cell 12:1425–1440

Borisjuk N, Sitailo L, Adler K, Malysheva L, Tewes A, Borisjuk L, Manteuffel R (1998) Calreticulin expression in plant cells: developmental regulation, tissue specificity and intracellular distribution. Planta 206:504–514

Bouché N, Scharlat A, Snedden W, Bouchez D, Fromm H (2002) A novel family of calmodulin-binding transcription activators in multicellular organisms. J Biol Chem 277:21851–21861

Bush DS (1995) Calcium regulation in plant cells and its role in signaling. Ann Rev Plant Physiol Plant Mol Biol 46:95–122

Chen F, Hayes PM, Mulrooney DM, Pan A (1994) Identification and characterization of cDNA clones encoding plant calreticulin in barley. Plant Cell 6:835–843

Chena X, Lina, WH, Wanga Y, Luana S, Xuea HW (2008) An inositol polyphosphate 5-phosphatase functions in PHOTOTROPIN1 signaling in Arabidopsis by altering cytosolic Ca^{2+} [W]. Plant Cell 20:353–366

Cheng SH, Willmann MR, Chen HC, Sheen J (2002) Calcium signaling through protein kinases. The *Arabidopsis* calcium-dependent protein kinase gene family. Plant Physiol 129:469–485

Clark GB, Roux SJ (1995) Annexins of plant cells. Plant Physiol 109:1133–1139

Coughlan SJ, Hastings C, Winfrey RW Jr (1997) Cloning and characterization of the calreticulin gene from *Ricinus communis* L. Plant Mol Biol 34:897–911

Denecke J, Carlsson LE, Vidal S, Hoglund AS, Ek B, van Zeijl MJ, Sinjorgo KM, Palva ET (1995) The tobacco homolog of mammalian calreticulin is present in protein complexes in vivo. Plant Cell 7:391–406

Denis V, Cyert MS (2002) Internal Ca^{2+} release in yeast is triggered by hypertonic shock and mediated by a TRP channel homologue. J Cell Biol 156:29–34

Dresselhaus T, Hagel C, Lorz H, Kranz E (1996) Isolation of a full-length cDNA encoding calreticulin from a PCR library of in vitro zygotes of maize. Plant Mol Biol 31:23–34

Eckardt NA (2001) A calcium-regulated gatekeeper in phloem sieve tubes. Plant Cell 13:989–992

Ehtesham NZ, Phan TN, Gaikwad A, Sopory SK, Tuteja N (1999) Calnexin from *Pisum sativum*: cloning of the cDNA and characterization of the encoded protein. DNA Cell Biol 18:853–862

Furch AC, Hafke JB, Schulz A, van Bel AJ (2007) Ca^{2+}-mediated remote control of reversible sieve tube occlusion in *Vicia faba*. J Exp Bot 58:2827–2838
Furuyama T, Dzelzkalns VA (1999) A novel calcium-binding protein is expressed in *Brassica pistils* and anthers late in flower development. Plant Mol Biol 39:729–737
Garciadeblas B, Benito B, Rodríguez-Navarro A (2001) Plant cells express several stress calcium ATPases but apparently no sodium ATPase. Plant Soil 235:181–192
Gardai SJ, McPhillips KA, Frasch SC, Janssen WJ, Starefeldt A, Murphy-Ullrich JE, Furch AC, Hafke JB, Schulz A, van Bel AJ (2007) Ca^{2+}-mediated remote control of reversible sieve tube occlusion in *Vicia faba*. J Exp Bot 58:2827–2838
Goode JH, Settlage SB, Wilson RF, Dewey RE (1995) Isolation of a calnexin homolog from developing soybean seeds. Plant Physiol 108:1341
Górecka KM, Trebacz K, Górecki R, Pikula S (2007a) Participation of annexin At1 in plant response to abiotic stress. Postepy Biochem 53:154–158
Gorecka KM, Thouverey C, Buchet R, Pikula S (2007b) Potential role of annexin AnnAt1 from *Arabidopsis thaliana* in pH-mediated cellular response to environmental stimuli. Plant Cell Physiol 48:792–803
Harada A, Shimazaki K (2007) Phototropins and blue light-dependent calcium signaling in higher plants. Photochem Photobiol 83:102–111
Harmon AC, Gribskov M, Gubrium E, Harper JF (2001) The CDPK superfamily of protein kinases. New Phytol 151:175–183
Harper JF (2001) Dissecting calcium oscillators in plant cells. Trends Plant Sci 6:395–397
Harper JF, Huang JF, Lloyd SJ (1994) Genetic identification of an auto inhibitor in CDPK, a protein kinase with a calmodulin like domain. Biochemistry 33:7278–7287
Hassan AM, Wesson C, Trumble WR (1995) Calreticulin is the major Ca^{2+}-storage protein in the endoplasmic reticulum of the pea plant (*Pisum sativum*). Biochem Biophys Res Commun 211:54–59
Hirschi K (2001) Vacuolar H^+/Ca^{2+} transport: who's directing the traffic? Trends Plant Sci 6:100–104
Huang L, Franklin AE, Hoffman NE (1993) Primary structure and characterization of an Arabidopsis thaliana calnexin-like protein. J Biol Chem 268:6560–6566
Hwang I, Sze H, Harper JF (2000) A calcium-dependent protein kinase can inhibit a calmodulin-stimulated Ca^{2+} pump (ACA2) located in the endoplasmic reticulum of *Arabidopsis*. Proc Natl Acad Sci U S A 97:6224–6229
Jeong JC, Shin D, Lee J, Kang CH, Baek D, Cho MJ, Kim MC, Yun DJ (2007) Isolation and characterization of a novel calcium/calmodulin-dependent protein kinase, AtCK, from Arabidopsis. Mol Cells 24:276–282
Jia X-Y, Xu C-Y, Jing R-L, Li R-Z, Mao X-G, Wang J-P, Chang X-P (2008) Molecular cloning and characterization of wheat calreticulin (CRT) gene involved in drought stressed responses. J Exp Bot 59:739–751
Kaplan B, Davydov O, Knight H, Galon Y, Knight MR, Fluhr R, Fromm H (2006) Rapid transcriptome changes induced by cytosolic Ca^{2+} transients reveal ABRE-related sequences as Ca^{2+}-responsive *cis* elements in Arabidopsis. Plant Cell 18:2733–2748
Katou S, Kuroda K, Seo S, Yanagawa Y, Tsuge T, Yamazaki M, Miyao A, Hirochika H, Ohashi Y (2007) A calmodulin-binding mitogen-activated protein kinase phosphatase is induced by wounding and regulates the activities of stress-related mitogen-activated protein kinases in rice. Plant Cell Physiol 48:332–344
Katiyar-Agarwal S, Zhu j, Kim K, Agarwal M, Fu X, Huang A, Zhu JK (2006) The plasma membrane Na+/H+ antiporter SOS1 interacts with RCD1 and functions in oxidative stress tolerance in Arabidopsis. Proc Natl Acad Sci U S A 103:18816–18821
Kim KN, Cheong YH, Gupta R, Luan S (2000) Interaction specificity of *Arabidopsis* calcineurine B-like calcium sensor and their target kinases. Plant Physiol 124:1844–1853
Knight H, Trewavas AJ, Knight MR (1996) Cold calcium signaling in *Arabidopsis* involves two cellular pools and a change in calcium signature after acclimation. Plant Cell 8:489–503
Knoblauch M, Peters WS, Ehlers K, Van Bel AJE (2001) Reversible calcium-regulated stopcocks in legume sieve tubes. Plant Cell 13:1221–1230

Knoblauch M, Noll GA, Müller T, Prüfer D, Schneider-Hüther I, Scharner D, Van Bel AJE, Peters WS (2003) ATP-independent contractile proteins from plants. Nat Mater 2:573–574
Kolukisaoglu U, Weinl S, Blazevic D, Batistic O, Kudla J (2004) Calcium sensors and their interacting protein kinases: genomics of the *Arabidopsis* and rice CBL–CIPK signaling networks. Plant Physiol 134:43–58
Kretsinger RH, Nockolds CE (1973) Carp muscle calcium-binding protein. II. Structure determination and general description. J Biol Chem 248:3313–3326
Kudla J, Xu Q, Harter K, Gruissem W, Luan S (1999) Genes for calcineurin B-like proteins in *Arabidopsis* are di.erentially regulated by stress signals. Proc Natl Acad Sci U S A 96:4718–4723
Kwiatkowski BA, Zielinkska-Kwiatkowska AG, Migdalski A, Kleczkowski LA, Wasilweska LD (1995) Cloning of two cDNAs encoding calnexin-like proteins from maize (*Zea mays*) leaves: identification of potential calcium-binding domains. Gene 65:219–222
Li Z, Komatsu S (2000) Molecular cloning and characterization of calreticulin, a calcium-binding protein involved in the regeneration of rice cultured suspension cells. Eur J Biochem 267:737–745
Li W, Llopis J, Whitney M, Zlokarnik G, Tsien RY (1998) Cell permeant caged InsP3 ester shows that Ca^{2+} spike frequency can optimize gene expression. Nature 392:936–941
Lim CO, Kim HY, Kim MG, Lee SI, Chung WS, Park SH, Hwang I, Cho MJ (1996) Expressed sequence tags of Chinese cabbage flower bud cDNA. Plant Physiol 111:577–588
Liu J, Zhu JK (1998) A calcium sensor homolog required for plant salt tolerance. Science 280:1943–1945
Liu HT, Li GL, Chang H, Sun DY, Zhou RG, Li B (2007) Calmodulin-binding protein phosphatase PP7 is involved in thermotolerance in Arabidopsis. Plant Cell Environ 30:156–64
Liu HT, Gao F, Li G-L, Han J-L, Liu De-L, Sun Da-Y, Zhou R-G (2008) The calmodulin-binding protein kinase 3 is part of heat shock signal transduction in *Arabidopsis thaliana*. Plant J 55:760–773
Lu YT, Hidaka H, Feldman LJ (1996) Characterization of a calcium/calmodulin protein kinase homologue from maize roots showing light-regulated gravitropism. Planta 199:18–24
Luan S, Kudla J, Rodríguez-Concepción M, Yalovsky S, Gruissem W (2002) Calmodulins and calcineurin B-like proteins: calcium sensors for specific signal response coupling in plants. Plant Cell 14:S389–S400
Mahajan S, Tuteja N (2005) Cold, salinity and drought stresses: an overview. Arch Biochem Biophys 444:139–158
Mahajan S, Sopoy SK, Tuteja N (2006a) CBL–CIPK paradigm: role in calcium and stress signaling in plants. Proc Indian Natl Sci Acad U S A 72:63–78
Mahajan S, Sopoy SK, Tuteja N (2006b) Cloning and characterization of CBL–CIPK signaling components from a legume (*Pisum sativum*) FEBS J 273:907–925
Mahajan S, Pandey GK, Tuteja N (2008) Calcium- and salt-stress signaling in plants: Shedding light on SOS pathway. Arch Biochem Biophys 471:146–158
Martinez-Noel G, Tognetti J, Nagaraj V, Wiemken A, Pontis H (2006) Calcium is essential for fructan synthesis induction mediated by sucrose in wheat. Planta 225:183–191
Menegazzi P, Guzzo F, Baldan B, Mariani P, Treves S (1993) Purification of calreticulin-like protein(s) from spinach leaves. Biochem Biophy Res Commun 190:130–1135
Moore CA, Bowen HC, Scrase-Field S, Knight MR, White PJ (2002) The deposition of suberin lamellae determines the magnitude of cytosolic Ca^{2+} elevations in root endodermal cells subjected to cooling. Plant J 30:457–466
Napier RM, Trueman S, Henderson J, Boyce JM, Hawes C, Fricker MD, Venis MA (1995) Purification, sequencing and functions of calreticulin from maize. J Exp Bot 46:1603–1613
Nardi MC, Feron R, Navazio L, Mariani P, Pierson E, Wolters-Arts M, Knuiman B, Mariani C, Derksen J (2006) Expression and localization of calreticulin in tobacco anthers and pollen tubes. Planta 223:1263–1271
Noll GA, Fontanellaz ME, Rüping B, Ashoub A, van Bel AJ, Fischer R, Knoblauch M, Prüfer D (2007) Spatial and temporal regulation of the forisome gene for1 in the phloem during plant development. Plant Mol Biol 65:285–294

Opas M, Szewczenko-Pawlikowski M, Jass GH, Mesaeli N, Michalak M (1996) Calreticulin modulates cellular adhesiveness via regulation of expression of vinculin. J Cell Biol 135:1–11
Ostwald TJ, MacLennan DH (1974) Isolation of a high affinity calcium-binding protein from sarcoplasmic reticulum. J Biol Chem 249:974–979
Pardo JM, Reddy MP, Yang S, Maggio A, Huh GH, Matsumoto T, Coca MA et al (1998) Stress signaling through Ca^{2+}/calmodulin-dependent protein phosphatase calcineurin mediates salt adaptation in plants. Proc Natl Acad Sci U S A 95:9681–9686
Peters WS, Knoblauch M, Warmann SA, Schnetter R, Shen AQ, Pickard WF (2007) Tailed forisomes of *Canavalia gladiata*: a new model to study Ca^{2+}-driven protein contractility. Ann Bot 100:101–109
Poovaiah BW, Reddy ASN (1993) Calcium and signal transduction in plants. Crit Rev Plant Sci 12:185–211
Price AH, Taylor A, Ripley SJ, Griffiths A, Trewavas AJ, Knight MR (1994) Oxidative signals in tobacco increase cytosolic calcium. Plant Cell 6:1301–1310
Rasi-Caldogno F, Carnelli A, DeMichelis MI (1995) Identification of the plasma membrane Ca^{2+} ATPase and of its autoinhibitory domain. Plant Physiol 108:105–113
Ritchie SM, Swanson SJ, Gilroy S (2002) From common signalling components to cell specific responses: insights from the cereal aleurone. Physiol Plant 115:342–351
Rudd JJ, Franklin-Tong VE (2001) Unravelling response-specificity in Ca^{2+} signalling pathways in plant cells. New Phytol 151:7–33
Sanders D, Pelloux J, Brownlee C, Harper JF (2002) Calcium at the crossroads of signaling. Plant Cell 14(Suppl):S401–S417
Sarwat M, Tuteja N (2007) Calnexin: a versatile calcium binding integral membrane-bound chaperone of endoplasmic reticulum. Calcium Binding Proteins 2:36–50
Tang R-H, Han S, Zheng H, Cook CW, Choi CS, Woerner TE, Jackson RB, Zhen-Ming Pei1 Z-M (2007) Coupling diurnal cytosolic Ca^{2+} oscillations to the CAS IP3 pathway in Arabidopsis. Science 315:1423–1426
Trewavas A (1999) Le calcium, c'est la vie: calcium makes waves. Plant Physiol 120:1–6
Tuteja N (2007) Mechanisms of high salinity tolerance in plants. Methods Enzymol 428:419–438
Tuteja N, Mahajan S (2007) Further characterization of calcineurin B-like protein and its interacting partner CBL-interacting protein kinase from *Pisum sativum*. Plant Signal Behav 2:358–361
Wang X (2001) Plant phospholipases. Annu Rev Plant Physiol Plant Mol Biol 52:211–231
Wyatt SE, Tsou PL, Robertson D (2002) Expression of the high capacity calcium-binding domain of calreticulin increases bioavailable calcium stores in plants. Transgen Res 11:1–10
Xiong L, Schumaker KS, Zhu JK (2002) Cell signaling during cold, drought, and salt stress. Plant Cell 14:S165–S183
Xiong TC, Bourque S, Lecourieux D, Amelot N, Grat S, Briere C, Mazars C, Pugin A, Ranjeva R (2006) Calcium signaling in plant cell organelles delimited by a double membrane. Biochim Biophys Acta 1763:1209–1215
Yang T, Poovaiah BW (2002) A calmodulin-binding/CGCG box DNA-binding protein family involved in multiple signaling pathways in plants. J Biol Chem 277:45049–45058
Yoo JH, Park CY, Kim JC, Heo WD, Cheong MS, Park HC, Kim MC, Moon BC, Choi MS, Kang YH, Lee JH, Kim HS, Lee SM, Yoon HW, Lim CO, Yun DJ, Lee SY, Chung WS, Cho MJ (2005) Direct interaction of a divergent CaM isoform and the transcription factor, MYB2, enhances salt tolerance in arabidopsis. J Biol Chem 280:3697–3706
Zhang L, Lu YT (2003) Calmodulin-binding protein kinases in plants. Trends Plant Sci 8:123–127
Zhang Y-Q, Rao R (2008) A spoke in the wheel: Calcium spikes disrupt yeast cell cycle. Cell Cycle 7:870–873.
Zielinski RE (1998) Calmodulin and calmodulin-binding proteins in plants. Annu Rev Plant Physiol Plant Mol Biol 49:697–725

MAPK Signaling

Miroslav Ovečka and Irene K. Lichtscheidl

Abstract Plant adaptation to environmental and endogenous factors is based on the effective reaction to signals perceived by specialized plant sensing systems, and generation of relevant responses. Passing of perceived extracellular signals to response-generating effectors requires effective intracellular signal transduction pathways. Mitogen-activated protein kinase (MAPK) cascades are conserved transducers of extracellular signals operating in higher plants. They are involved in the regulation of diverse cellular processes occurring during the whole ontogenetic plant life and ranging from normal cell proliferation to stress-inducing plant-to-environment interactions. The complexity of plant MAPK signaling networks is described in the light of recent findings achieved in plant response to biotic factors, abiotic stress signals, and morphogenic events relevant to plant growth and development.

1 Introduction

Plant growth and development are under the continuous influence of different environmental and endogenous factors during the whole period of ontogenesis that plants must effectively react to. Inability to avoid and prevent their negative influence generally leads to devastating effects on basic metabolism, transport processes, membranes, and cell structure. Such evolutionary force provoked the evolution of a number of

M. Ovečka (✉)
Institute of Botany, Slovak Academy of Sciences, Dubravska cesta 14, 845 23, Bratislava, Slovak Republic
and
Instituto de Agrobiotecnología, Universidad Pública de Navarra
Gobierno de Navarra and Consejo Superior de Investigaciones Científicas, Ctra. de Mutilva s/n, 31192, Mutilva Baja, Navarra, Spain
e-mail: botuove@savba.sk

I.K. Lichtscheidl
Institution of Cell imaging and Ultrastructure Research, University of Vienna, Althanstrasse 14, 1090, Vienna, Austria

F. Baluška and S. Mancuso (eds.), *Signaling in Plants,* Signaling and Communication in Plants,
DOI: 10.1007/978-3-540-89228-1_3, © Springer-Verlag Berlin Heidelberg 2009

different response mechanisms that protect plants at the molecular, cellular, organ, plant, and population levels. Cellular responses induced by environmental stimuli are composed of a cascade of events, starting with perception and molecular recognition of the external stimulus, followed by generation of a signal and its transmittance via specific signaling pathways, and leading further to the activation of defined effectors responsible for the generation of appropriate molecular reactions (Soropy and Munshi 1998; Møller and Chua 1999). The results of these fast and specific reactions are defense, protection, or altered development of the plant.

One line of signal transduction mechanisms is modulated by protein phosphorylation and dephosphorylation that is controlled by protein kinases and protein phosphatases. This fast and effective signaling system fulfills all requirements needed to interconnect the sensing system with gene expression processes in coordination with the physiological status of the cell. One of the most conserved and best characterized protein kinase signaling pathways is mitogen-activated protein kinase (MAPK) cascade. Upon signal perception, it mediates the transmission of those signals that control numerous cellular processes, including cell division, differentiation, and responses to environmental stresses.

The study of signaling in plants is a fast developing area of plant biology, supplying constantly new exciting data to the incomplete picture of MAPK-based signal transduction that we have so far. Comprehensive analysis of the state of the art is summarized in reviews published in the last few years (Bögre et al. 2000; Wrzaczek and Hirt 2001; Tena et al. 2001; Jonak et al. 2002; Šamaj et al. 2004a, b; Nakagami et al. 2005; Colcombet and Hirt 2008). Our aim is to briefly summarize some recent achievements in the field of plant MAPK signaling that shed more light on stress-related and morphological adaptations in plant development.

2 Principle of the MAPK Signaling Cascade

Activation of MAPKs is spatially and temporally performed in modules. MAPK modules are universal transducers of extracellular signals in all eukaryotic cells. The basic module of a MAPK pathway is composed of three functionally interlinked protein kinases (MAPK; MAPK kinase, MAPKK; and MAPKK kinase, MAPKKK). MAPKs have been categorized within a large family of serine/threonine protein kinases, depending on the amino acids that are phosphorylated during activation. The activation of MAPK requires phosphorylation of both threonine and tyrosine residues in a so-called T-X-Y activation loop in subdomain VIII of the catalytic kinase domain. The dual mode of activation is unique and phosphorylation of both tyrosine and threonine residues is required for the full activation of MAPKs. MAPK phosphorylation is catalyzed by a dual-specificity MAPKK. The MAPKKs are activated by phosphorylation of their threonine and serine residues by MAPKKKs (Seger and Krebs 1995; Robinson and Cobb 1997).

After perception of a signal at the plasma membrane, the cytoplasmic phase of the signaling pathway includes transmittance, intensification, and targeting of the signal. MAPKs can activate a number of proteins with regulatory function. The substrate specificity of the activated MAPK pathway as well as the existence of several synergistic and/or antagonistic MAPK pathways activated upon different external cues mediate effective cell responses. They are specific for signals from the activated MAPK pathway and may involve activation of transcription factors, cytoplasmic enzymes, and/or cytoskeletal proteins. Cells hence have efficient control of gene expression, metabolism, cytoskeletal structure, and dynamics, as well as of other cell components with regulatory functions (Garrington and Johnson 1999; Bögre et al. 2000; Jonak et al. 2002; Nakagami et al. 2005).

The mechanisms of the specificity in different cell types and in response to different stimuli are partially based on the proper localization and accessibility of MAPKs within the cell. A special class of interacting proteins (anchoring and scaffold proteins) brings together specific kinases for selective activation, and they also control the subcellular localization of signaling complexes within the cell (Garrington and Johnson 1999). The scaffolding mechanism is responsible for the specificity of MAPK signaling by creating multienzyme complexes composed of several signal molecules in certain compartments. This physical grouping favors rapid passage of the signal through the cascade and prevents unwanted crosstalk (Whitmarsh and Davis 1998). In addition, the scaffold proteins can promote the pathway specificity by presenting preferred substrates in locally high concentration, and by discriminating between single pathways that share common signaling components (Harris et al. 2001; Nakagami et al. 2005).

The termination of the MAPK activated period is an integral part of a signal transduction wave triggered by one stimulus. The inactivation of MAPKs is catalyzed by several types of protein phosphatases: serine/threonine-specific phosphoprotein phosphatases and metal-ion-dependent protein phosphatases, phosphotyrosine phosphatases, and dual-specificity phosphatases (Farkas et al. 2007). Such dephosphorylation of MAPKs is necessary for resetting the pathway to the basal level, when a new input can activate the pathway again. This feedback inhibition of the signaling cascade delineates the MAPK activation usually as a transient signal transduction process (Luan 1998; Brown et al. 1997). Serine/threonine protein phosphatases reversibly dephosphorylate MAPKs at their serine and threonine residues. They are categorized in two groups: type 1 (PP-1) and type 2 (PP-2A, PP-2B, PP-2C), according to their substrate specificity, ion requirements, and sensitivity to inhibitors (Soropy and Munshi 1998; Cohen 1989).

With the completion and analysis of the *Arabidopsis* genome, a more complex view of MAPK functioning, homology, and nomenclature in plants is possible (Wrzaczek and Hirt 2001; MAPK Group 2002; Hamel et al. 2006). In addition, completion of the genome sequences for rice and poplar makes it possible to examine phylogenetic relationships of *Arabidopsis*, poplar, and rice MAPK genes. On the basis of this project, 20 *Arabidopsis* members, 21 poplar homologs, and 15 rice homologs were found at the level of MAPKs. At the level of MAPKKs, the list contains ten *Arabidopsis* members, nine poplar homologs, and eight rice homologs (Hamel et al. 2006; Sterck et al. 2005; Liu and Xue 2007; Agrawal et al. 2003).

3 MAPK Activation by Pathogenic Attacks

For interaction of plants with pathogens, plant cell surface receptors play a pivotal role. Sensing of pathogens is based on the detection of pathogenic molecules, either so-called pathogen-associated molecular patterns (PAMPs) or microbe-associated molecular patterns (Tena et al. 2001; Nakagami et al. 2005). This recognizing step is followed by the activation of a range of events leading to alteration of gene expression, where MAPK cascades function as a converging module (Nühse et al. 2000; Asai et al. 2002; Nurnberger et al. 2004).

Considerable progress in functional studies of the MAPK involvement in plant pathogen response comes from studies with bacterial or yeast elicitors. An example is the activation of MAPKs in *Arabidopsis* after treatment with bacterial flagellin peptide. Identification of somatic embryogenesis receptor kinase (SERK) 3/brassinosteroid-associated kinase (BAK) 1, a member of the SERK family (Russinova et al. 2004; Shiu and Bleecker 2001), shed more light on the perception mechanism. *Arabidopsis* AtSERK3/BAK1 is required for early PAMP responses because it rapidly forms an elicitor-induced complex with the receptor for the bacterial flagellin peptide flg22, called "flagellin sensing 2" (FLS2). In mutated *At-SERK3/BAK1* plants responses to flg22 are considerably reduced (Heese et al. 2007). Specific physical interaction of flg22 with the FLS2 protein has been demonstrated by immunoprecipitation with FLS2-specific antibodies and chemical cross-linking with a radiolabeled derivative of the flg22 in *Arabidopsis* (Chinchilla et al. 2006). AtSERK3/BAK1 belongs to the family of receptor-like kinases, that contain, in addition to extracellular domains and transmembrane span regions, a cytoplasmic domain possessing conserved kinase motifs (Shiu and Bleecker 2001; Shiu et al. 2004). AtSERK3/BAK1 thus may interlink signal perception of PAMP with downstream signal transduction components.

Upon flg22 treatment, transient activation of AtMPK4, AtMPK6, and AtMPK3 was observed. The transient expression system allowed for identification of upstream MAPKKs (AtMKK1, AtMKK4, and AtMKK5) and an upstream MAPKKK (AtMEKK1) (Asai et al. 2002; Nühse et al. 2000; Mészáros et al. 2006). The activation of this MAPK module is induced by the flagellin receptor FLS2, leucine-rich repeat receptor kinase. The downstream targets of activated AtMPK6 and AtMPK3 are WRKY transcription factors (Asai et al. 2002). Interestingly, in addition to signaling functions based on the phosphorylation kinase activity, a structural role for MEKK1 in pathogenesis-associated signaling was suggested, which would be independent of its protein kinase activity (Suarez-Rodriguez et al. 2007). MEKK1 is involved also in regulation of redox homeostasis in *Arabidopsis*; MEKK1-deficient plants accumulate reactive oxygen species (ROS) and are altered in hormone signaling (Nakagami et al. 2006).

The role of the MKK3 pathway in pathogen defense has been documented using *Pseudomonas syringae* (Dóczi et al. 2007). Gene expression analysis revealed a basal expression level in nonattacked plants. Upon infection by *Pseudomonas syringae* pv. *tomato* DC3000 the expression of MKK3 was strongly induced, which was not the case after application of abiotic stresses (Dóczi et al. 2007).

Triggering of defense-induced MAPK signaling events has been described also in plant-*Agrobacterium* interactions. *Agrobacterium*-mediated T-DNA transfer induces the activation of the *Agrobacterium*-induced MAPK MPK3 which directly acts on the transcription factor VIP1 and causes its relocation from the cytoplasm to the nucleus for the expression of defense genes (Djamei et al. 2007). In this case, however, the plant defense response seems to be overruled by agrobacteria for their own advantage: agrobacteria use VIP1 as a T-DNA shuttle to import their genes into the host nucleus.

All these data indicate the importance of MAPK signaling for the activation of plant defense responses against pathogenic bacteria and fungi (Ramonell et al. 2005; Zeidler et al. 2004; Zipfel et al. 2004). However, the evolutionary efforts of plants to develop effective defense mechanisms leading to plant innate immunity was accompanied by effective efforts of bacteria to avoid them. Pathogenic bacteria utilize distinct type III virulence effectors that can help them to manipulate host MAPK signaling cascades and to promote their own pathogenicity (Shan et al. 2007). Some of them, recently identified, such as AvrPto and AvrPtoB from *Pseudomonas syringae*, are potent suppressors of MAPK activation. It was shown in *Arabidopsis* that AvrPto and AvrPtoB suppressed specifically MAPK activation and early defense signaling events by flg22 upstream of MAPKKK. Hence, the activation of involved MAPKs, MPK3 and MPK6, by flg22 was completely blocked (He et al. 2006).

4 Involvement of MAPKs in Abiotic Stress Responses

The general utilization of plant MAPK cascades as signal transduction pathways for various stresses (Nakagami et al. 2005) reveals the exciting phenomenon that some MAPK cascade components are activation-specific to particular stress factors, while others are "promiscuous." An obvious example is the parallelism and cross-talk of biotic and abiotic signaling pathways; MAPK modules can function as a bridge interconnecting upstream signaling steps with appropriate downstream effectors with high selectivity or flexibility. Accumulating data suggest that hormone and ROS signaling pathways play key roles in the crosstalk between biotic and abiotic stress signaling (Apel and Hirt 2004; Torres and Dangl 2005; Fujita et al. 2006). MPK6, for instance, is involved in the production of ethylene (Liu and Zhang 2004), it is activated by oxidative stress (Xiong and Yang 2003), together with MPK3 it is activated in the ROS signaling pathway (Rentel et al. 2004), and a MPK4/MPK6 cascade has been shown to function during cold and salt stress signaling (Teige et al. 2004; Ichimura et al. 2000).

Temperature stress, such as low and high temperature, leads to direct inhibition of metabolic activity. For sensing of cold stress membrane rigidification may be sufficient to increase Ca^{2+} concentration in the cytosol, hence changing cytoskeleton stability and boosting the abscisic acid (ABA) and ROS production. Effective cold signaling, however, requires the activation of MAPK cascades (Sangwan et al. 2002; Zhu et al. 2007; Fujita et al. 2006). MKK2 was shown to be activated by cold stress in *Arabidopsis thaliana* together with the downstream components MPK4 and MPK6;

MKK2 overexpression constitutively upregulated MPK4 and MPK6 activity, resulting in salt and freezing tolerance, while *mkk2* mutant plants were salt- and cold-stress-hypersensitive (Teige et al. 2004). In addition to cold stress signaling, temperature-responsive MAPKs are involved in the defense against heat stress. In alfalfa, a heat-shock-activated MAPK has been identified with the capacity to be activated at 37°C. At 25°C it is normally inactive, but interestingly it was activated when cells were transferred from 4 to 25°C. Stress-activated MAPK (SAMK), another alfalfa MAPK that is normally activated by cold (Jonak et al. 1996), was activated when cells were transferred from 37 to 25°C (Sangwan et al. 2002; Sangwan and Dhindsa 2002). This clearly indicates that the relative temperature shift rather than the absolute temperature generates a signal for activation of the responsive pathway.

Plant responses to other abiotic stresses such as water stress, osmotic stress, or wounding are closely related to the signaling roles of plant hormones ABA and jasmonic acid (JA) (Denekamp and Smeekens 2003). T-DNA insertional mutation of the *Arabidopsis MKK9* gene gives no phenotype in control conditions; however, the mutant is salt- and mannitol-insensitive. Because germination of *mkk9* mutant plants is insensitive to low concentrations of ABA, it is apparent that MKK9 acts as a negative regulator of the abiotic stress response (Alzwiy and Morris 2007).

AtMPK1 and AtMPK2 are two MAPKs responding to mechanical injury. They are also activated by the stress signals ABA and H_2O_2 in combination with wounding, while JA is able to activate them in the absence of wounding, and the induction of AtMPK1/AtMPK2 kinase activity by wounding and JA was not prevented in the JA-insensitive *coi1* mutant (Ortiz-Masia et al. 2007).

Activation of alfalfa stress-induced MAPK (SIMK) by moderate osmotic stress was documented (Munnik et al. 1999). Yeast two-hybrid screening identified an upstream MAPKK, SIMKK. SIMKK was shown to activate SIMK in response to salt stress (Kiegerl et al. 2000). Both SIMK and SAMK are also activated by wounding. SIMK activation by wounding was shown to be closely correlated with the wound-induced expression of MP2C, the *Medicago* type 2C protein phosphatase; MP2C directly inactivated SIMK after a period of activation, thus resetting the SIMK signaling pathway (Meskiene et al. 2003). It was clearly demonstrated that MP2C is a MAPK phosphatase that directly inactivates SIMK but not the wound-activated SAMK (Meskiene et al. 1998, 2003). Both SIMK and SAMK, however, can be activated by the constitutively active SIMKK in the absence of external stimuli (Meskiene et al. 2003; Kiegerl et al. 2000; Cardinale et al. 2002). Similarly, *Arabidopsis* serine/threonine phosphatase AP2C1, which is highly similar to alfalfa MP2C, inactivates the stress-responsive MAPKs MPK4 and MPK6. AP2C1 phosphatase influences the regulation of defense responses and levels of stress hormones in *Arabidopsis*. *ap2c1* mutant plants exhibited elevated JA levels in response to wounding, while AP2C1-overexpressing plants were compromised in the synthesis of wound-induced ethylene and in the defense response to the pathogens. Further, it was proved that AP2C1/MPK4 and AP2C1/MPK6 are coupled in functional complexes when coexpressed in protoplasts. These data indicate specific and functional interactions of AP2C1 with MPK4 and MPK6 (Schweighofer et al. 2007).

Plants depend on water uptake. Nutrients from the rhizosphere are taken up by plant roots together with water in the form of solutes. However, essential ions can be limiting factors in nonphysiological concentrations, which triggers developmental changes and subsequent root adaptations. The question is which signaling mechanisms mediate such adaptations. Unbalanced status of ions in the root environment can modulate the effectiveness of nutrient uptake by regulation of ion channels in the plant root cells. Ion channels are indispensable for nutrient uptake and for the balance of turgor pressure. Sensing and signaling of any changes in nutrient balance can involve production of "morphogenic" compounds such as ROS that have been shown to play a role in root responses to some nutrient deprivation, and the expression of several nutrient-responsive genes.

An additional field in the signaling of plant abiotic stresses has been opened owing to the problem of environmental pollution with excess of heavy metals. Transduction of metal-induced signals by MAPK pathways was clearly established; in *Medicago*, copper or cadmium ions activated four distinct MAPKs. Among them, SIMK was very rapidly activated by copper. This activation was specifically mediated by the upstream kinase SIMKK (Jonak et al. 2004). Similar responses to copper ions have been demonstrated in rice, where three different MAPKs were activated, including OsMPK6, the ortholog of SIMK; both cadmium and copper induced MAPK activation in rice (Yeh et al. 2003, 2007). Pretreatment with hydroxyl radical scavengers abolished cadmium- and copper-induced MAPK activation, indicating that MAPK pathways involved in those heavy metal stresses operate through ROS-generating systems (Yeh et al. 2007). Rapid and transient activation of a protein kinase with a molecular mass of 58 kDa is induced by toxic concentrations of aluminum in cell suspension cultures of *Coffea arabica* L. A MAPK nature of the protein kinase was proved by immunolocalization of cell extracts with anti pTEpY antibodies recognizing specifically an active form of MAPK (Arroyo-Serralta et al. 2005). Rapid activation of a 48-kDa MAPK-like protein by aluminum application to the root apex was reported also in wheat (Osawa and Matsumoto 2001). A similar effect with the activation of a 40- and 42-kDa myelin basic protein kinase has been elicited by zinc in rice roots; kinase immunoblot analysis proved the specificity to MAPKs. Experiments with inhibitors showed the involvement of ROS and the independence from the phosphoinositide 3-kinase related cellular signaling system (Lin et al. 2005).

From these data it is apparent that MAPK activation is a specific reaction of root cells to perception of heavy metal stress signals from the environment, while ROS-generating and/or utilizing systems share the activation. This reinforces the scenario that activation of MAPKs by heavy metals is related to activation of oxidative stress components. Thus, ROS become interesting players within the complex plant signaling networks.

5 Morphogenetic Roles of MAPKs

Organization of the plant body is driven by the precise regulation of cell morphogenesis, the process that controls all basic plant development phases by determining the shape and the size of individual cells. Deterministic processes of plant "shaping" are maintained at the cellular level through regulation of cell division and cell expansion (Dolan and Roberts 1995; Jürgens 1996). Recent studies have revealed a potential involvement of MAPK cascades in the control of morphogenesis, particularly in cell division and pattern formation. Cytokinesis, the last stage of cell division, depends on the correct formation of the cell plate. This process is guided by the phragmoplast. Limiting factors during the phragmoplast lateral expansion are the correct turnover of phragmoplast microtubules (MTs), their disassembly in the middle, and their reuse for polymerization in growing leading edges. Functional analysis of MAPKs in alfalfa, *Arabidopsis*, and tobacco revealed their role in the regulation of cytokinesis:

A tobacco MAPKKK, nucleus- and phragmoplast-localized protein kinase 1 (NPK1) (Banno et al. 1993), displays cell-specific intracellular relocation in dividing cells, with specific accumulation at the equator layer of the phragmoplast during cytokinesis (Nishihama and Machida 2001). Overexpression of a kinase-negative mutant version of NPK1 induced the formation of multinucleate cells with incomplete cell plates. Hence, it has been shown that NPK1 regulates the lateral expansion of the cell plate and the phragmoplast (Nishihama et al. 2001), and that NPK1 is indispensable for proper termination of the cell division process. The activity and the proper localization of NPK1 are regulated by phragmoplast-associated kinesin-like proteins NACK1 and NACK2 (NPK1-activating kinesins) (Nishihama et al. 2002).

NPK1 is an upstream MAPKKK member of the so called NACK-PQR pathway, a MAPK cascade consisting of NQK1/NtMEK1 (MAPKK) and NRK1/NTF6 (MAPK) (Nishihama et al. 2002; Soyano et al. 2003; Krysan et al. 2002). Importantly, this MAPK cascade is rapidly inactivated by the depolymerization of phragmoplast MTs (Soyano et al. 2003). A significant interlinking role in this story has been found for cell-division-related MT-associated proteins (MAPs). One of them is MAP65, a 65-kDa MAP with MT-bundling activity purified from tobacco BY-2 cells (Jiang and Sonobe 1993). Because phosphorylation of NtMAP65-1a by the MAPK NRK1 was documented in tobacco (Sasabe et al. 2006), the MAP65 family became one of the key targets of activated MAPKs in cytokinesis (Sasabe and Machida 2006). When the NACK-PQR pathway is inactivated, MAP65 bundles MTs and thus stabilizes phragmoplast structures. Once the NACK-PQR pathway has been activated in the phragmoplast and MAP65 has been phosphorylated, its MT-bundling activity is suppressed. The resulting MT instability promotes their turnover and thus accelerates an expansion of the phragmoplast (Sasabe and Machida 2006).

The members of the ANP family of *Arabidopsis* MAPKKKs are another example for positive regulators of cell division (Krysan et al. 2002) with cytokinesis-dependent activation (Bögre et al. 1999; Calderini et al. 1998). Furthermore, activation during cytokinesis and phragmoplast-specific localization were described for the MAPKs alfalfa MMK3 and tobacco NTF6 (Bögre et al. 1999; Calderini et al. 1998).

Tobacco NPK1 is a good example for the interlacing of plant responses to external and endogenous signals; in addition to its involvement in cytokinesis, its altered expression by silencing interfered with the functioning of some disease-resistance genes (Jin et al. 2002). The same is the case for the stress-induced MAPK SIMK of *Medicago* that is expressed in the root meristem. The majority of SIMK is localized in the nucleus of root interphase cells (Baluška et al. 2000). With the progress of the cell cycle SIMK relocation occurs; after onset of mitosis SIMK escapes from the nucleus and associates with the spindle poles in early prometaphase. Interestingly, salt stress, known to trigger specifically the activity of SIMK, also induces such relocation to MT structures such as the preprophase band and the phragmoplast (Baluška et al. 2000). Tissue-specific and cell-cycle-dependent intracellular distribution of SIMK in root meristematic cells of *Medicago sativa* can be directly relevant to biological functions of SIMK in root development. For instance, SIMK was redistributed from nuclei to the cytoplasm during the root hair formation process; it is accumulated within outgrowing bulges and at tips of emerging root hairs. During redistribution, SIMK is activated (Šamaj et al. 2002). The polar localization of SIMK in root hairs raises the question of its possible function as a molecular transducer in tip growth (Šamaj et al. 2004a, b).

Recent developmental studies have shown an even broader involvement of MAPK signaling pathways in plant morphogenesis. The impact of this functional signaling segment in plant ontogeny was documented in stomata. Development of stomatal cell complexes is regulated at the genetic level, but is also deeply influenced by environmental signals. After the MAPKKK gene *YODA* had been identified as the main player in the establishment of guard cell identity and pattern formation (Bergmann et al. 2004), a complete MAPK signaling cascade of these key regulators (YDA-MKK4/MKK5-MPK3/MPK6) was identified (Wang et al. 2007). Identification of the pathway brought a complex view of how cell fate specification during stomatal development, and during asymmetric cell division in general, might be regulated by intercellular signaling. MAPK cascades can be involved in the restriction of asymmetric cell division frequency, in maintaining polarity, and in coordination of cell fate specification (Wang et al. 2007). MAPKKK YODA, for instance, is responsible for determination of cell fates in the basal linage during embryonic development. Loss-of-function mutation *yda* suppresses elongation of the zygote and the extraembryonic suspensor is not developed properly. The opposite effect has been caused by gain-of-function mutation when an overdeveloped suspensor suppresses embryonic development (Lukowitz et al. 2004).

The complex role of MPK6 in multiple developmental processes has been shown in parallel studies. Plants homozygous for null *MPK6* mutation displayed reduced male fertility and abnormal anther development. Additional abnormalities were observed during seed production: *mpk6* plants contained embryos abnormally grown out of their seed coats. MPK6 thus appears to be involved in the regulation of multiple aspects of plant development (Bush and Krysan 2007).

Involvement of MAPK cascades in molecular switching mechanisms during early embryonal development is suggested also from a recent study proving novel overlapping functions of MPK3 and MPK6 in the regulation of the ovule integument development. Genetic experiments provided strong evidence that the sterile phenotype in plants is due to the loss of function of MPK6 and the partial loss of function of MPK3 (Wang et al. 2008). Overlapping expression patterns in the developing ovules support the expectation that MPK3 and MPK6 have overlapping functions in regulating ovule development. Because essential functions documented for both

MPK3 and MPK6 were observed in ovule integument development, but cell division in other parts of the double mutant plants was normal, the authors suggest that MPK3 and MPK6 might be involved in transducing cell-division-promoting signals either from embryo sac or from neighboring sporophytic cells (Wang et al. 2008).

Recent studies have demonstrated that both MPK3 and MPK6 have a broader influence on multiple developmental programs, such as embryo, stomata, inflorescence, and anther development (Bush and Krysan 2007; Wang et al. 2007). In addition to facilitating developmental signals, these two kinases are known to participate also in different stress-responsive signaling pathways, such as in response to the bacterial elicitor flg22 (Asai et al. 2002). This is another indication that MAPKs might be convergent points in pathways with different MAPKKKs and MAPKKs related to different signals.

Another crucial aspect in plant development is the regulation of polar auxin transport-based processes. Phenotypical analysis of *Arabidopsis thaliana bud1* (bushy and dwarf1) mutant showed reduction in lateral root formation, simplification in venation patterns, temperature-dependent shortening of the hypocotyl, and a faster curvature reaction to changed gravitropic vector. Molecular genetic analysis determined the *bud1* background as a result of increased expression of *MKK7*. Assay with radiolabeled indole-3-acetic acid showed that the increased expression of MKK7 in *bud1* caused a deficiency in auxin transport, while antisense transgenic plants with repressed MKK7 expression enhanced auxin transport (Dai et al. 2006). Such comparative analyses proved that MKK7 in *Arabidopsis thaliana* has a negative relationship to polar auxin transport.

These few examples clearly indicate how interconnection of the developmental program with gene expression patterns can be modulated by signaling components. Control of both processes, gene expression pattern and its direct developmental outputs, has a high priority and thus finding versatile MAPK signaling networks within the ontogenetic program has logical justification.

6 Signaling of Hormones and Oxidative Stresses

Hormone signaling pathways are integral regulatory components of the plant development program and interact extensively with others during plant growth, development, and adaptation to a plethora of biotic and abiotic stresses. Evidence for an ethylene-activated MAPK signal transduction pathways has been experimentally proved; ethylene responses in plants are regulated by a MAPK module comprising the CTR1 protein, a negative regulator of ethylene responses, which has sequence similarity to the Raf family of MAPKKK (Kieber et al. 1993; Huang et al. 2003). Upon treatment with aminocyclopropane-1-carboxylic acid (ACC), two *Medicago* MAPKs, MMK3 and SIMK, were activated. ACC-induced activation of SIMK and MMK3 was specifically mediated by the upstream kinase SIMKK. In transgenic *Arabidopsis* plants overexpressing SIMKK, AtMPK6 activation was constitutive and followed by ethylene-induced target gene expression (Ouaked et al. 2003). Thus, SIMKK has been established as a positive regulator of ethylene responses.

Investigations of MAPK signaling in mediation of auxin response, tested at the level of auxin-responsive gene expression, showed an interesting crosstalk between activation of auxin-regulated transcription and other MAPK-mediated signaling cascades. Expression of the constitutively active form of NPK1, the tobacco MAPKKK, specifically activates the MAPK pathway, but leads also to suppression of auxin-induced gene expression in transfected mesophyll leaf protoplasts (Kovtun et al. 1998). Similarly, *Arabidopsis* ANP1, an ortholog of tobacco NPK1, mediates H_2O_2 signal transduction under oxidative stress by activation of two stress-related MAPKs, AtMPK3 and AtMPK6. In parallel, the same MAPKKK inhibits auxin response (Kovtun et al. 2000).

The ABA signaling network consists of the activation of AtMPK3 (Lu et al. 2002) or AtMPK6 (Liu et al. 2008) depending on the nature of the stimulus. In addition, AtMPK6 can be activated through upstream MKK3 in the JA signaling network (Takahashi et al. 2007). MAPK signaling cascades are also targets of NO signals, as has been documented by MAPK activation after application of NO donors in *Arabidopsis* (Clarke et al. 2000) and tobacco (Kumar and Klessig 2000).

ROS are linked to the activation of several stress-responsive MAPK pathways in plants (Apel and Hirt 2004; Kovtun et al. 2000; Rentel et al. 2004; Romeis et al. 1999; Yuasa et al. 2001). An important question in the effort to resolve a general scheme of ROS participation in signaling events is where ROS members are produced. In plant stress reactions to signals such as pathogenesis or wounding, the place defined for ROS generation was the apoplast (Luan et al. 2002; Overmyer et al. 2003). Controversially, biotic or abiotic stress reactions provoked intracellular ROS production in chloroplasts, peroxisomes, and mitochondria (Datt et al. 2003; Yao et al. 2002). Concerning stress reactions with apoplastic ROS accumulation, NADPH oxidases are intensively studied. Impaired ROS production in the double mutant of *AtrbohD* and *AtrbohF*, *Arabidopsis* guard-cell-expressed NADPH oxidase catalytic subunit genes, prevents also ABA-induced cellular reactions. However, stomatal closing and activation of plasma membrane Ca^{2+}-permeable channels are rescued by application of exogenous H_2O_2 (Kwak et al. 2003). The role of the AtrbohD and AtrbohF NADPH oxidase catalytic subunits was established also in pathogen-induced ROS production in *Arabidopsis*, because ROS production after infection was reduced in *atrbohD*/*atrbohF* double mutants (Torres et al. 2002).

The function of ROS-mediated plasma membrane Ca^{2+} channel activation correlates with the localization of Ca^{2+} influx. A single-channel examination revealed that hyperpolarization-activated Ca^{2+} channels have significantly higher activity in some types of cells, such as in root hairs and in cells of the root elongation zone, than in mature cells (Demidchik et al. 2007). NADPH oxidase (RHD2/AtrbohC, *Arabidopsis* respiratory burst oxidase homolog), is an enzyme catalyzing ROS production in tip-growing plant cells such as root hairs. It has been proposed that activated RHD2/AtrbohC enzymes produce ROS, which in turn activate the hyperpolarization-activated Ca^{2+} channels in the tip. These channels are responsible for generating the tip-focused Ca^{2+} influx (Foreman et al. 2003). A major role of polarized ROS production is therefore the maintenance of polarity in tip-growing cells through facilitating calcium influx that is involved in the modulation of the dynamic state of the actin cytoskeleton.

It was shown that *Arabidopsis* serine/threonine protein kinase OXI1 (oxidative stress inducible 1), belonging to the AGC family of protein kinases, is activated upon ROS treatment. In addition, OXI1 is required for activation of the downstream MAPKs, MPK3 and MPK6, in response to ROS (Rentel et al. 2004). Interestingly, OXI1 is also activated by phosphatidic acid (PA) (Anthony et al. 2004). Phospholipase D and PA have been implicated in H_2O_2-induced activation of MAPK cascades (Zhang et al. 2003). Available data indicate that phospholipase D produces PA after activation by developmental and environmental signals. PA activates NADPH oxidase that produces ROS and ROS activates certain protein kinases (Zhang et al. 2005).

Recent data have provided more direct evidence for ROS-dependent activation of different MAPK signaling pathways. Regulation of a variety of processes based on equilibrated ROS homeostasis in *Arabidopsis thaliana* has been studied in respect to MEKK1-dependent signaling specificity. A complex experimental approach clearly indicates that MEKK1 mediates ROS-induced MPK4 signaling. Immunoprecipitation studies by in vitro kinase assays showed strong activation of MPK3, MPK4, and MPK6 by H_2O_2 treatment in wild-type plants, while H_2O_2-induced activation of MPK4 was lost in a *mekk1* mutant (Nakagami et al. 2006). A MAPKKK most closely related to *Arabidopsis* MEKK1 is oxidative stress-activated MAPKKK1 (OMTK1) in alfalfa. In activating tests for different stresses, OMTK1 was activated by H_2O_2, while MMK3, another alfalfa MAPK, was specifically activated in the cell death process (Nakagami et al. 2004).

Correlation of interlaced activities of MAPKs with ROS has been shown for MEKK1; under the influence of different biotic and abiotic stress inputs, MEKK1 is involved in the activation of MPK3, MPK4, and MPK6 (Asai et al. 2002; Teige et al. 2004). For the regulation of vegetative plant development, MEKK1 in addition plays an important role; *mekk1* mutant plants exhibit a dwarf phenotype leading to lethality after true leaves have been started. Moreover, *mekk1* mutant plants have shorter root hairs and fewer lateral roots than wild-type plants (Nakagami et al. 2006). This is consistent with studies showing a great impact of ROS on root development processes and a specific distribution of superoxide ($O_2^{\cdot-}$) and hydrogen peroxide (H_2O_2) in the *Arabidopsis* root tip; the apoplast of the cell elongation zone contains increased amounts of $O_2^{\cdot-}$, while the differentiation zone and the root hair formation zone are enriched in H_2O_2. Interestingly, when the concentrations of $O_2^{\cdot-}$ and H_2O_2 individually were decreased, antagonistic effects on root elongation were observed (Dunand et al. 2007).

Another function of ROS can be found in salt stress reactions. When plants are stressed by high salt concentrations, primary aspects such as osmotic stress and ion imbalances are followed by a secondary oxidative stress caused by the generation of ROS. In line with general salt stress responses, increased plasma membrane endocytosis and intracellular ROS production within endosomes were observed (Leshem et al. 2007). Production of ROS is also one of the major mechanisms of plant defense responses to pathogens (Lamb and Dixon 1997) and during plant-microbe symbiotic interactions (Santos et al. 2001; Ramu et al. 2002; Levine et al. 1996; Peleg-Grossman et al. 2007).

Plant defense reactions against pathogens include rapid programmed cell death. Activation of the SIPK/Ntf4/WIPK pathway, a general cascade found in tobacco, induces cell death phenotypes with loss of membrane potential, electrolyte leakage, and rapid dehydration (Liu et al. 2007). An upstream kinase of the SIPK/Ntf4/WIPK pathway is NtMEK2 (Ren et al. 2006; Yang et al. 2001). Studies using a constitutively active mutant of NtMEK2 provided more clear evidence for the role of SIPK, Ntf4, and WIPK in cell death, because activation of endogenous SIPK, Ntf4, and WIPK led to similar cell death in the absence of the pathogen (Jin et al. 2003; Ren et al. 2006; Yang et al. 2001; Yoshioka et al. 2003). Importantly, loss of membrane potential in plants with a constitutively active mutant form of NtMEK2 was associated with the generation of ROS (Liu et al. 2007). SIPK/Ntf4/WIPK cascade has been shown to be activated rapidly in perception of most of the stresses tested so far (reviewed in Mizoguchi et al. 1997; Tena et al. 2001; MAPK Group 2002; Nakagami et al. 2005; Pedley and Martin 2005).

7 Conclusions

In principle, the MAPK signaling cascade is a network of interacting proteins, interconnecting sensing components at the cell surface with the effectors in the cell interior by a sequence of phosphorylation steps. Plants are continuously faced with different extracellular cues, and thus signaling pathways must be able to integrate diverse signals and to keep high selectivity for co-ordinated cellular functions. Extracellular cues may produce stimulus- or tissue-specific responses by utilization of one particular MAPK module. On the other hand, some components of MAPK modules have the ability to bridge several unrelated pathways. Thus, the mechanism regulated by phosphorylation ensures spatial and temporal substrate specificity and increases the sensitivity of the MAPK-based signal transduction. Determination of the molecular players and events occurring in cell signaling pathways is the key to our understanding of the principle of how plants cope with a changing environment. Although intensive effort will be needed to answer many principal questions in the future, the recent burst of new methodological approaches conveys hope. Techniques such as the determination of protein kinase activities, searching for protein kinase substrates, and mapping of phosphorylation sites in a large-scale dimension will make it possible to study plant signaling processes from a systems biology point of view.

Acknowledgements This work was supported by a grant no. APVV-0432–06 from the Grant Agency APVV and by a grant no. 02/0149/08 from the Grant Agency VEGA, Bratislava, Slovakia, by a grant of the Hochschuljubiläumsstiftung der Stadt Wien and by a grant no. L433-B17 from the Austrian FWF.

References

Agrawal GK, Iwahashi H, Rakwal R (2003) Rice MAPKs. Biochem Biophys Res Commun 302:171–180

Alzwiy IA, Morris PC (2007) A mutation in the *Arabidopsis* MAP kinase kinase 9 gene results in enhanced seedling stress tolerance. Plant Sci 173:302–308

Anthony RG, Henriques R, Helfer A, Mészáros T, Rios G, Testerink C, Munnik T, Deák M, Koncz C, Bögre L (2004) A protein kinase target of a PDK1 signalling pathway is involved in root hair growth in *Arabidopsis*. EMBO J 23:572–581

Apel K, Hirt H (2004) Reactive oxygen species: metabolism, oxidative stress, and signal transduction. Annu Rev Plant Biol 55:373–399

Arroyo-Serralta GA, Kú-González Á, Hernández-Sotomayor SM, Aguilar JJZ (2005) Exposure to toxic concentrations of aluminium activates a MAPK-like protein in cell suspension cultures of *Coffea arabica*. Plant Physiol Biochem 43:27–35

Asai T, Tena G, Plotnikova J, Willmann MR, Chiu W-L, Gomez-Gomez L, Boller T, Ausubel FM, Sheen J (2002) MAP kinase signalling cascade in *Arabidopsis* innate immunity. Nature 415:977–983

Baluška F, Ovečka M, Hirt H (2000) Salt stress induced changes in amounts and localisation of the mitogen-activated protein kinase SIMK in alfalfa roots. Protoplasma 212:262–267

Banno H, Hirano K, Nakamura T, Irie K, Nomoto S, Matsumoto K, Machida Y (1993) NPK1, a tobacco gene that encodes a protein with a domain homologous to yeast BCK1, STE11, and Byr2 protein kinases. Mol Cell Biol 13:4745–4752

Bergmann DC, Lukowitz W, Somerville CR (2004) Pattern controlled by a MAPKK kinase. Science 304:1494–1497

Bögre L, Calderini O, Binarova P, Mattauch M, Till S, Kiegerl S, Jonak C, Pollaschek C, Barker P, Huskisson NS, Hirt H, Heberle-Bors E (1999) A MAP kinase is activated late in plant mitosis and becomes localised to the plane of cell division. Plant Cell 11:101–113

Bögre L, Meskiene I, Heberle-Bors E, Hirt H (2000) Stressing the role of MAP kinases in mitogenic stimulation. Plant Mol Biol 43:705–718

Brown GC, Hoek JB, Kholodenko BN (1997) Why do protein kinase cascades have more then one level? Trends Biochem Sci 22:288

Bush SM, Krysan PJ (2007) Mutational evidence that the *Arabidopsis* MAP kinase MPK6 is involved in anther, inflorescence, and embryo development. J Exp Bot 58:2181–2191

Calderini O, Bögre L, Vicente O, Binarova P, Heberle-Bors E, Wilson C (1998) A cell cycle regulated MAP kinase with a possible role in cytokinesis in tobacco cells. J Cell Sci 111:3091–3100

Cardinale F, Meskiene I, Ouaked F, Hirt H (2002) Convergence and divergence of stress-induced mitogen-activated protein kinase signaling pathways at the level of two distinct mitogen-activated protein kinase kinases. Plant Cell 14:1–10

Chinchilla D, Bauer Z, Regenass M, Boller T, Felix G (2006) The *Arabidopsis* receptor kinase FLS2 binds flg22 and determines the specificity of flagellin perception. Plant Cell 18:465–476

Clarke A, Desikan R, Hurst RD, Hancock JT, Neill ST (2000) NO way back: nitric oxide and programmed cell death in *Arabidopsis thaliana* suspension cultures. Plant J 4:667–677

Cohen P (1989) The structure and regulation of protein phosphatases. Annu Rev Biochem 58:453–508

Colcombet J, Hirt H (2008) *Arabidopsis* MAPKs: a complex signalling network involved in multiple biological processes. Biochem J 413:217–226

Dai Y, Wang H, Li B, Huang J, Liu X, Zhou Y, Mou Z, Li J (2006) Increased expression of MAP KINASE KINASE7 causes deficiency in polar auxin transport and leads to plant architectural abnormality in *Arabidopsis*. Plant Cell 18:308–320

Datt JF, Pellinen R, Beeckman T, Van De Cotte B, Langebartels C, Kangasjervi J, Inze D, Van Breusegem F (2003) Changes in hydrogen peroxide homeostasis trigger an active cell death process in tobacco. Plant J 33:621–629

Demidchik V, Shabala VS, Davies JM (2007) Spatial variation in H_2O_2 response of *Arabidopsis thaliana* root epidermal Ca^{2+} flux and plasma membrane Ca^{2+} channels. Plant J 49:377–386

Denekamp M, Smeekens SC (2003) Integration of wounding and osmotic stress signals determines the expression of the AtMYB102 transcription factor gene. Plant Physiol 132:1415–1423

Djamei A, Pitzschke A, Nakagami H, Rajh I, Hirt H (2007) Trojan horse strategy in *Agrobacterium* transformation: Abusing MAPK defense signalling. Science 318:453–456

Dóczi R, Brader G, Pettkó-Szandtner A, Rajh I, Djamei A, Pitzschke A, Teige M, Hirt H (2007) The *Arabidopsis* mitogen-activated protein kinase kinase MKK3 is upstream of group C mitogen-activated protein kinases and participates in pathogen signalling. Plant Cell 19:3266–3279

Dolan L, Roberts K (1995) Plant development: pulled up by the roots. Curr Opin Genet Dev 5:432–438

Dunand C, Crèvecoeur M, Penel C (2007) Distribution of superoxide and hydrogen peroxide in *Arabidopsis* root and their influence on root development: possible interaction with peroxidises. New Phytol 174:332–341

Farkas I, Dombrádi V, Miskei M, Szabados L, Koncz C (2007) *Arabidopsis* PPP family of serine/threonine phosphatises. Trends Plant Sci 12:169–176

Foreman J, Demidchik V, Bothwell JHF, Mylona P, Miedema H, Torresk MA, Linstead P, Costa S, Brownlee C, Jonesk JDG, Davies JM, Dolan L (2003) Reactive oxygen species produced by NADPH oxidase regulate plant cell growth. Nature 422:442–446

Fujita M, Fujita Y, Noutoshi Y, Takahashi F, Narusaka Y, Yamaguchi-Shinozaki K, Shinozaki K (2006) Crosstalk between abiotic and biotic stress responses: a current view from the points of convergence in the stress signaling networks. Curr Opin Plant Biol 9:436–442

Garrington TP, Johnson GL (1999) Organization and regulation of mitogen-activated protein kinase signaling pathway. Curr Opin Cell Biol 11:211–218

Hamel L-P, Nicole M-C, Sritubtim S, Morency M-J, Ellis M, Ehlting J, Beaudoin N, Barbazuk B, Klessig D, Lee J, Martin G, Mundy J, Ohashi Y, Scheel D, Sheen J, Xing T, Zhang S, Seguin A, Ellis BE (2006) Ancient signals: comparative genomics of plant MAPK and MAPKK gene families. Trends Plant Sci 11:192–198

Harris K, Lamson RE, Nelson B, Hughes TR, Marton MJ, Roberts CJ, Boone C, Pryciak PM (2001) Role of scaffolds in MAP kinase pathway specificity revealed by custom design of pathway-dedicated signaling proteins. Curr Biol 11:1815–1824

He P, Shan L, Lin N-C, Martin GB, Kemmerling B, Nürnberger T, Sheen J (2006) Specific bacterial suppressors of MAMP signaling upstream of MAPKKK in *Arabidopsis* innate immunity. Cell 125:563–575

Heese A, Hann DR, Gimenez-Ibanez S, Jones AME, He K, Li J, Schroeder JI, Peck SC, Rathjen JP (2007) The receptor-like kinase SERK3/BAK1 is a central regulator of innate immunity in plants. Proc Natl Acad Sci U S A 104:12217–12222

Huang Y, Li H, Hutchison CE, Laskey J, Kieber JJ (2003) Biochemical and functional analysis of CTR1, a protein kinase that negatively regulates ethylene signaling in *Arabidopsis*. Plant J 33:221–233

Ichimura K, Mizoguchi T, Yoshida R, Yuasa T, Shinozaki K (2000) Various abiotic stresses rapidly activate *Arabidopsis* MAP kinases ATMPK4 and ATMPK6. Plant J 24:655–665

Jiang CJ, Sonobe S (1993) Identification and preliminary characterization of a 65 kDa higher-plant microtubule-associated protein. J Cell Sci 105:891–901

Jin H, Axtell MJ, Dahlbeck D, Ekwenna O, Zhang S, Staskawicz B, Baker B (2002) NPK1, an MEKK1-like mitogen-activated protein kinase kinase kinase, regulates innate immunity and development in plants. Dev Cell 3:291–297

Jin H, Liu Y, Yang K-Y, Kim CY, Baker B, Zhang S (2003) Function of a mitogen-activated protein kinase pathway in N-gene mediated resistance in tobacco. Plant J 33:719–731

Jonak C, Kiegerl S, Ligterink W, Barker PJ, Huskisson NS, Hirt H (1996) Stress signaling in plants: a mitogen-activated protein kinase pathway is activated by cold and drought. Proc Natl Acad Sci U S A 93:11274–11279

Jonak C, Ökrész L, Bögre L, Hirt H (2002) Complexity, cross talk and integration of plant MAP kinase signalling. Curr Opin Plant Biol 5:415–424

Jonak C, Nakagami H, Hirt H (2004) Heavy metal stress. Activation of distinct mitogen-activated protein kinase pathways by copper and cadmium. Plant Physiol 136:3276–3283

Jürgens G (1996) Cell division and morphogenesis in angiosperm embryogenesis. Semin Cell Dev Biol 7:867–872

Kieber JJ, Rothenberg M, Roman G, Feldmann KA, Ecker JR (1993) CTR1, a negative regulator of the ethylene response pathway in *Arabidopsis*, encodes a member of the raf family of protein kinases. Cell 72:427–441

Kiegerl S, Cardinale F, Siligan C, Gross A, Baudouin E, Liwosz A, Eklöf S, Till S, Bögre L, Hirt H, Meskiene I (2000) SIMKK, a mitogen-activated protein kinase (MAPK) kinase, is a specific activator of the salt stress-induced MAPK, SIMK. Plant Cell 12:2247–2258

Kovtun Y, Chiu W-L, Zeng W, Sheen J (1998) Suppression of auxin signal transduction by a MAPK cascade in higher plants. Nature 395:716–720

Kovtun Y, Chiu W-L, Tena G, Sheen J (2000) Functional analysis of oxidative stress-activated mitogen-activated protein kinase cascade in plants. Proc Natl Acad Sci U S A 97:2940–2945

Krysan PJ, Jester PJ, Gottwald JR, Sussman MR (2002) An *Arabidopsis* mitogen-activated protein kinase kinase kinase gene family encodes essential positive regulators of cytokinesis. Plant Cell 14:1109–1120

Kumar D, Klessig DF (2000) Differential induction of tobacco MAP kinases by the defence signals nitric oxide, salicylic acid, ethylene and jasmonic acid. Mol Plant Microbe Interact 13:347–351

Kwak JM, Mori IC, Pei Z-M, Leonhardt N, Torres MA, Dangl JL, Bloom RE, Bodde S, Jones JDG, Schroeder JI (2003) NADPH oxidase AtrbohD and AtrbohF genes function in ROS-dependent ABA signaling in *Arabidopsis*. EMBO J 22:2623–2633

Lamb C, Dixon RA (1997) The oxidative burst in plant disease resistance. Annu Rev Plant Physiol Plant Mol Biol 48:251–275

Leshem Y, Seri L, Levine A (2007) Induction of phosphatidylinositol 3-kinase-mediated endocytosis by salt stress leads to intracellular production of reactive oxygen species and salt tolerance. Plant J 51:185–197

Levine A, Pennell R, Alvarez M, Palmer R, Lamb CJ (1996) Calcium-mediated apoptosis in a plant hypersensitive disease resistance response. Curr Biol 6:427–437

Lin C-W, Chang H-B, Huang H-J (2005) Zinc induces mitogen-activated protein kinase activation mediated by reactive oxygen species in rice roots. Plant Physiol Biochem 43:963–968

Liu PF, Chang WC, Wang YK, Chang HY, Pan RL (2008) Signaling pathways mediating the suppression of *Arabidopsis thaliana* Ku gene expression by abscisic acid. Biochim Biophys Acta 1779:164–174

Liu Q, Xue Q (2007) Computational identification and phylogenetic analysis of the MAPK gene family in Oryza sativa. Plant Physiol Biochem 45:6–14

Liu Y, Zhang S (2004) Phosphorylation of 1-aminocyclopropane-1-carboxylic acid synthase by MPK6, a stress-responsive mitogen-activated protein kinase, induces ethylene biosynthesis in *Arabidopsis*. Plant Cell 16:3386–3399

Liu Y, Ren D, Pike S, Pallardy S, Gassmann W, Zhang S (2007) Chloroplast-generated reactive oxygen species are involved in hypersensitive response-like cell death mediated by a mitogen-activated protein kinase cascade. Plant J 51:941–954

Lu M, Han H, Guevara-Garcia A, Fedoroff NV (2002) Mitogen-activated protein kinase signaling in postgermination arrest of development by abscisic acid. Proc Natl Acad Sci U S A 99:15812–15817

Luan S (1998) Protein phosphatases and signaling cascades in higher plants. Trends Plant Sci 3:271–275

Luan S, Kudla J, Rodriguez-Concepcion M, Yalovsky S, Gruissem W (2002) Calmodulins and calcineurin B-like proteins: calcium sensors for specific signal response coupling in plants. Plant Cell 14:S389–S400

Lukowitz W, Roeder A, Parmenter D, Somerville C (2004) A MAPKK Kinase gene regulates extra-embryonic cell fate in *Arabidopsis*. Cell 116:109–119

MAPK Group (2002) Mitogen-activated protein kinase cascades in plants: a new nomenclature. Trends Plant Sci 7:301–308

Meskiene I, Bögre L, Glaser W, Balog J, Brandstotter M, Zwerger K, Ammerer G, Hirt H (1998) MP2C, a plant protein phosphatase 2C, functions as a negative regulator of mitogen-activated protein kinase pathways in yeast and plants. Proc Natl Acad Sci U S A 95:1938–1943

Meskiene I, Baudouin E, Schweighofer A, Liwosz A, Jonak C, Rodriguez PL, Jelinek H, Hirt H (2003) Stress-induced protein phosphatase 2C is a negative regulator of a mitogen-activated protein kinase. J Biol Chem 278:18945–18952

Mészáros T, Helfer A, Hatzimasoura E, Magyar Z, Serazetdinova L, Rios G, Bardóczy V, Teige M, Koncz C, Peck S Bögre L (2006) The *Arabidopsis* MAP kinase kinase MKK1 participates in defence responses to the bacterial elicitor flagellin. Plant J 48:485–498

Mizoguchi T, Ichimura K, Shinozaki K (1997) Environmental stress response in plants: the role of mitogen-activated protein kinases. Trends Biotechnol 15:15–19

Møller SG, Chua N-H (1999) Interaction and intersections of plant signaling pathways. J Mol Biol 293:219–234

Munnik T, Ligterink W, Meskiene I, Calderini O, Beyerly J, Musgrave A, Hirt H (1999) Distinct osmo-sensing protein kinase pathways are involved in signalling moderate and severe hyper-osmotic stress. Plant J 20:381–388

Nakagami H, Kiegerl S, Hirt H (2004) OMTK1, a novel MAPKKK, channels oxidative stress signaling through direct MAPK interaction. J Biol Chem 279:26959–26966

Nakagami H, Pitzschke A, Hirt H (2005) Emerging MAP kinase pathways in plant stress signalling. Trends Plant Sci 10:339–346

Nakagami H, Soukupová H, Schikora A, Žárský V, Hirt H (2006) A mitogen-activated protein kinase kinase kinase mediates reactive oxygen species homeostasis in *Arabidopsis*. J Biol Chem 281:38697–38704

Nishihama R, Machida Y (2001) Expansion of the phragmoplast during plant cytokinesis: a MAPK pathway may MAP it out. Curr Opin Plant Biol 4:507–512

Nishihama R, Ishikawa M, Araki S, Soyano T, Asada T, Machida Y (2001) The NPK1 mitogen-activated protein kinase kinase kinase is a regulator of cell-plate formation in plant cytokinesis. Gen Dev 15:352–363

Nishihama R, Soyano T, Ishikawa M, Araki S, Tanaka H, Asada T, Irie K, Ito M, Terada M, Banno H, Yamazaki Y, Machida Y (2002) Expansion of the cell plate in plant cytokinesis requires a kinesin-like protein/MAPKKK complex. Cell 109:87–99

Nühse TS, Peck SC, Hirt H, Boller T (2000) Microbial elicitors induce activation and dual phosphorylation of the *Arabidopsis thaliana* MAPK 6. J Biol Chem 275:7521–7526

Nurnberger T, Brunner F, Kemmerling B, Piater L (2004) Innate immunity in plants and animals: striking similarities and obvious differences. Immunol Rev 198:249–266

Ortiz-Masia D, Perez-Amador MA, Carbonell J, Marcote MJ (2007) Diverse stress signals activate the C1 subgroup MAP kinases of *Arabidopsis*. FEBS Lett 581:1834–1840

Osawa H, Matsumoto H (2001) Possible involvement of protein phosphorylation in aluminum-responsive malate efflux from wheat root apex. Plant Physiol 126:411–420

Ouaked F, Rozhon W, Lecourieux D, Hirt H (2003) A MAPK pathway mediates ethylene signaling in plants. EMBO J 22:1282–1288

Overmyer K, Brosche M, Kangasjarvi J (2003) Reactive oxygen species and hormonal control of cell death. Trends Plant Sci 8:335–342

Pedley KF, Martin GB (2005) Role of mitogen-activated protein kinases in plant immunity. Curr Opin Plant Biol 8:541–547

Peleg-Grossman S, Volpin H, Levine A (2007) Root hair curling and *Rhizobium* infection in *Medicago truncatula* are mediated by phosphatidylinositide-regulated endocytosis and reactive oxygen species. J Exp Bot 58:1637–1649

Ramonell K, Berrocal-Lobo M, Koh S, Wan J, Edwards H, Stacey G, Somerville S (2005) Loss-of-function mutations in chitin responsive genes show increased susceptibility to the powdery mildew pathogen *Erysiphe cichoracearum*. Plant Physiol 138:1027–1036

Ramu SK, Peng HM, Cook DR (2002) Nod factor induction of reactive oxygen species production is correlated with expression of the early nodulin gene Rip1 in *Medicago truncatula*. Mol Plant Microbe Interact 15:522–528

Ren D, Yang K-Y, Li G, Liu Y, Zhang S (2006) Activation of Ntf4, a tobacco MAPK, during plant defense response and its involvement in hypersensitive response-like cell death. Plant Physiol 141:1482–1493

Rentel MC, Lecourieux D, Ouaked F, Usher SL, Petersen L, Okamoto H, Knight H, Peck SC, Grierson CS, Hirt H, Knight MR (2004) OXI1 kinase is necessary for oxidative burst-mediated signalling in *Arabidopsis*. Nature 427:858–861

Robinson MJ, Cobb MH (1997) Mitogen-activated protein kinase pathways. Curr Opin Cell Biol 9:180–186

Romeis T, Piedras P, Zhang S, Klessig DF, Hirt H, Jones JD (1999) Rapid avr9- and Cf-9-dependent activation of MAP kinases in tobacco cell cultures and leaves. Convergence of resistance gene, elicitor, wound, and salicylate responses. Plant Cell 11:273–287

Russinova E, Borst J-W, Kwaaitaal M, Caño-Delgado A, Yin Y, Chory J, Vries SC (2004) Heterodimerization and endocytosis of *Arabidopsis* brassinosteroid receptors BRI1 and AtSERK3 (BAK1). Plant Cell 16:3216–3229

Šamaj J, Ovečka M, Hlavačka A, Lecourieux F, Meskiene I, Lichtscheidl I, Lenart P, Salaj J, Volkmann D, Bögre L, Baluška F, Hirt H (2002) Involvement of the mitogen-activated protein kinase SIMK in regulation of root hair tip growth. EMBO J 21:3296–3306

Šamaj J, Baluška F, Hirt H (2004a) From signal to cell polarity: mitogen-activated protein kinases as sensors and effectors of cytoskeleton dynamicity. J Exp Bot 55:189–198

Šamaj J, Baluška F, Menzel D (2004b) New signalling molecules regulating root hair tip growth. Trends Plant Sci 9:217–220

Sangwan V, Dhindsa RS (2002) In vivo and in vitro activation of temperature-responsive plant map kinases. FEBS Lett 531:561–564

Sangwan V, Orvar BL, Beyerly J, Hirt H, Dhindsa RS (2002) Opposite changes in membrane fluidity mimic cold and heat stress activation of distinct plant MAP kinase pathways. Plant J 31:629–638

Santos R, Herouart D, Sigaud S, Touati D, Puppo A (2001) Oxidative burst in alfalfa-*Sinorhizobium meliloti* symbiotic interaction. Mol Plant Microbe Interact 14:86–89

Sasabe M, Machida Y (2006) MAP65: a bridge linking a MAP kinase to microtubule turnover. Curr Opin Plant Biol 9:563–570

Sasabe M, Soyano T, Takahashi Y, Sonobe S, Igarashi H, Itoh TJ, Hidaka M, Machida Y (2006) Phosphorylation of NtMAP65–1 by a MAP kinase down-regulates its activity of microtubule bundling and stimulates progression of cytokinesis of tobacco cells. Genes Dev 20:1004–1014

Seger R, Krebs EG (1995) The MAPK signaling cascade. FASEB J 9:726–735

Shan L, He P, Sheen J (2007) Intercepting host MAPK signaling cascades by bacterial type III effectors. Cell Host Microbe 1:167–174

Shiu S-H, Bleecker AB (2001) Receptor-like kinases from Arabidopsis form a monophyletic gene family related to animal receptor kinases. Proc Natl Acad Sci U S A 98:10763–10768

Shiu S-H, Karlowski WM, Pan R, Tzeng Y-H, Mayer KFX, Li W-H (2004) Comparative analysis of the receptor-like kinase family in *Arabidopsis* and rice. Plant Cell 16:1220–1234

Schweighofer A, Kazanaviciute V, Scheikl E, Teige M, Doczi R, Hirt H, Schwanninger M, Kant M, Schuurink R, Mauch F, Buchala A, Cardinale F, Meskiene I (2007) The PP2C-type phosphatase AP2C1, which negatively regulates MPK4 and MPK6, modulates innate immunity, jasmonic acid, and ethylene levels in *Arabidopsis*. Plant Cell 19:2213–2224

Soropy SK, Munshi M (1998) Protein kinases and phosphatases and their role in cellular signaling in plants. Crit Rev Plant Sci 17:245–318

Soyano T, Nishihama R, Morikiyo K, Ishikawa M, Machida Y (2003) NQK1/NtMEK1 is a MAPKK that acts in the NPK1 MAPKKK-mediated MAPK cascade and is required for plant cytokinesis. Genes Dev 17:1055–1067

Sterck L, Rombauts S, Jansson S, Sterky F, Rouzé P, Van de Peer Y (2005) EST data suggest that poplar is an ancient polyploidy. New Phytol 167:165–170

Suarez-Rodriguez MC, Adams-Phillips L, Liu Y, Wang H, Su SH, Jester PJ, Zhang S, Bent AF, Krysan PJ (2007) MEKK1 is required for flg22-induced MPK4 activation in *Arabidopsis* plants. Plant Physiol 143:661–669

Takahashi F, Yoshida R, Ichimura K, Mizoguchi T, Seo S, Yonezawa M, Maruyama K, Yamaguchi-Shinozaki K, Shinozaki K (2007) The mitogen-activated protein kinase cascade MKK3-MPK6 is an important part of the jasmonate signal transduction pathway in *Arabidopsis*. Plant Cell 19:805–818

Teige M, Scheikl E, Eulgem T, Doczi R, Ichimura K, Shinozaki K, Dangl JL, Hirt H (2004) The MKK2 pathway mediates cold and salt stress signaling in *Arabidopsis*. Mol Cell 15:141–152

Tena G, Asai T, Chiu W-L, Sheen J (2001) Plant mitogen-activated protein kinase signaling cascades. Curr Opin Plant Biol 4:392–400

Torres MA, Dangl JL (2005) Functions of the respiratory burst oxidase in biotic interactions, abiotic stress and development. Curr Opin Plant Biol 8:397–403

Torres MA, Dangl JL, Jones JDG (2002) *Arabidopsis* gp91phox homologues AtrbohD and AtrbohF are required for accumulation of reactive oxygen intermediates in the plant defense response. Proc Natl Acad Sci U S A 99:517–522

Wang H, Ngwenyama N, Liu Y, Walker JC, Zhang S (2007) Stomatal development and patterning are regulated by environmentally responsive mitogen-activated protein kinases in *Arabidopsis*. Plant Cell 19:63–73

Wang H, Liu Y, Bruffett K, Lee J, Hause G, Walker JC, Zhang S (2008) Haplo-insufficiency of MPK3 in MPK6 mutant background uncovers a novel function of these two MAPKs in *Arabidopsis* ovule development. Plant Cell 20:602–613

Whitmarsh AJ, Davis R (1998) Structural organization of MAP kinase signaling modules by scaffold proteins in yeast and mammals. Trends Biochem Sci 23:481–485

Wrzaczek M, Hirt H (2001) Plant MAP kinase pathways: how many and what for? Biol Cell 93:81–87

Xiong L, Yang Y (2003) Disease resistance and abiotic stress tolerance in rice are inversely modulated by an abscisic acid-inducible mitogen-activated protein kinase. Plant Cell 15:745–759

Yang K-Y, Liu Y, Zhang S (2001) Activation of a mitogen-activated protein kinase pathway is involved in disease resistance in tobacco. Proc Natl Acad Sci U S A 98:741–746

Yao N, Tada Y, Sakamoto M, Nakayashiki H, Park P, Tosa Y, Mayama S (2002) Mitochondrial oxidative burst involved in apoptotic response in oats. Plant J 30:567–579

Yeh C-M, Hung W-C, Huang H-J (2003) Copper treatment activates mitogen-activated protein kinase signalling in rice. Physiol Plant 119:392–399

Yeh C-M, Chien P-S, Huang H-J (2007) Distinct signalling pathways for induction of MAP kinase activities by cadmium and copper in rice roots. J Exp Bot 58:659–671

Yoshioka H, Numata N, Nakajima K, Katou S, Kawakita K, Rowland O, Jones JDG, Doke N (2003) *Nicotiana benthamiana* gp91^{phox} homologs NbrbohA and NbrbohB participate in H2O2 accumulation and resistance to *Phytophthora infestans*. Plant Cell 15:706–718

Yuasa T, Ichimura K, Mizoguchi T, Shinozaki K (2001) Oxidative stress activates ATMPK6, an *Arabidopsis* homologue of MAP kinase. Plant Cell Physiol 42:1012–1016

Zeidler D, Zahringer U, Gerber I, Dubery I, Hartung T, Bors W, Hutzler P, Durner J (2004) Innate immunity in *Arabidopsis thaliana*: lipopolysaccharides activate nitric oxide synthase (NOS) and induce defense genes. Proc Natl Acad Sci U S A 101:15811–15816

Zhang W, Wang C, Qin C, Wood T, Olafsdottir G, Wang X (2003) Phospholipase Dδ and phosphatidic acid decrease H_2O_2-induced cell death in *Arabidopsis*. Plant Cell 15:2285–2295

Zhang W, Yu L, Zhang Y, Wang X (2005) Phospholipase D in the signaling networks of plant response to abscisic acid and reactive oxygen species. Biochim Biophys Acta 1736:1–9

Zhu J, Dong C-H, Zhu J-K (2007) Interplay between cold-responsive gene regulation, metabolism and RNA processing during plant cold acclimation. Curr Opin Plant Biol 10:290–295

Zipfel C, Robatzek S, Navarro L, Oakeley EJ, Jones JD, Felix G, Boller T (2004) Bacterial disease resistance in *Arabidopsis* through flagellin perception. Nature 428:764–767

Integrated Nod Factor Signaling in Plants

Sibylle Hirsch and Giles Oldroyd

Abstract Many legumes enter symbiotic relationships with rhizobial bacteria to acquire the macronutrient nitrogen. Bacteria reduce nitrogen to ammonia, a form accessible for the plant, in plant-derived organs, called "nodules." The symbiotic interaction is initiated by the release of the bacterial signal Nod factor into the rhizosphere and its recognition by plant roots. The perception of Nod factor in the plasma membrane induces a signaling pathway that uses calcium as a secondary messenger. Genetic analysis in legume species such as *Medicago truncatula* and *Lotus japonicus* revealed many components which are essential for the Nod factor induced signaling pathway. We describe the structural specificity of Nod factor recognition and the genes involved in the activation and perception of calcium oscillations during Nod factor signaling. In addition, the processes that lead to the initiation of nodule organogenesis following Nod factor signaling are briefly discussed.

1 Introduction

One of the most important examples of a beneficial symbiosis in the biosphere is the symbiotic interaction between nitrogen-fixing soil bacteria and their plant hosts. Nitrogen fixation takes place in unique organs predominantly associated with roots, called "nodules." The formation of nodules provides an oxygen-limited environment that is suitable for the activity of the oxygen-inhibited bacterial enzyme nitrogenase. Nitrogen-fixing bacteria convert atmospheric nitrogen into ammonia, which is absorbed by the plant to meet its nutritional needs. As nitrogen is usually lacking from many soils around the world, biological nitrogen fixation has a significant impact on global agriculture. Indeed, early agricultural systems invariantly used co-cultivation or rotation of a legume with a cereal crop to both enhance soil fertility and provide a balance of protein-rich and carbohydrate-rich foods (e.g., soybean/rice in Asia, bean/maize in the Americas, and lentil, chickpea/wheat, barley in Mesopotamia).

S. Hirsch and G. Oldroyd (✉)
Department of Disease and Stress Biology, John Innes Centre, Norwich, NR4 7UH, UK
e-mail: Sibylle.hirsch@bbsrc.ac.uk

F. Baluška and S. Mancuso (eds.), *Signaling in Plants*, Signaling and Communication in Plants, 71
DOI: 10.1007/978-3-540-89228-1_4, © Springer-Verlag Berlin Heidelberg 2009

Nitrogen-fixing root nodule symbioses occur in two major associations: the legume–*Rhizobium* association and the actinorhiza–*Frankia* association. The actinorhiza –*Frankia* symbiosis differs from the interaction of legumes with rhizobia in several morphological and cytological aspects (Pawlowski and Bisseling 1996). However, both legume and actinorhizal symbioses are initiated by an exchange of signals, a process, that has been well studied for the legume–*Rhizobium* interaction but is poorly understood for the actinorhizal symbiosis. The legume–*Rhizobium* association is initiated by the release of plant-made compounds known as flavonoids from the plant root into the rhizosphere and their recognition by rhizobia. Upon recognition of flavonoids, *nod* genes of rhizobia are induced and these produce lipochitooligosaccharide signal molecules called "Nod factors" (Denarie et al. 1996; Long 1996; Spaink 2000). Rhizobial infection of the host plant and formation of the nodule primordia depend on the appropriate recognition of Nod factor by the host legume.

Bacterial infection can occur by both intracellular and intercellular mechanisms. The intracellular mechanism is used in most root nodule symbioses and generally starts with the attachment of rhizobia to the root hairs of legumes. This results in the induction of root hair curling and in the entrapment of the bacteria within a curled root hair, a structure also known as an infection pocket. The bacteria induce the hydrolysis of the plant cell wall and this coupled with plasma membrane invagination forms a tubular structure, the infection thread. Bacteria enter the plant root through the infection threads which grow towards the root cortex. Concomitant with this infection process, Nod factor triggers the formation of the nodule primordia by inducing cell divisions in the root cortex. After the infection thread reaches the nodule primordia, rhizobia are released inside the plant cell and enclosed by a plant-derived membrane. Within this structure the rhizobial bacteria differentiate into bacteroids, and these coupled with the surrounding plant membrane are referred to as "symbiosomes," the cellular structures capable of nitrogen fixation (Newcomb 1981). Intracellular invasion occurs in the majority of legume hosts and in the model legumes *Medicago truncatula* and *Lotus japonicus*. During intercellular infection, which is observed in many tropical legumes, the bacteria enter via cracks in the epidermis that can form where lateral or adventitious roots emerge (Chandler et al. 1982; Goormachtig et al. 2004; James et al. 1992). Although the two infection processes differ in many ways, they both depend on Nod factor and its signal transduction pathway in the plant host (Cullimore et al. 2001; D'Haeze et al. 2000; Oldroyd 2001).

Nod factor is a key player in the coordination of nodulation and therefore we will focus our attention on Nod factor, its perception, and the induction of the Nod factor signaling cascade in legumes.

2 Nod Factor Production in Bacteria

To establish a functional root nodule symbiosis, rhizobial bacteria have to recognize and respond to the presence of the legume host. This is realized by the chemical interplay of flavonoid compounds released from the legume root and bacterially derived Nod factor, a potent signal to the legume host.

2.1 *Flavonoids and* nod *Gene Induction*

Flavonoids are plant secondary metabolites that are synthesized via the phenylpropanoid pathway. They are diverse molecules with a wide range of structures and function. For decades flavones and isoflavones, subclasses of flavonoids, have been known to play a role in nodulation (Peters et al. 1986). Approximately 30 flavonoids which act as inducers of *nod* genes in rhizobia have been isolated (Begum et al. 2001; Cooper 2007; Hungria et al. 1992; Smit et al. 1992). *nod* genes, which are predominantly required for Nod factor synthesis, are mediated by NodD transcriptional regulators. NodDs associate with promoter elements of *nod* genes to activate their transcription (Fisher and Long 1993). It has long been discussed whether flavonoids form a complex with NodD at the DNA, but no evidence for a direct interaction between the two molecules has yet been found. However, it has been observed that flavonoids stimulate an increase in DNA binding of NodD1 to *nod* gene promoters (Peck et al. 2006). Recent work indicates the further importance of flavonoids for nodulation once the bacteria have entered the plant root and exogenous flavonoids are no longer available (Subramanian et al. 2007). In *M. truncatula* and soybean it has been shown that the suppression of endogenous flavonoid production abolishes nodulation (Subramanian et al. 2006; Wasson et al. 2006).

Besides the induction of *nod* gene expression, flavonoids elicit strong chemotactic and growth responses to concentrate rhizobia at the root surface (Rolfe 1988; Stougaard 2000). In addition, the activation of genes expressing type III secretion proteins and rhamnose-rich lipopolysaccharides requires flavonoids (Kobayashi et al. 2004). Furthermore, flavonoids have been proposed to have a role much later in nodule development by regulating auxin transport during the initiation of the nodule primordium (Wasson et al. 2006). Thus, flavonoids may play multiple roles during the legume–*Rhizobium* symbiosis with the induction of Nod factor signal production as a key function.

2.2 *Nod Factor Synthesis, Structure, and Specificity*

Nod factors are lipochitooligosaccharides comprising a backbone of β-1–4-linked *N*-acetyl-D-glucosamine residues with N-linked acyl groups attached to the nonreducing end (Fig. 1). Nod factors show a wide variation in their structure; they differ in the length of the backbone, the saturation of the fatty acid residue, and the number and types of substituent groups (Denarie et al. 1996; Spaink 2000). The additional modifications to the basic molecule can include sulfuryl, methyl, mannosyl, carbamoyl, acetyl, fucosyl, and arabinosyl moieties. The structure of the Nod factor molecule varies between different rhizobial strains. This diversity in Nod factor structure is often the basis for the specificity of interaction between certain bacterial strains and legume species.

As outlined earlier, Nod factor production is a function of the rhizobial *nod* genes. Each rhizobial strain has a characteristic array of *nod* genes. The *nodABC* genes are required for the production of the Nod factor backbone (the N-acylated

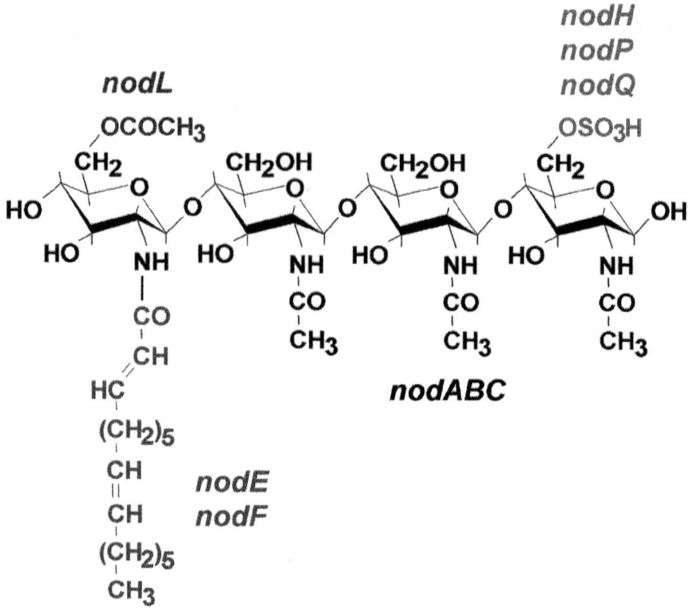

Fig. 1 Structure of Nod factor from *Sinorhizobium meliloti*. The Nod factor backbone of β-1–4-linked *N*-acetyl-D-glucosamine residues, which requires *nodABC*, and its linked *N*-acyl group is common for all rhizobial strains, while the number and types of substituent groups can vary. Nod factor from *S. meliloti* carries an *O*-sulfate group at the reducing end which requires *nodHPQ*. The appropriate attachment of the *N*-acyl group is *nodEF*-dependent and the attachment of the *O*-acetyl group at the nonreducing end depends on *nodL*

chitin) (Fig. 1). It was shown that *nodABC* can be functionally exchanged between rhizobial strains (Denarie et al. 1996), indicating that those genes are not involved in defining the bacterial host range. For the *Medicago* symbiont, *Sinorhizobium meliloti*, *nodE*, *nodF*, *nodL*, *nodH*, *nodP*, and *nodQ* are host-specific *nod* genes. *S. meliloti* Nod factor carries an *O*-sulfate group at the reducing glucosamine residue, which is a combined function of *nodH*, *nodP*, and *nodQ* (Fig. 1) (Roche et al. 1991; Schwedock and Long 1990). *S. meliloti* Nod factor can only induce nodulation in its *Medicago* hosts when this sulfate is present. The lack of the host-specific *nod* genes can alter the host range of *S. meliloti* (Denarie et al. 1996). For example, *nodH* mutants, which lack the *O*-sulfate group, lose their ability to nodulate *Medicago sativa*, but gain the activity on *Vicia sativa* (Faucher et al. 1989).

S. meliloti nodE and *nodF* regulate the attachment of the appropriate $C_{16:2}$ *N*-acyl group, while *nodL* is required for the *O*-acetyl modification (Fig. 1) (Demont et al. 1993; Spaink et al. 1991). Mutations in *nodE*, *nodF*, and *nodL* lead to delayed and reduced nodulation in *S. meliloti* hosts *Medicago* spp., but still elicit morphological responses associated with nodulation (Ardourel et al. 1994; Debelle et al. 1986; Swanson et al. 1987). *nodF*/*nodL* double mutants showed a more severe phenotype: they were unable to penetrate into legume hosts and to form infection threads (Ardourel et al. 1994). However, these mutants were able to activate cortical cell division, indicating that the Nod factor structural requirements are more stringent for bacterial entry than for the induction of the nodule primordium (Ardourel et al. 1994). This work and work from other laboratories uncoupled bacterial infection from the activation of earlier Nod factor responses, suggesting that two specificities for Nod factor recognition might exist: low-stringency recognition that is necessary for earlier stages of the symbiosis and high-stringency recognition necessary for bacterial infection (Firmin et al. 1993; Geurts et al. 1997; Oldroyd and Downie 2004; Spaink et al. 1991; Walker and Downie 2000).

3 Nod Factor Recognition in the Plant

Nod factor released from rhizobial bacteria is required to activate several plant processes associated with nodulation (Downie and Walker 1999). Genetic approaches have been used to identify genes involved in the induction of these plant responses.

3.1 Receptor-Like Kinases as Candidates for the Nod Factor Receptor

To identify receptors in legumes responsible for the perception of Nod factor, mutant screens have been performed. Legume mutants lacking all Nod factor induced plant responses were identified in *L. japonicus, M. truncatula*, and *Pisum sativum*. In *L. japonicus* two genes, *LjNFR1* and *LjNFR5*, were found to be essential for Nod factor perception (Fig. 2) (Madsen et al. 2003; Radutoiu et al. 2003). Both genes encode transmembrane receptor-like kinases with two to three extracellular lysine motifs (LysM) and an intracellular serine/threonine protein kinase domain. While LjNFR1 contains an apparently functional serine/threonine kinase domain (Huse and Kuriyan 2002; Schenk and Snaar-Jagalska 1999), *LjNFR5* encodes a kinase lacking the activation loop that usually regulates kinase activity (Madsen et al. 2003). This leads to the hypothesis that LjNFR5 forms a receptor complex with LjNFR1 which contributes the kinase activation domain (Radutoiu et al. 2003). LysM motifs present in the extracellular domains of both receptor-like kinases have previously been identified to bind peptidoglycans (Bateman and Bycroft 2000). The binding occurs to the *N*-acetylglucosamine-*N*-acetylmureine backbone, which is chemically similar to the Nod factor backbone (Steen et al. 2003); but the direct binding of Nod factor to LjNFR1 and LjNFR5 LysM domains remains to be demonstrated. However, recent work revealed that the specificity of Nod factor recognition is mediated by LjNFR1 and LjNFR5 (Radutoiu et al. 2007). The expression of both receptor-like kinases from *L. japonicus* in *M. truncatula* and *Lotus filicaulis* extends their host range to include the *L. japonicus* symbiont *Mesorhizobium loti*. Furthermore, domain swaps and amino acid substitutions demonstrated the importance of the LysM domains, especially the LysM2 domain of LjNFR5 in Nod factor specificity (Radutoiu et al. 2007).

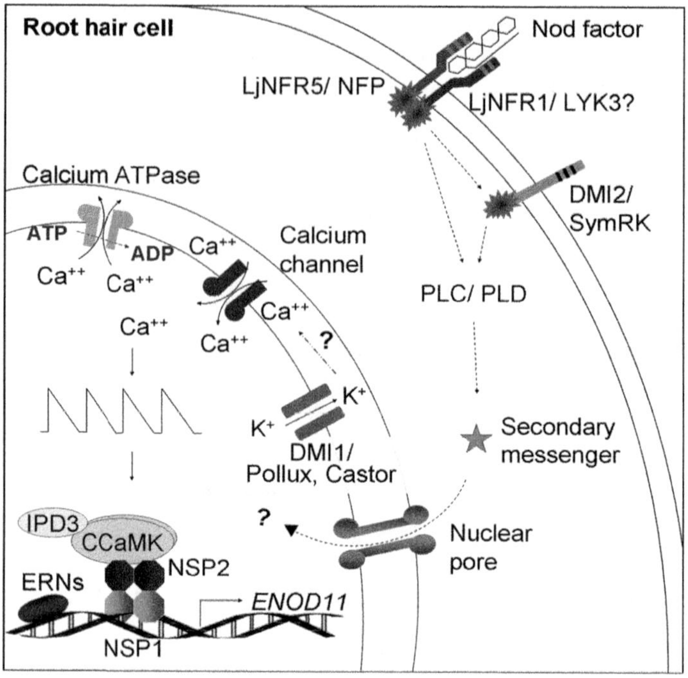

Fig. 2 Nod factor signaling in the legume-*Rhizobium* symbiosis. Nod factor recognition requires two receptor-like kinases, LjNFR1 (LYK3?) and LjNFR5/NFP, consisting of extracellular LysM domains and an intracellular kinase domain. A second type of receptor-like kinase is involved in Nod factor signaling: DMI2/SymRK, which contains extracellular leucine-rich repeat motifs. Phospholipase C, phospholipase D, and a secondary messenger are proposed to link Nod factor recognition to calcium spiking in the nucleus. The activation of calcium spiking also depends on components of the nuclear pore complex, NUP85 and NUP133, and up to two potassium channels, DMI1/Castor, Pollux, on the nuclear membrane. The calcium signal is presumed to be decoded and transduced by CCaMK, NSP1, NSP2, and ERNs. *PLC* phospholipase C, *PLD* phospholipase D, *CCaMK* calcium/calmodulin-dependent protein kinase, *ENOD11* early nodulation gene

Characterization of *M. truncatula* and *P. sativum* mutants has led to the identification of the *LjNFR5* orthologues *MtNFP* and *PsSYM10*, respectively (Amor et al. 2003; Arrighi et al. 2006; Madsen et al. 2003; Walker et al. 2000). Loss-of-function mutations in *MtNFP* and *PsSYM10* caused the loss of all Nod factor inducible plant responses, similar to *Ljnfr5*. Several homologues of *LjNFR1* exist in *M. truncatula* and mutations or RNA interference of two of these, *LYK3* and *LYK4*, show specific defects in *S. meliloti* infection, but not in the induction of earlier Nod factor responses (Limpens et al. 2003; Smit et al. 2007). Further analysis of a weak *lyk3* allele revealed that LYK3 is involved in the recognition of the *O*-acetyl modification of the *S. meliloti* Nod factor (Fig. 1). This work provides evidence in *M. truncatula* for the previously proposed two-receptor model that was based on genetic work in *S. meliloti* (Ardourel et al. 1994): *MtNFP* provides low-stringency recognition for early Nod factor responses, while *LYK3* provides high-stringency recognition that is required to allow *S. meliloti* infection.

3.2 The Role of Lectin Nucleotide Phosphohydrolase in Nod Factor Recognition

During the past few years much less attention had been paid to a possible alternative or additional candidate for Nod factor perception. In 1999 a lectin was isolated from roots of the legume *Dolichos biflorus* (Quinn and Etzler 1987) that catalyzes the hydrolysis of phosphoanhydride bonds of nucleoside diphosphates and triphosphates (Etzler et al. 1999). It was therefore called "lectin nucleotide phosphohydrolase" (LNP). LNP binds to Nod factors and Nod factor binding induces its phosphohydrolase activity in vitro. An antiserum to LNP of *D. biflorus* and *Glycine max* inhibits root hair deformation and nodule formation, suggesting that this protein has a function in the legume–*Rhizobium* symbiosis (Day et al. 2000; Etzler et al. 1999). Moreover, the expression of an LNP gene in *G. max* is induced by rhizobia, while LNP overexpression enhances nodulation (Day et al. 2000; McAlvin and Stacey 2005). Although these data highlight the relevance of LNP in nodulation, a role for LNP as a Nod factor receptor remains controversial.

4 The Role of Calcium in Nod Factor Signaling

The calcium ion is an important component in a diverse array of signaling pathways in plants. It is a ubiquitous intracellular secondary messenger that has been found to increase upon many physiological stimuli. Oscillations in calcium ion concentrations have been observed in stomatal guard cells (McAinsh et al. 1995) and have been shown to specifically regulate stomatal closure (Allen et al. 2001). Calcium ions also play a pivotal role in Nod factor signaling (Fig. 2). Application of Nod factor induces two separable calcium responses: a rapid influx of calcium ions and oscillations in cytosolic calcium around the nucleus.

With use of ion-specific microelectrodes an influx of calcium ions into *M. sativa* root hair cells could be shown after Nod factor addition (Felle et al. 1999a, b). The calcium ion influx is followed by the efflux of chloride ions within a few seconds, leading to Nod factor induced membrane depolarization (Felle et al. 1998). To rebalance the membrane charge, cation channels are activated, allowing the efflux of potassium ions and the repolarization of the membrane (Felle et al. 1998). Calcium ion influx associated with the root hair tip has also been observed in *Phaseolus vulgaris*, *P. sativum*, and *M. truncatula* root hair cells (Cardenas et al. 1999; Shaw and Long 2003; Walker et al. 2000), suggesting that it is a common feature in the legume–*Rhizobium* symbiosis. Calcium ion influx occurs within 1 min after the application of Nod factor at a minimum concentration of 1–10 nM (Felle et al. 1998; Shaw and Long 2003).

In addition to the calcium flux response, Nod factor induces oscillations in cytosolic calcium ion concentrations associated with the nucleus, called "calcium spiking" (Fig. 2). This was first shown in *M. sativa* root hair cells using calcium-sensitive reporter dyes (Ehrhardt et al. 1996). Since then calcium spiking has been observed in *P. vulgaris, L. japonicus, M. truncatula*, and *P. sativum* root hair cells following the addition of Nod factor (Cardenas et al. 1999; Harris et al. 2003; Oldroyd et al. 2001; Wais et al. 2000; Walker et al. 2000). A calcium spike always shows a characteristic shape: a sudden increase in calcium levels followed by a more gradual decline (Meyer and Stryer 1988). In contrast to the Nod factor induced calcium flux, calcium spiking starts with a lag of approximately 10 min after Nod factor addition. The reason for this lag between Nod factor addition and calcium spiking induction has yet to be explained. Interestingly, calcium spiking appears at a minimum Nod factor concentration of 1–10 pM, a roughly 1,000-fold lower concentration than that required for induction of the calcium flux (Oldroyd et al. 2001; Shaw and Long 2003). In addition, calcium spiking is activated by Nod-factor-like chitin oligomers without the activation of the calcium flux (Walker et al. 2000). These discoveries indicate that for the induction of calcium spiking a relatively low stringency of Nod factor perception is required, while a much higher concentration and specific Nod factor structures are needed to elicit the calcium flux. This correlates with previous work indicating a high stringency of Nod factor recognition required for bacterial infection, and a model has been proposed which suggests a role of the calcium flux in the activation of infection thread formation (Miwa et al. 2006). This model predicts that upon the first contact between the legume root and the rhizobia the perception of a low Nod factor concentration initiates calcium spiking and subsequently promotes the accumulation of bacteria inside the infection foci. Once rhizobia are trapped in the root hair curl, the Nod factors should accumulate to high concentrations and therefore activate calcium flux, which could then lead to the initiation of the infection thread. In contrast to the observed lag phase using calcium-sensitive dyes, in this model the calcium flux occurs later than the calcium spiking response. Indeed, it has been shown that root hair cells initially induced to spike were able to show a calcium ion influx when Nod factor levels were secondarily raised (Shaw and Long 2003).

Pharmacological and genetic research indicate a function for calcium spiking in the activation of early Nod factor induced responses such as gene expression (Charron et al. 2004; Wais et al. 2000; Walker et al. 2000). It has been widely suggested that the nature of the calcium oscillations encodes information to define the outcome of the downstream response (Allen et al. 2001; Evans et al. 2001; Oldroyd and Downie 2006).

5 Genes Involved in the Activation of Calcium Spiking

Calcium spiking plays a crucial role in the Nod factor induced signaling pathway; it probably acts as a secondary messenger to transduce the Nod factor signal. Mutant screens have revealed several components required for the Nod factor dependent activation of calcium spiking. In addition to the LysM receptor-like kinases (see Sect. 3.1), a leucine-rich repeat (LRR) receptor-like kinase (Endre et al. 2002; Stracke et al. 2002), up to two putative cation channels (Ane et al. 2004; Edwards et al. 2007; Imaizumi-Anraku et al. 2005; Riely et al. 2007), and two nucleoporins

(Kanamori et al. 2006; Saito et al. 2007) are necessary for the induction of calcium spiking (Wais et al. 2000). Biochemical and pharmacological studies further indicate that multiple phospholipid signaling pathways are linked to Nod factor signaling (Charron et al. 2004; den Hartog et al. 2001; Engstrom et al. 2002; Sun et al. 2007).

5.1 LRR Receptor-Like Kinase

In addition to the LysM receptor-like kinases, the Nod factor signal transduction pathway requires a second type of receptor-like kinase that contains three LRR motifs in the extracellular domain (Fig. 2). The importance of this LRR receptor-like kinase for nodulation has been characterized in *Sesbania rostrata* (*SrSymRK*), *M. sativa* (*MsNORK*), *P. sativum* (*PsSym19*), *M. truncatula* (*DMI2*), and *L. japonicus* (*LjSymRK*) (Capoen et al. 2005; Endre et al. 2002; Stracke et al. 2002). In addition, the orthologues of this receptor-like kinase have been shown to be critical for nodulation in the actinorhizal plants *Casuarina glauca* (*CgSymRK*) and *Datisca glomerata* (*DgSymRK*), indicating genetic mechanisms shared between the two types of root nodule symbioses (Gherbi et al. 2008; Markmann et al. 2008). This LRR receptor-like kinase is essential for the activation of calcium spiking and early nodulin gene expression, but is not required for the activation of the calcium flux (Catoira et al. 2000; Miwa et al. 2006; Shaw and Long 2003; Wais et al. 2000). In contrast, the LysM receptor-like kinases are necessary for all Nod factor induced responses including the calcium flux, indicating a bifurcation in the Nod factor signaling pathway immediately downstream of the LysM receptor kinases, with one branch leading to calcium spiking and one branch leading to the calcium flux. In addition, partial suppression of *DMI2* and *SrSymRK* using RNA interference revealed a function for these genes at later stages of the symbiosis, during bacterial release and symbiosome formation (Capoen et al. 2005; Limpens et al. 2005). Consistent with these observations, the receptor-like kinase is expressed in roots prior to infection and is induced in the preinfection zone of *M. truncatula* nodules (Bersoult et al. 2005; Limpens et al. 2005).

5.2 The Role of Ion Channels in Nodulation Signaling

Calcium spiking is associated with the nucleus and it has been proposed that the nuclear envelope and the endoplasmic reticulum may act as the internal stores for this calcium response. A number of genes have been identified that are necessary for Nod factor induced calcium spiking and potentially function in regulating the release of calcium from these stores (Miwa et al. 2006; Shaw and Long 2003; Wais et al. 2000). In *L. japonicus* there are two highly homologous genes, *CASTOR* and *POLLUX*, encoding for predicted integral membrane proteins (Imaizumi-Anraku et al. 2005). In *M. truncatula* and *P. sativum* only one such gene, *DMI1* and *SYM8*, respectively, was identified (Fig. 2) (Ane et al. 2004; Edwards et al. 2007). *DMI1* and *SYM8* are orthologous and are more closely related to *POLLUX* than they are to *CASTOR*. DMI1 has been reported to localize to the periphery of nuclei of root cells (Riely et al. 2007). DMI1, SYM8, CASTOR, and POLLUX show weak homology to MthK, a calcium-gated potassium channel from *Methanobacterium thermoautotrophicum* (Jiang et al. 2002). Observations using *sym8* and a structural model of the SYM8 channel indicate that SYM8/DMI1 may form a cation channel (but not a calcium channel) which could be opened by the binding of a signaling ligand (Edwards et al. 2007). SYM8/DMI1 may allow cation flow across the nuclear membrane, thus changing the membrane polarization, and this could regulate the opening of an as yet unidentified calcium channel. The hypothesis that SYM8/DMI1 regulates calcium channels is supported by the discovery that DMI1 interferes with calcium release from internal endoplasmic reticulum stores in yeast (Peiter et al. 2007).

5.3 Two Nucleoporins Required for Calcium Spiking

Two additional genes essential for Nod factor activation of calcium spiking have been identified in *L. japonicus*, *NUP133* and *NUP85*, and they both encode nucleoporins (Fig. 2) (Kanamori et al. 2006; Saito et al. 2007). Nucleoporins are components of the nuclear pore that consists of more than 30 proteins and function in the transport of large proteins and RNAs (Suntharalingam and Wente 2003). Nuclear pore complexes are additionally necessary for the appropriate localization of proteins to the inner nuclear membrane (Suntharalingam and Wente 2003). A NUP133-green fluorescent protein fusion localizes to the nuclear rim of root and root hair cells (Kanamori et al. 2006), indicating that NUP133 and NUP85 are members of a nuclear pore complex in legumes. Therefore, we might hypothesize that both nucleoporins are involved in the activation of calcium spiking by the possible transport and localization of calcium or other cation channels, such as SYM8/DMI1. However, further research will be required to clarify the role of NUP133 and NUP85 in early symbiotic signal transduction.

5.4 Phospholipase C and Phospholipase D

During the past few years a function for phospholipids in early Nod factor signaling has been discovered (Fig. 2). Phospholipase C, which is activated by heterotrimeric G-proteins, hydrolyzes plasma membrane phospholipid phosphatidylinositol 4,5-bisphosphate into inositol 1,4,5-trisphosphate and diacylglycerol. Inositol 1,4,5-trisphosphate, in turn, can lead to the release of calcium from intracellular stores, such as the endoplasmatic reticulum. Mastoparan, a tetradecapeptide, activates heterotrimeric G-proteins and so artificially induces effector enzymes downstream of G-proteins. It has been shown that mastoparan treatment activates calcium spiking, nodulin gene expression, and root hair deformation, indicating a role of G-proteins in Nod factor signaling (den Hartog et al. 2001; Pingret et al. 1998; Sun et al. 2007). Nodulin gene expression triggered by mastoparan requires the calcium/calmodulin-dependent protein kinase CCaMK (see Sect. 6.1), but does not require DMI1 and DMI2 (Charron et al. 2004; Sun et al. 2007). This suggests that mastoparan activates a component of the Nod factor signaling pathway downstream of DMI1/DMI2 and upstream of calcium spiking and CCaMK. Additionally, phospholipase C and phospholipase D were both found to be activated by Nod factor (den Hartog et al. 2001; den Hartog et al. 2003) and their requirement in signal transduction has been revealed by pharmacological studies. Root hair deformation and nodulin expression were inhibited by application of neomycin, an inhibitor of phospholipase C (Charron et al. 2004; den Hartog et al. 2001; Pingret et al. 1998). Cyclopiazonic acid and U-73122, inhibitors of type IIA calcium pumps and phospholipase C, abolish Nod factor induced calcium spiking (Engstrom et al. 2002). *n*-Butanol and cyclopiazonic acid inhibit both nodulin gene expression and mastoparan-induced calcium spiking (Charron et al. 2004; Sun et al. 2007). These pharmacological studies support the requirement of phospholipase C, phospholipase D, and calcium pumps as components in Nod factor-induced calcium signaling.

6 Genes Involved in the Perception of Calcium Spiking

Recent discoveries have deciphered components in the Nod factor signaling pathway which link Nod factor perception to the pivotal calcium spiking response. In addition, a number of components of the Nod factor signaling pathway have been defined that function in the perception and transduction of the calcium spiking signal. A calcium/calmodulin-dependent protein kinase and at least three transcription factors are likely to link calcium to the activation of nodulation gene expression.

6.1 The Function of a Calcium/Calmodulin-Dependent Protein Kinase

A calcium/calmodulin-dependent protein kinase (CCaMK) required for the legume–*Rhizobium* symbiosis has been identified in *M. truncatula* (*DMI3*), *L. japonicus* (*LjCCaMK*), and *P. sativum* (*PsSYM9*) (Levy et al. 2004; Mitra et al. 2004). Analogous to proteins that function upstream of calcium spiking, CCaMK is necessary for Nod factor induced gene expression, but is not necessary for the activation of calcium spiking (Catoira et al. 2000; Wais et al. 2000). This indicates that CCaMK plays a role in signaling downstream of the calcium oscillations (Fig. 2). The CCaMK protein consists of a serine/threonine kinase domain, followed by a calmodulin binding domain and three calcium binding EF-hand domains. It belongs to a family of plant-specific proteins, but its calmodulin binding domain is similar to the mammalian CaMKII which has the ability to decode calcium spiking (De Koninck and Schulman 1998). In general, plant chimeric calcium/calmodulin dependent protein kinases are dually regulated: free calcium ions, which bind to the EF-hand domains, induce autophosphorylation and thereby enhance the binding of calcium complexed with calmodulin. The binding of calmodulin in turn promotes substrate phosphorylation (Gleason et al. 2006; Patil et al. 1995; Ramachandiran et al. 1997; Takezawa et al. 1996). An additional domain important for the function of CCaMK is the autoinhibitory domain which negatively regulates kinase activity. Specific removal of this domain leads to gene expression and spontaneous nodule formation in the absence of rhizobia or Nod factor (Gleason et al. 2006). A point mutation in the autophosphorylation site of *LjCCaMK* also resulted in spontaneous nodulation when expressed from its native promoter (Tirichine et al. 2006), implying that CCaMK is a central regulator for nodule organogenesis. A newly isolated protein, IPD3, has been shown to interact directly with *M. truncatula* CCaMK in yeast and in planta (Messinese et al. 2007). IPD3, which lacks homology to functionally characterized proteins, might act downstream of CCaMK in the Nod factor induced signaling pathway.

6.2 A Suite of Transcription Factors Transduce the Signal Downstream of CCaMK

NSP1 and NSP2 are two putative GRAS domain transcriptional regulators which function in Nod factor signaling downstream of calcium spiking (Kalo et al. 2005; Oldroyd and Long 2003; Smit et al. 2005; Wais et al. 2000). Both GRAS domain proteins have been identified in *M. truncatula* and *L. japonicus* (*LjNSP1*, *LjNSP2*) (Heckmann et al. 2006). The GRAS family of putative transcriptional regulators is found throughout the plant kingdom and these proteins have diverse roles in plant development (Bolle 2004). Although NSP1 and NSP2 both encode GRAS family members they are not very similar (17% identity, 32% similarity), indicating different functions in the Nod factor signaling pathway. *M. truncatula nsp1* and *nsp2* show defects in the activation of the nodule primordium, bacterial infection, and Nod factor induced gene expression and show reduced root hair deformation (Catoira et al. 2000; Oldroyd and Long 2003). The constitutive CCaMK construct did not autoactivate gene expression in the *nsp2* mutant, but showed partial activation in the *nsp1* mutant (Gleason et al. 2006). These observations place NSP1 and NSP2 downstream of CCaMK and indicate that the calcium signal is transduced by CCaMK through the two GRAS domain regulators (Fig. 2). Recently, it has been shown that NSP1 and NSP2 homodimerize and heterodimerize with each other and that NSP1 associates with the promoter of the early nodulation gene *ENOD11* (Hirsch et al., unpublished results). The fact that NSP2 also heterodimerizes with the kinase domain of CCaMK indicates a direct activation of the GRAS protein complex by CCaMK and links calcium spiking to the induction of nodulation gene expression through these transcription factors.

Like NSP1 and NSP2, ERN1, an ethylene response factor (ERF) transcription factor, is required for nodulation and acts downstream of CCaMK (Middleton et al. 2007). ERN1 and two close homologues, ERN2 and ERN3, have been identified to associate with the *ENOD11* promoter close to the NSP1 binding site (Fig. 2) (Andriankaja et al. 2007). Although this might indicate a direct heterodimerization between NSP1 and the ERN proteins, no such interaction has yet been described. While ERN1 and ERN2 function as transcriptional activators, ERN3 represses ERN1/ERN2-dependent transcriptional activation of *ENOD11* (Andriankaja et al. 2007). Another protein which negatively regulates *ENOD11* expression is NIN (Marsh et al. 2007; Schauser et al. 1999). *NIN* encodes a protein with transmembrane domains, but also a putative DNA binding domain. NIN negatively regulates the spatial pattern of *ENOD11* expression 24 h after Nod factor treatment (Marsh et al. 2007). Despite this apparent negative regulation of early nodulation gene expression, genetic studies have revealed that NIN is a positive regulator of nodulation and bacterial infection. Thus, NIN is likely to act in the activation of gene expression at later stages of nodule organogenesis.

7 The Arbuscular Mycorrhizal Symbiosis

More than 80% of terrestrial plants form beneficial symbioses with arbuscular mycorrhizal (AM) fungi. The AM symbiosis is thought to have evolved more than 400 million years ago and is therefore one of the oldest symbioses entered into by plants. The AM fungi enhance the uptake of macronutrients such as organic nitrogen and phosphate through the plant root (Harrison 1999; Hodge et al. 2001). Interestingly, nodulation and mycorrhizal signaling pathways share several common features, including the activation of calcium oscillations and the induction of a common set of signaling genes. At least seven genes that play a role in both types of symbioses have been identified so far. The LRR receptor-like kinase (Sect. 5.1) and the putative ion channels (Sect. 5.2) are dually involved in the activation of calcium spiking in nodulation and arbuscular mycorrhization (Catoira et al. 2000; Kosuta et al. 2008). CCaMK and IPD3 (Sect. 6.1), which act downstream of calcium spiking, are also shared signaling components, while NSP1 and NSP2 (Sect. 6.2) are not (Catoira et al. 2000; Messinese et al. 2007). The fact that root nodule and AM symbioses harbor conserved early-signaling genes supports the model that nodulation genes have been recruited from the more ancient AM symbiosis.

8 Cytokinin, a Positive Regulator of Nodulation

Hormones which possess a general role in plant organogenesis have also been shown to be important for nodule formation. Cytokinin, an adenine-derived signaling molecule, is a positive regulator of nodule organogenesis. The exogenous cytokinin application to legume roots activates Nod-factor-like responses such as cortical cell division and early nodulin gene expression (Cooper and Long 1994; Fang and Hirsch 1998; Mathesius et al. 2000). With use of a cytokinin response gene from *Arabidopsis thaliana* cytokinin levels could be measured in transgenic *L. japonicus* root tissue during plant development and symbiotic interactions with rhizobia. The cytokinin response regulator was expressed in curled root hairs and nodule primordia after rhizobia inoculation (Lohar et al. 2004), indicating that cytokinins play a role in the legume–*Rhizobium* symbiosis. In support of this, suppression of the cytokinin receptor by RNA interference led to cytokinin-insensitive roots, which showed a strong reduction in nodule formation in *M. truncatula* (Gonzalez-Rizzo et al. 2006). A loss-of-function mutation in the cytokinin receptor gene of *L. japonicus*, *LHK1*, also dramatically reduced nodule formation (Murray et al. 2007). *lhk1* mutants showed a block in the activation of nodule primordia, but a hyperinfection phenotype with a large number of infection threads that lost directionality during their growth in the root cortex. Furthermore, a gain-of-function mutation in *LHK1* resulted in spontaneous nodule formation in the absence of rhizobia or Nod factor and hypersensitivity to exogenous cytokinin (Tirichine et al. 2007). These observations highlight the importance of cytokinins in nodule organogenesis, but also indicate that they are dispensable for bacterial infection in the epidermis.

9 Conclusions

Genetic and biochemical studies have defined many components of the Nod factor signaling pathway. It has become clear that calcium oscillations, one of the earliest measurable plant responses, play a central role in nodulation. Two receptor-like kinases with extracellular LysM domains and one receptor-like kinase with an extracellular LRR domain are essential for the activation of calcium oscillations. The necessity of three receptor-like kinases at the plasma membrane suggests that a phoshorylation cascade is involved in the initiation of Nod factor signaling. Two cation channel(s), nucleoporins, and phospholipase C/phospholipase D are also required for the induction of calcium spiking, but their exact functions remain to be defined. Although the requirement of calcium channels and calcium pumps for internal calcium stores is critical for calcium spiking, the proteins fulfilling these functions in Nod factor signaling have not yet been identified. A calcium/calmodulin-dependent protein kinase, CCaMK, which acts downstream of calcium oscillations is an obvious candidate for decoding the calcium signal. Its dual mode of calcium binding might enable the protein to decipher the information encoded in the amplitude and frequency of the calcium spikes. Recent advances indicate that the calcium signal is directly linked to nodulation gene expression through two GRAS domain proteins and a suite of ERF transcription factors acting downstream of CCaMK. Cytokinin is a key regulator of nodule organogenesis and changes in cytokinin levels are likely to be a target for Nod factor signaling.

References

Allen GJ, Chu SP, Harrington CL, Schumacher K, Hoffman T, Tang YY, Grill E, Schroeder JI (2001) A defined range of guard cell calcium oscillation parameters encodes stomatal movements. Nature 411:1053–1057

Amor BB, Shaw SL, Oldroyd GED, Maillet F, Penmetsa RV, Cook D, Long SR, Denarie J, Gough C (2003) The NFP locus of *Medicago truncatula* controls an early step of Nod factor signal transduction upstream of a rapid calcium flux and root hair deformation. Plant J 34:495–506

Andriankaja A, Boisson-Dernier A, Frances L, Sauviac L, Jauneau A, Barker DG, de Carvalho-Niebel F (2007) AP2-ERF transcription factors mediate Nod factor dependent Mt ENOD11 activation in root hairs via a novel cis-regulatory motif. Plant Cell 19:2866–2885

Ane JM, Kiss GB, Riely BK, Penmetsa RV, Oldroyd GED, Ayax C, Levy J, Debelle F, Baek JM, Kalo P et al (2004) *Medicago truncatula* DMI1 required for bacterial and fungal symbioses in legumes. Science 303:1364–1367

Ardourel M, Demont N, Debelle FD, Maillet F, Debilly F, Prome JC, Denarie J, Truchet G (1994) *Rhizobium meliloti* lipooligosaccharide nodulation factors - different structural requirements for bacterial entry into target root hair-cells and induction of plant symbiotic developmental responses. Plant Cell 6:1357–1374

Arrighi JF, Barre A, Ben Amor B, Bersoult A, Soriano LC, Mirabella R, de Carvalho-Niebel F, Journet EP, Gherardi M, Huguet T et al (2006) The *Medicago truncatula* lysine motif-receptor-like kinase gene family includes NFP and new nodule-expressed genes. Plant Physiol 142:265–279

Bateman A, Bycroft M (2000) The structure of a LysM domain from E-coli membrane-bound lytic murein transglycosylase D (MltD). J Mol Biol 299:1113–1119

Begum AA, Leibovitch S, Migner P, Zhang F (2001) Specific flavonoids induced nod gene expression and pre-activated nod genes of *Rhizobium leguminosarum* increased pea (*Pisum sativum* L.) and lentil (*Lens culinaris* L.) nodulation in controlled growth chamber environments. J Exp Bot 52:1537–1543

Bersoult A, Camut S, Perhald A, Kereszt A, Kiss GB, Cullimore JV (2005) Expression of the *Medicago truncatula* DMI2 gene suggests roles of the symbiotic nodulation receptor kinase in nodules and during early nodule development. Mol Plant Microbe Interact 18:869–876

Bolle C (2004) The role of GRAS proteins in plant signal transduction and development. Planta 218:683–692

Capoen W, Goormachtig S, De Rycke R, Schroeyers K, Holsters M (2005) SrSymRK, a plant receptor essential for symbiosome formation. Proc Natl Acad Sci U S A 102:10369–10374

Cardenas L, Feijo JA, Kunkel JG, Sanchez F, Holdaway-Clarke T, Hepler PK, Quinto C (1999) Rhizobium Nod factors induce increases in intracellular free calcium and extracellular calcium influxes in bean root hairs. Plant J 19:347–352.

Catoira R, Galera C, de Billy F, Penmetsa RV, Journet EP, Maillet F, Rosenberg C, Cook D, Gough C, Denarie J (2000) Four genes of Medicago truncatula controlling components of a Nod factor transduction pathway. Plant Cell 12:1647–1665

Chandler MR, Date RA, Roughley RJ (1982) Infection and root-nodule development in stylosanthes species by Rhizobium. J Exp Bot 33:47–57

Charron D, Pingret JL, Chabaud M, Journet EP, Barker DG (2004) Pharmacological evidence that multiple phospholipid signaling pathways link rhizobium nodulation factor perception in *Medicago truncatula* root hairs to intracellular responses, including Ca^{2+} spiking and specific ENOD gene expression. Plant Physiol 136:3582–3593

Cooper JB, Long SR (1994) Morphogenetic rescue of *Rhizobium meliloti* nodulation mutants by trans-zeatin secretion. Plant Cell 6:215–225

Cooper JE (2007) Early interactions between legumes and rhizobia: disclosing complexity in a molecular dialogue. J Appl Microbiol 103:1355–1365

Cullimore JV, Ranjeva R, Bono JJ (2001) Perception of lipo-chitooligosaccharidic Nod factors in legumes. Trends Plant Sci 6:24–30

Day RB, McAlvin CB, Loh JT, Denny RL, Wood TC, Young ND, Stacey G (2000) Differential expression of two soybean apyrases, one of which is an early nodulin. Mol Plant Microbe Interact 13:1053–1070

Debelle F, Rosenberg C, Vasse J, Maillet F, Martinez E, Denarie J, Truchet G (1986) Assignment of symbiotic developmental phenotypes to common and specific nodulation (Nod) genetic-loci of *Rhizobium meliloti*. J Bacteriol 168:1075–1086.

De Koninck P, Schulman H (1998) Sensitivity of CaM kinase II to the frequency of Ca^{2+} oscillations. Science 279:227–230

Demont N, Debelle F, Aurelle H, Denarie J, Prome JC (1993) Role of the *Rhizobium meliloti*-Nodf and Node genes in the biosynthesis of lipo-oligosaccharidic nodulation factors. J Biol Chem 268:20134–20142

Denarie J, Debelle F, Prome JC (1996) Rhizobium lipo-chitooligosaccharide nodulation factors: signaling molecules mediating recognition and morphogenesis. Annu Rev Biochem 65:503–535

den Hartog M, Musgrave A, Munnik T (2001) Nod factor-induced phosphatidic acid and diacylglycerol pyrophosphate formation: a role for phospholipase C and D in root hair deformation. Plant J 25:55–65

den Hartog M, Verhoef N, Munnik T (2003) Nod factor and elicitors activate different phospholipid signaling pathways in suspension-cultured alfalfa cells. Plant Physiol 132:311–317

D'Haeze W, Mergaert P, Prome JC, Holsters M (2000) Nod factor requirements for efficient stem and root nodulation of the tropical legume *Sesbania rostrata*. J Biol Chem 275:15676–15684

Downie JA, Walker SA (1999) Plant responses to nodulation factors. Curr Opin Plant Biol 2:483–489

Edwards A, Heckmann AB, Yousafzai F, Duc G, Downie JA (2007) Structural implications of mutations in the pea SYM8 symbiosis gene, the DMI1 ortholog, encoding a predicted ion channel. Mol Plant Microbe Interact 20:1183–1191

Ehrhardt D, Wais R, Long SR (1996) Calcium spiking in alfalfa root hairs responding to *Rhizobium meliloti* nodulation signals. Plant Physiol 111:61002–61002

Endre G, Kereszt A, Kevei Z, Mihacea S, Kalo P, Kiss GB (2002) A receptor kinase gene regulating symbiotic nodule development. Nature 417:962–966

Engstrom EM, Ehrhardt DW, Mitra RM, Long SR (2002) Pharmacological analysis of nod factor-induced calcium spiking in *Medicago truncatula*. Evidence for the requirement of type IIA calcium pumps and phosphoinositide signaling. Plant Physiol 128:1390–1401

Etzler ME, Kalsi G, Ewing NN, Roberts NJ, Day RB, Murphy JB (1999) A Nod factor binding lectin with apyrase activity from legume roots. Proc Natl Acad Sci U S A 96:5856–5861

Evans NH, McAinsh MR, Hetherington AM (2001) Calcium oscillations in higher plants. Curr Opin Plant Biol 4:415–420

Fang YW, Hirsch AM (1998) Studying early nodulin gene ENOD40 expression and induction by nodulation factor and cytokinin in transgenic alfalfa. Plant Physiol 116:53–68

Faucher C, Camut S, Denarie J, Truchet G (1989) The Nodh and Nodq host range genes of *Rhizobium meliloti* behave as avirulence genes in R-leguminosarum Bv Viciae and determine changes in the production of plant-specific extracellular signals. Mol Plant Microbe Interact 2:291–300

Felle HH, Kondorosi E, Kondorosi A, Schultze M (1998) The role of ion fluxes in Nod factor signalling in *Medicago sativa*. Plant J 13:455–463

Felle HH, Kondorosi E, Kondorosi A, Schultze M (1999a) Elevation of the cytosolic free Ca^{2+} is indispensable for the transduction of the nod factor signal in alfalfa. Plant Physiol 121:273–279

Felle HH, Kondorosi E, Kondorosi A, Schultze M (1999b) Nod factors modulate the concentration of cytosolic free calcium differently in growing and non-growing root hairs of *Medicago sativa* L. Planta 209:207–212

Firmin JL, Wilson KE, Carlson RW, Davies AE, Downie JA (1993) Resistance to nodulation of cv Afghanistan peas Is overcome by Nodx, which mediates an O-acetylation of the *Rhizobium leguminosarum* lipo-oligosaccharide nodulation factor. Mol Microbiol 10:351–360

Fisher RF, Long SR (1993) Interactions of Nodd at the Nod box - Nodd binds to 2 distinct sites on the same face of the helix and induces a bend in the DNA. J Mol Biol 233:336–348
Geurts R, Heidstra R, Hadri AE, Downie JA, Franssen H, van Kammen A, Bisseling T (1997) Sym2 of pea is involved in a nodulation factor-perception mechanism that controls the infection process in the epidermis. Plant Physiol 115:351–359
Gherbi H, Markmann K, Svistoonoff S, Estevan J, Autran D, Giczey G, Auguy F, Peret B, Laplaze L, Franche C et al (2008) SymRK defines a common genetic basis for plant root endosymbioses with arbuscular mycorrhiza fungi, rhizobia, and Frankia bacteria. Proc Natl Acad Sci U S A 105:4928–4932
Gleason C, Chaudhuri S, Yang TB, Munoz A, Poovaiah BW, Oldroyd GED (2006) Nodulation independent of rhizobia induced by a calcium-activated kinase lacking autoinhibition. Nature 441:1149–1152
Gonzalez-Rizzo S, Crespi M, Frugier F (2006) The *Medicago truncatula* CRE1 cytokinin receptor regulates lateral root development and early symbiotic interaction with *Sinorhizobium meliloti*. Plant Cell 18:2680–2693
Goormachtig S, Capoen W, James EK, Holsters M (2004) Switch from intracellular to intercellular invasion during water stress-tolerant legume nodulation. Proc Natl Acad Sci U S A 101:6303–6308
Harris JM, Wais R, Long SR (2003) Rhizobium-induced calcium spiking in *Lotus japonicus*. Mol Plant Microbe Interact 16:335–341
Harrison MJ (1999) Molecular and cellular aspects of the arbuscular mycorrhizal symbiosis. Annu Rev Plant Physiol Plant Mol Biol 50:361–389
Heckmann AB, Lombardo F, Miwa H, Perry JA, Bunnewell S, Parniske M, Wang TL, Downie JA (2006) *Lotus japonicus* nodulation requires two GRAS-domain regulators, one of which is functionally conserved in a non-legume. Plant Physiol 142:1739–1750
Hodge A, Campbell CD, Fitter AH (2001). An arbuscular mycorrhizal fungus accelerates decomposition and acquires nitrogen directly from organic material. Nature 413:297–299
Hungria M, Johnston AWB, Phillips DA (1992) Effects of flavonoids released naturally from bean (*Phaseolus vulgaris*) on Nodd-regulated gene-transcription in *Rhizobium leguminosarum* Bv Phaseoli. Mol Plant Microbe Interact 5:199–203
Huse M, Kuriyan J (2002) The conformational plasticity of protein kinases. Cell 109:275–282
Imaizumi-Anraku H, Takeda N, Charpentier M, Perry J, Miwa H, Umehara Y, Kouchi H, Murakami Y, Mulder L, Vickers K et al (2005) Plastid proteins crucial for symbiotic fungal and bacterial entry into plant roots. Nature 433:527–531
James EK, Sprent JI, Sutherland JM, McInroy SG, Minchin FR (1992) The structure of nitrogen-fixing root-nodules on the aquatic mimosoid legume *Neptunia plena*. Ann Bot 69:173–180
Jiang YX, Lee A, Chen JY, Cadene M, Chait BT, MacKinnon R (2002) Crystal structure and mechanism of a calcium-gated potassium channel. Nature 417:515–522
Kalo P, Gleason C, Edwards A, Marsh J, Mitra RM, Hirsch S, Jakab J, Sims S, Long SR, Rogers J et al (2005) Nodulation signaling in legumes requires NSP2, a member of the GRAS family of transcriptional regulators. Science 308:1786–1789
Kanamori N, Madsen LH, Radutoiu S, Frantescu M, Quistgaard EMH, Miwa H, Downie JA, James EK, Felle HH, Haanig LL et al (2006) A nucleoporin is required for induction of Ca^{2+} spiking in legume nodule development and essential for rhizobial and fungal symbiosis. Proc Natl Acad Sci U S A 103:359–364
Kobayashi H, Graven YN, Broughton WJ, Perret X (2004) Flavonoids induce temporal shifts in gene-expression of nod-box controlled loci in Rhizobium sp NGR234. Mol Microbiol 51:335–347
Kosuta S, Hazledine S, Sun J, Miwa H, Morris RJ, Downie JA, Oldroyd GED (2008) Differential and chaotic calcium signatures in the symbiosis signaling pathway of legumes. Proc Natl Acad Sci U S A 105:9823–9828
Levy J, Bres C, Geurts R, Chalhoub B, Kulikova O, Duc G, Journet EP, Ane JM, Lauber E, Bisseling T et al (2004) A putative Ca^{2+} and calmodulin-dependent protein kinase required for bacterial and fungal symbioses. Science 303:1361–1364
Limpens E, Franken C, Smit P, Willemse J, Bisseling T, Geurts R (2003) LysM domain receptor kinases regulating rhizobial Nod factor-induced infection. Science 302:630–633

Limpens E, Mirabella R, Fedorova E, Franken C, Franssen H, Bisseling T, Geurts R (2005) Formation of organelle-like N-2-fixing symbiosomes in legume root nodules is controlled by DMI2. Proc Natl Acad Sci U S A 102:10375–10380

Lohar DP, Schaff JE, Laskey JG, Kieber JJ, Bilyeu KD, Bird DM (2004) Cytokinins play opposite roles in lateral root formation, and nematode and Rhizobial symbioses. Plant J 38:203–214

Long SR (1996) Rhizobium symbiosis: nod factors in perspective. Plant Cell 8:1885–1898

Madsen EB, Madsen LH, Radutoiu S, Olbryt M, Rakwalska M, Szczyglowski K, Sato S, Kaneko T, Tabata S, Sandal N, Stougaard J (2003) A receptor kinase gene of the LysM type is involved in legume perception of rhizobial signals. Nature 425:637–640

Markmann K, Giczey G, Parniske M (2008) Functional adaptation of a plant receptor- kinase paved the way for the evolution of intracellular root symbioses with bacteria. PLOS Biol 6:e68

Marsh JF, Rakocevic A, Mitra RM, Brocard L, Sun J, Eschstruth A, Long SR, Schultze M, Ratet P, Oldroyd GE (2007) *Medicago truncatula* NIN is essential for Rhizobium-independent nodule organogenesis induced by autoactive CCaMK. Plant Physiol 144:324–335

Mathesius U, Charon C, Rolfe BG, Kondorosi A, Crespi M (2000) Temporal and spatial order of events during the induction of cortical cell divisions in white clover by *Rhizobium leguminosarum* bv. trifolii inoculation or localized cytokinin addition. Mol Plant Microbe Interact 13:617–628

McAinsh MR, Webb AAR, Taylor JE, Hetherington AM (1995) Stimulus-induced oscillations in guard-cell cytosolic-free calcium. Plant Cell 7:1207–1219

McAlvin CB, Stacey G (2005) Transgenic expression of the soybean apyrase in *Lotus japonicus* enhances nodulation. Plant Physiol 137:1456–1462

Messinese E, Mun JH, Yeun LH, Jayaraman D, Rouge P, Barre A, Lougnon G, Schornack S, Bono JJ, Cook DR, Ane JM (2007) A novel nuclear protein interacts with the symbiotic DMI3 calcium- and calmodulin-dependent protein kinase of *Medicago truncatula*. Mol Plant Microbe Interact 20:912–921

Meyer T, Stryer L (1988) Molecular model for receptor-stimulated calcium spiking. Proc Natl Acad Sci U S A 85:5051–5055

Middleton PH, Jakab J, Penmetsa RV, Starker CG, Doll J, Kalo P, Prabhu R, Marsh JF, Mitra RM, Kereszt A et al (2007) An ERF transcription factor in *Medicago truncatula* that is essential for Nod factor signal transduction. Plant Cell 19:1921–1234

Mitra RM, Gleason CA, Edwards A, Hadfield J, Downie JA, Oldroyd GED, Long SR (2004) A Ca^{2+}/calmodulin-dependent protein kinase required for symbiotic nodule development: gene identification by transcript-based cloning. Proc Natl Acad Sci U S A 101:4701–4705

Miwa H, Sun J, Oldroyd GE, Downie JA (2006) Analysis of Nod-factor-induced calcium signaling in root hairs of symbiotically defective mutants of *Lotus japonicus*. Mol Plant Microbe Interact 19:914–923

Murray JD, Karas BJ, Sato S, Tabata S, Amyot L, Szczyglowski K (2007) A cytokinin perception mutant colonized by Rhizobium in the absence of nodule organogenesis. Science 315:101–104

Newcomb W (1981) Nodule development. In: Gibson AH, Newton WE (eds) Current perspectives in nitrogen fixation, Elsevier/North-Holland Biomedical Press, sole distributors for the USA and Canada, Elsevier North-Holland (Amsterdam, New York, New York), pp 303–304

Oldroyd GE, Downie JA (2006) Nuclear calcium changes at the core of symbiosis signalling. Curr Opin Plant Biol 9:351–357

Oldroyd GED (2001) Dissecting symbiosis: developments in Nod factor signal transduction. Ann Bot 87:709–718

Oldroyd GED, Downie JA (2004) Calcium, kinases and nodulation signalling in legumes. Nat Rev Mol Cell Biol 5:566–576

Oldroyd GED, Long SR (2003) Identification and characterization of nodulation-signaling pathway 2, a gene of *Medicago truncatula* involved in Nod factor signaling. Plant Physiol 131:1027–1032

Oldroyd GED, Mitra RM, Wais RJ, Long SR (2001) Evidence for structurally specific negative feedback in the Nod factor signal transduction pathway. Plant J 28:191–199

Patil S, Takezawa D, Poovaiah BW (1995) Chimeric plant calcium/calmodulin-dependent protein-kinase gene with a neural visinin-like calcium-binding domain. Proc Natl Acad Sci U S A 92:4897–4901

Pawlowski K, Bisseling T (1996) Rhizobial and actinorhizal symbioses: what are the shared features? Plant Cell 8:1899–1913

Peck MC, Fisher RF, Long SR (2006) Diverse flavonoids stimulate NodD1 binding to nod gene promoters in *Sinorhizobium meliloti*. J Bacteriol 188:5417–5427

Peiter E, Sun J, Heckmann AB, Venkateshwaran M, Riley BK, Otegui MS, Edwards A, Freshour G, Hahn MG, Cook DR et al (2007) The *Medicago truncatula* DMI1 protein modulates cytosolic calcium signaling. Plant Physiol 145:192–203

Peters NK, Frost JW, Long SR (1986) A plant flavone, luteolin, induces expression of *Rhizobium meliloti* nodulation genes. Science 233:977–980

Pingret JL, Journet EP, Barker DG (1998) Rhizobium nod factor signaling: evidence for a G protein-mediated transduction mechanism. Plant Cell 10:659–671

Quinn JM, Etzler ME (1987) Isolation and characterization of a lectin from the roots of *Dolichos biflorus*. Arch Biochem Biophys 258:535–544

Radutoiu S, Madsen LH, Madsen EB, Felle HH, Umehara Y, Gronlund M, Sato S, Nakamura Y, Tabata S, Sandal N, Stougaard J (2003) Plant recognition of symbiotic bacteria requires two LysM receptor-like kinases. Nature 425:585–592

Radutoiu S, Madsen LH, Madsen EB, Jurkiewicz A, Fukai E, Quistgaard EM, Albrektsen AS, James EK, Thirup S, Stougaard J (2007) LysM domains mediate lipochitin-oligosaccharide recognition and Nfr genes extend the symbiotic host range. EMBO J 26:3923–3935

Ramachandiran S, Takezawa D, Wang W, Poovaiah BW (1997) Functional domains of plant chimeric calcium/calmodulin-dependent protein kinase: regulation by autoinhibitory and visinin-like domains. J Biochem 121:984–990

Riely BK, Lougnon G, Ane JM, Cook DR (2007) The symbiotic ion channel homolog DMI1 is localized in the nuclear membrane of *Medicago truncatula* roots. Plant J 49:208–216

Roche P, Debelle F, Maillet F, Lerouge P, Faucher C, Truchet G, Denarie J, Prome JC (1991) Molecular-basis of symbiotic host specificity in *Rhizobium meliloti* - Nodh and Nodpq genes encode the sulfation of lipo-oligosaccharide signals. Cell 67:1131–1143

Rolfe BG (1988) Flavones and isoflavones as inducing substances of Legume nodulation. Biofactors 1:3–10

Saito K, Yoshikawa M, Yano K, Miwa H, Uchida H, Asamizu E, Sato S, Tabata S, Imaizumi-Anraku H, Umehara Y et al (2007) NUCLEOPORIN85 is required for calcium spiking, fungal and bacterial symbioses, and seed production in *Lotus japonicus*. Plant Cell 19:610–624

Schauser L, Roussis A, Stiller J, Stougaard J (1999) A plant regulator controlling development of symbiotic root nodules. Nature 402:191–195

Schenk PW, Snaar-Jagalska BE (1999) Signal perception and transduction: the role of protein kinases. Biochim Biophys Acta 1449:1–24

Schwedock J, Long SR (1990) Atp sulphurylase activity of the Nodp and Nodq gene products of *Rhizobium meliloti*. Nature 348:644–647

Shaw SL, Long SR (2003) Nod factor elicits two separable calcium responses in *Medicago truncatula* root hair cells. Plant Physiol 131:976–984

Smit G, Puvanesarajah V, Carlson RW, Barbour WM, Stacey G (1992) *Bradyrhizobium japonicum* Nodd1 can be specifically induced by soybean flavonoids that do not induce the Nodyabcsuij operon. J Biol Chem 267:310–318

Smit P, Raedts J, Portyanko V, Debelle F, Gough C, Bisseling T, Geurts R (2005) NSP1 of the GRAS protein family is essential for rhizobial Nod factor-induced transcription. Science 308:1789–1791

Smit P, Limpens E, Geurts R, Fedorova E, Dolgikh E, Gough C, Bisseling T (2007) Medicago LYK3, an entry receptor in Rhizobial Nod factor signaling. Plant Physiol 145:183–191

Spaink HP (2000) Root nodulation and infection factors produced by rhizobial bacteria. Annu Rev Microbiol 54:257–288

Spaink HP, Sheeley DM, Vanbrussel AAN, Glushka J, York WS, Tak T, Geiger O, Kennedy EP, Reinhold VN, Lugtenberg BJJ (1991) A Novel highly unsaturated fatty-acid moiety of lipo-oligosaccharide signals determines host specificity of Rhizobium. Nature 354:125–130

Steen A, Buist G, Leenhouts KJ, El Khattabi M, Grijpstra F, Zomer AL, Venema G, Kuipers OP, Kok J (2003) Cell wall attachment of a widely distributed peptidoglycan binding domain is hindered by cell wall constituents. J Biol Chem 278:23874–23881

Stougaard J (2000) Regulators and regulation of legume root nodule development. Plant Physiol 124:531–540

Stracke S, Kistner C, Yoshida S, Mulder L, Sato S, Kaneko T, Tabata S, Sandal N, Stougaard J, Szczyglowski K, Parniske M (2002) A plant receptor-like kinase required for both bacterial and fungal symbiosis. Nature 417:959–962

Subramanian S, Stacey G, Yu O (2006) Endogenous isoflavones are essential for the establishment of symbiosis between soybean and Bradyrhizobium japonicum. Plant J 48:261–273

Subramanian S, Stacey G, Yu O (2007) Distinct, crucial roles of flavonoids during legume nodulation. Trends Plant Sci 12:282–285

Sun J, Miwa H, Downie JA, Oldroyd GE (2007) Mastoparan activates calcium spiking analogous to Nod factor-induced responses in *Medicago truncatula* root hair cells. Plant Physiol 144:695–702

Suntharalingam M, Wente SR (2003) Peering through the pore: nuclear pore complex structure, assembly, and function. Dev Cell 4:775–789

Swanson JA, Tu JK, Ogawa J, Sanga R, Fisher RF, Long SR (1987) Extended region of nodulation genes in *Rhizobium meliloti* 1021 1. Phenotypes of Tn5 insertion mutants. Genetics 117:181–189

Takezawa D, Ramachandiran S, Paranjape V, Poovaiah BW (1996) Dual regulation of a chimeric plant serine threonine kinase by calcium and calcium calmodulin. J Biol Chem 271:8126–8132

Tirichine L, James EK, Sandal N, Stougaard J (2006) Spontaneous root-nodule formation in the model legume *Lotus japonicus*: a novel class of mutants nodulates in the absence of rhizobia. Mol Plant Microbe Interact 19:373–382

Tirichine L, Sandal N, Madsen LH, Radutoiu S, Albrektsen AS, Sato S, Asamizu E, Tabata S, Stougaard J (2007) A gain-of-function mutation in a cytokinin receptor triggers spontaneous root nodule organogenesis. Science 315:104–107

Wais RJ, Galera C, Oldroyd G, Catoira R, Penmetsa RV, Cook D, Gough C, Denarie J, Long SR (2000) Genetic analysis of calcium spiking responses in nodulation mutants of *Medicago truncatula*. Proc Natl Acad Sci U S A 97:13407–13412

Walker SA, Downie JA (2000) Entry of *Rhizobium leguminosarum* bv. viciae into root hairs requires minimal nod factor specificity, but subsequent infection thread growth requires nodO or nodE. Mol Plant Microbe Interact 13:754–762

Walker SA, Viprey V, Downie JA (2000) Dissection of nodulation signaling using pea mutants defective for calcium spiking induced by Nod factors and chitin oligomers. Proc Natl Acad Sci U S A 97:13413–13418

Wasson AP, Pellerone FI, Mathesius U (2006) Silencing the flavonoid pathway in *Medicago truncatula* inhibits root nodule formation and prevents auxin transport regulation by rhizobia. Plant Cell 18:1617–1629

Physiological Roles of Cyclic Nucleotide Gated Channels in Plants

Tal Sherman and Hillel Fromm

Abstract Plant cyclic nucleotide gated channels (CNGCs) are diverse in their temporal and spatial expression patters, their subcellular membrane localization, their involvement in signaling cascades, and their physiological roles. We describe the latest developments in understanding the functions of plant CNGCs based on their expression profiles and the phenotypic analysis of CNGC mutants, and engineered plants with ectopic expression of intact or mutant CNGCs. While this summary provides some clues to the functions of specific plant CNGCs, it emphasizes the lack of comprehension of the roles and modes of action of most plant CNGCs. Deciphering these is a great challenge to plant biologists.

1 Introduction

Cyclic nucleotide gated channels (CNGCs) were initially discovered in mammalian vertebrate photoreceptors (Fesenko et al. 1985) and olfactory epithelium (Nakamura and Gold 1987). The first plant CNGC was discovered by Schuurink et al. (1998) while screening the *Hordeum vulgare* aleurone complementary DNA expression library using an *Arabidopsis* calmodulin (CaM) conjugated to horseradish peroxidase as a probe. The basic structure of CNGCs is conserved amongst mammalians and plants (six transmembrane domains, pore between the fifth and sixth transmembrane domains, N- and C-terminal tails protruding into the cytoplasm). However, the CaM binding domain of the mammalian CNGC is located at the N-terminus, while that of plants is located at the C-terminus and coincides with the cyclic nucleotide binding domain (CNBD) (Kohler et al. 1999; Arazi et al. 2000).

T. Sherman and H. Fromm (✉)
Department of Plant Sciences, Faculty of Life Sciences, Tel Aviv University, Tel Aviv 69978, Israel
e-mail: Hillelf@post.tau.ac.il

F. Baluška and S. Mancuso (eds.), *Signaling in Plants,* Signaling and Communication in Plants,
DOI: 10.1007/978-3-540-89228-1_5, © Springer-Verlag Berlin Heidelberg 2009

Since the discovery of CNGCs in plants, a wealth of data regarding plant CNGCs has been accumulated. The *Arabidopsis* Genome Initiative revealed a family of 20 CNGCs (Maser et al. 2001), while other plant-genome projects, including rice, corn, cotton, grape, and moss, revealed many homologs of these proteins. Structural aspects of plant CNGCs were reviewed in the past. Hua et al. (2003) analyzed the special features of the channel's pore selectivity filter and this was further reviewed by Talke et al. (2003). The CNBD was discussed in a review by Kaplan et al. (2007). Maser et al. (2001) classified the 20 *Arabidopsis* CNGCs into groups I, II, III, IVA, and IVB. Group I includes CNGC1, CNGC3, CNGC10, CNGC11, CNGC12, and CNGC13. Group II includes CNGC5, CNGC6, CNGC7, CNGC8, and CNGC9. Group III includes CNGC14, CNGC15, CNGC16, CNGC17, and CNGC18. Group IVA includes CNGC19 and CNGC20.Group IVB includes CNGC2 and CNGC4. On the basis of amino acid sequences of the *Arabidopsis* CNGCs, a phylogenetic tree was constructed (Maser et al. 2001). Similar phylogenetic trees were constructed on the basis of channel pore (Talke et al. 2003) and CNBD (Kaplan et al. 2007) sequences. Here we address the physiological roles of plant CNGCs by means of their expression patterns (reporter genes, microarray data, etc.) and phenotypic analysis of mutants and genetically engineered plants (knockout, knock down, and overexpression).

2 Expression Patterns of CNGCs

Clues to reveal the functions of the different CNGCs could be obtained by determining their temporal and spatial expression patters. Various approaches for analyzing the expression patterns of these channels were tested. Reporter gene fusions, transcript profiling, and immunodetection enabled researchers to localize CNGCs in specific organs and cell types. Another way to evaluate gene expression is using DNA microarrays and online databases (data-mining tools) that collect microarray data such as the *Arabidopsis* eFP Browser (Winter et al. 2007), SUBA (Heazlewood et al. 2007), and GENEVESTIGATOR (Zimmermann et al. 2004). Data sets based on expressed sequence tags such as MPSS (Brenner et al. 2000) can be used for this purpose as well (Talke et al. 2003). To learn about the expression pattern of all 20 *Arabidopsis thaliana* CNGCs (AtCNGCs), we used the GENEVESTIGATOR analysis tool. Figures 1 and 2 show the output results from the Meta-Analyzer platform in which the input included all the above-mentioned genes. The expression profile was analyzed by development (Fig. 1) and anatomy (Fig. 2) using all 2,620 ATH1 DNA chips available from this GENEVESTIGATOR version (similar results were obtained using GENEVESTIGATOR V3 with 3,110 DNA chips).

Looking at Fig. 1, it is noticeable that most AtCNGCs are expressed between the "rosette" and "silique" stages, with an emphasis on "young rosette," "young flower," and "silique." It is interesting that some of the genes grouped together by the expression profiles generated by GENEVESTIGATOR coincide with the groups of Maser's phylogenetic tree (Maser et al. 2001). For example, *CNGC7* and *CNGC8*, both belonging to group II, show high expression in the "silique" stage. Group III's

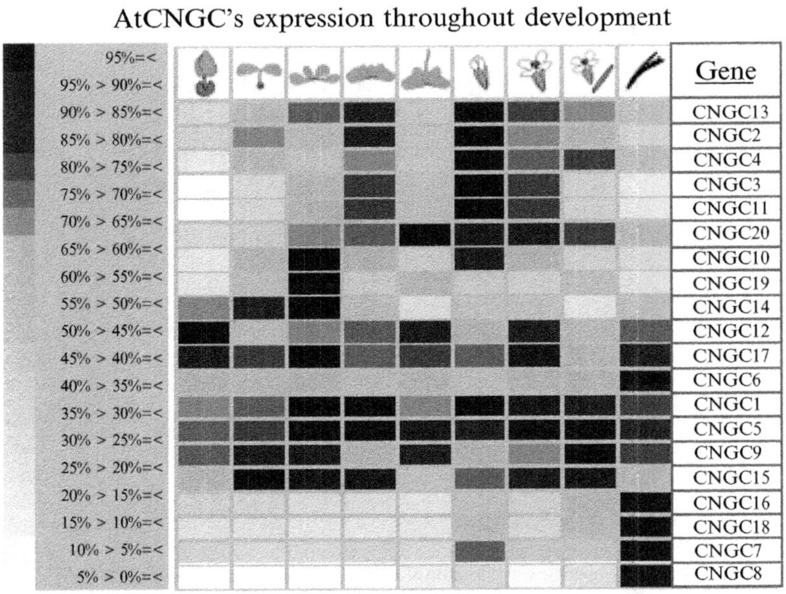

Fig. 1 Gene expression profiling by development of all 20 *Arabidopsis thaliana* cyclic nucleotide gated channels (AtCNGCs) using the GENEVESTIGATOR Meta-Analyzer tool. The results are given in heat map codes. A color legend is given at the *left*. The symbols from *left* to *right* represent germinated seed, seedling, young rosette, developed rosette, bolting, young flower, developed flower, flowers and siliques, and siliques

CNGC16 and *CNGC18* are also highly expressed in the "silique" stage. Another pair from group III is *CNGC14* and *CNGC17*, which are located on the same branch in this group and are both strongly expressed in the "young rosette" stage. Group I includes *CNGC1*, *CNGC10*, *CNGC13*, *CNGC3*, *CNGC11*, and *CNGC12*; in Fig. 1, all of them with the exception of *CNGC12*, have strong expression in the "young flower" stage. *CNGC2* and *CNGC4* that appear together in group IVB also have strong expression in the "young flower" stage.

A general overview of the anatomy-based analysis in Fig. 2 reveals four major hot spots: six hot spots in the "pollen" column, nine spots in the "cauline leaf" column, the same nine appear also in the "senescence leaf" column, and four genes get high scores in the "hypocotyl," "xylem," and "cork" columns. Comparison of the genes according to their phylogenetic groups with their anatomy expression profiles revealed them to be congruent. *CNGC7* and *CNGC8* are together with *CNGC16* and *CNGC18* in "pollen." *CNGC1*, *CNGC13*, *CNGC3*, *CNGC11*, and *CNGC12* are all together in "senesence leaf," and, except for *CNGC13*, all of these genes get high values in "cauline leaf" as well. All of the above data can lead one to speculate on a possible connection between the phylogenetic relationship and function in the AtCNGC family. Since data regarding CNGC expression is available from various experimental approaches, we can now compare such published results with those obtained from the above-mentioned analysis by GENEVESTIGATOR.

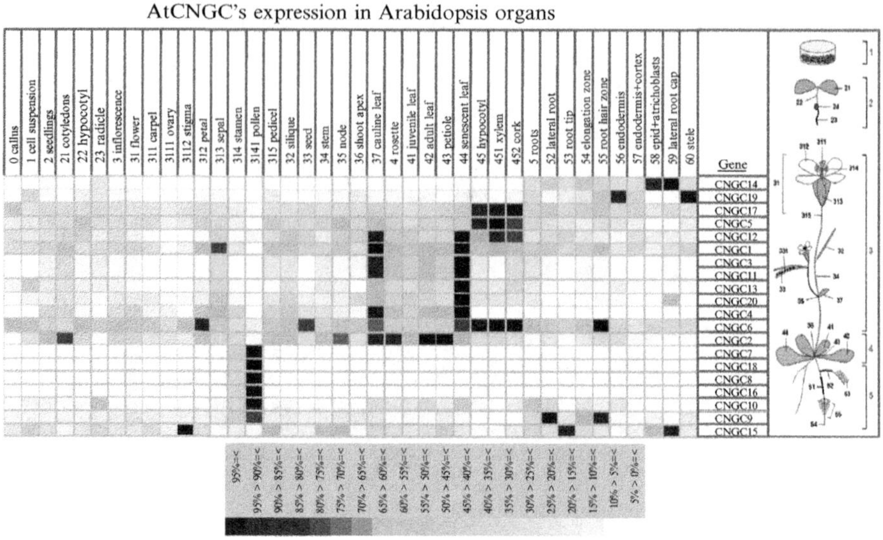

Fig. 2 Gene expression profiling by anatomy of all AtCNGCs using the GENEVESTIGATOR Meta-Analyzer tool. The results are given in heat map codes. A color legend is given at the *left*. The different plant organs are indicated in the *upper row* by numbers that also appear in the drawing on the *right*

With use of the cell eFP Browser based on the SUBA database, all AtCNGCs were analyzed. All proteins were predicted to localize to the plasma membrane (PM) like their mammalian homologs, except for CNGC20, which was localized to the chloroplast (data not shown). NtCBP4 antibodies were used to detect the protein in the PM of tobacco leaves and roots from wild-type and NtCBP4 over-expressing tobacco calli (Arazi et al. 1999). CNGC1, the *Arabidopsis* NtCBP4 homolog, was immunodetected in the *Arabidopsis* microsomal membrane root fraction but not in membrane proteins from leaves of 3-week-old seedlings (Ma et al. 2006). AtCNGC3 was localized to the PM using the green fluorescent protein (GFP) reporter gene in onion epidermal cells and in *Arabidopsis* leaf protoplasts (Gobert et al. 2006). Transmission electron microscopy was used to detect immunogold-anti-CNGC10 along the PM of *Arabidopsis* leaf cells (Borsics et al. 2007). In later work from the same group, GFP was used to localize AtCNGC5 and AtCNGC10. Both proteins were visualized in the PM of *Arabidopsis* leaf protoplast (Christopher et al. 2007). Quantitative real-time PCR of leaves and roots revealed that AtCNGC5 is expressed equally in roots and leaves, while AtCNGC10 is preferably expressed in roots. This was verified at the protein level by immunoprecipitation of proteins from roots and leaves using anti-AtCNGC5 and anti-AtCNGC10 antibodies. AtCNGC10 was visualized using transmission electron microscopy in sections of *Arabidopsis* root tip cells. The protein was detected in the PM of root meristematic cells, root columella cells, and border tip cells (Christopher et al. 2007).

The *CNGC2* putative promoter was cloned upstream of the *GUS* reporter gene to learn about its expression profile during plant development (Kohler et al. 2001). GUS staining in cotyledons was strong, while stems showed weak staining. Leaf GUS expression was weak and became stronger when the plant began to flower; no GUS staining was detected in roots. GUS expression in flowers was noticeable from floral stage 15 according to Smyth et al. (1990) in sepals, petals, stamens, and in dehiscent siliques at the valve–replum junction. Northern blot analysis using RNA extracted from different organs of 6-week-old plants revealed expression in inflorescence, stem, and leaf, while no transcript was detected in roots. Northern blot analysis also identified accumulation of CNGC2 transcripts during the progression of leaf senescence.

CNGC2 was found to be important for plants growing in calcified environments (Chan et al. 2003). When grown in the presence of 20 mM $CaCl_2$ in growth medium or watered with 10 mM $CaCl_2$ in soil, *cngc2* vegetative and reproductive development were severely impaired. A recent follow-up DNA microarray-based analysis, evaluated gene expression in the wild type and *cngc2* in response to high Ca^{2+} concentrations (Chan et al. 2008). These researchers found that the transcript profile of wild-type plants grown on high external Ca^{2+} resembled the profile of *cngc2* plants grown on control medium. Pathogenesis-related genes PR5 (At75040) and PR1 (At2g14610) were upregulated in the *cngc2* mutant grown on control media. This is in agreement with the evidence that the *cngc2* mutant contains high salicylic acid (SA) levels (Clough et al. 2000). Another interesting gene upregulated in the same experiment was ACA13 (At3g22910), a PM Ca^{2+} ATPase. Since this gene was upregulated when wild-type plants were challenged with high levels of Na^+(Maathuis et al. 2003), the authors believe this strengthens the hypothesis that *cngc2* has defects in sensing or maintaining Ca^{2+} homeostasis. When looking at the transcript profile of genes downregulated in *cngc2* grown on high external Ca^{2+}, another Ca^{2+} transporter appears; CAX1 (At2g38170), a tonoplast Ca^{2+}/H^+ antiporter. This gene was also downregulated in the *cngc2* mutant under control conditions. Knockout *cax1* plants exhibit multiple developmental defects, including reduction in root length and lateral root number, retardation in flowering time, shorter inflorescence length, and shorter and fewer secondary shots (Cheng et al. 2003). All of these *cax1* phenotypes led Cheng et al. (2003) to suspect that auxin levels are altered in the *cax1* mutant, and indeed this was confirmed through a series of experiments including root elongation assay in the presence of auxin and *IAA28:GUS* expression analysis in the *cax1* background. Interestingly, two auxin-responsive genes (At1g29430 and At5g27780) were downregulated in the *cngc2* mutant in control conditions and their expression was further decreased by high Ca^{2+}. In contrast, a protein kinase (Atg04500) whose transcript levels were increased twofold in control conditions was upregulated by high Ca^{2+} concentrations.

The phylogenetically closest member of *AtCNGC2* is *AtCNGC4* (Maser et al. 2001; Kaplan et al. 2007). Plants harboring a promoter::GUS fusion for this gene were analyzed in a pathogenicity assay, inoculated with an avirulent *Xanthomonas* strain. The results showed that GUS expression was upregulated 8–10 h after inoculation and peaked at 24 h after inoculation. In contrast, *CNGC2* promoter::GUS lines that were used as a control in this experiment showed constitutive GUS expression in leaves, and this expression was downregulated in the inoculation site 24 h after inoculation (Balague et al. 2003).

Promoter::GUS fusion was also employed to study *AtCNGC3* expression during development (Gobert et al. 2006). According to the GUS assay, *CNGC3* expression initiates in the seed prior to germination and is maintained in the root of the young seedling with low expression in cotyledons. GUS expression in roots is limited to the epidermis and cortex cells and no staining can be detected in the stellar regions. Gobert et al. (2006) noticed that there is a difference in the expression pattern between plants that were grown on agar plates compared with those that were grown in soil. Leaves of plants grown in soil showed significantly lower staining than leaves of plants grown on agar plates, indicating that expression of this gene is affected by environmental conditions.

A recent study of transporters in the *Arabidopsis* male gametophyte found several CNGCs (*CNGC7*, *CNGC8*, *CNGC16*, and *CNGC18*) specifically or preferentially expressed in pollen (Bock et al. 2006). Indeed, two new papers ascribe to AtCNGC18 a pivotal role in maintaining a Ca^{2+} gradient in the tip of germinating pollen grains (Chang et al. 2007; Frietsch et al. 2007). Chang et al. also confirmed that all four proteins are expressed in pollen. However, interestingly, through the of fusing the coding sequence of each CNGC to the yellow fluorescent protein under the *LAT52* promoter it was shown that CNGC7, CNGC8, and CNGC16 localized to inner membranes, whereas CNGC18 was the only protein to localize to the PM. This is the first report of CNGCs not localized to the PM. Frietsch et al (2007) used GFP fused to CNGC18 and found it localized to the PM as well. Using the *LAT52* promoter::GUS reporter within their *cngc18-1* T-DNA knockout line, Frietsch et al. (2007) were able to distinguish *cngc18-1* pollen from wild-type pollen. Using this tool, they showed that *cngc18-1* germinating pollen did not grow into the transmitting tract of wild-type pistils, linking male sterility to the CNGC knockout. Transgenic plants harboring the *CNGC18* promoter:GUS construct showed GUS expression only in pollen throughout floral stages 12–14 (Frietsch et al. 2007).

Interestingly, the data retrieved from the GENEVESTIGATOR query was not in full agreement with the data that have already been published. Although GENEVESTIGATOR made correct predictions about all four pollen-specific CNGCs, it failed to detect any considerable expression of *CNGC10* in the root. No significant score is visible for *CNGC3* root as well. Another surprise is the relatively low expression of *CNGC1* in roots. Since *cngc1* phenotypes are attributed to metal ion uptake and root functions (Sunkar et al. 2000; Ma et al. 2006) combined with the fact that *CNGC1* expression was detected in roots (Ma et al. 2006), it is puzzling that the highest score for this gene is in senescent and cauline leaves. All phenotypes and expression patterns are summarized in Table 1.

Table 1 CNGC-associated phenotypes and relevant expression profiles

Gene name	Knock-out, knock-down, and over expression phenotypes	Expression	Reference
AtCNGC1	AtCNGC1 T-DNA knock-out seedlings were more tolerant to Pb^{2+} [(1)] and contained less Ca^{2+} than WT [(2)]. *cncg1* showed reduced curvature in gravistimulated roots, the elongation zone of these gravistimulated roots contained lower NO, and the growth rate of the primary root in this mutant was greater. The *cngc1* roots caused greater acidification of the growing media than WT roots [(2)].	AtCNGC1 was immunodetected in roots only	[(1)](Sunkar et al., 2000) [(2)](Ma, 2006) [(3)](Arazi et al., 1999)
NtCBP4	Plants over-expressing NtCBP4 exhibited improved tolerance to Ni^{2+}, which was associated with reduced Ni^{2+} accumulation, and hypersensitivity to Pb^{2+}, associated with enhanced Pb^{2+} accumulation[(3)]. Seedlings expressing the NtCBP4 truncated C-terminal protein lacking the CaMBD, showed improved tolerance to Pb^{2+} with attenuated accumulation of this metal[(1)]	NtCBP4 protein was immunodetected in membrane fractions from leaves and roots [(3)]	
AtCNGC2, DND1	The *cngc2* mutant contains high level of SA, constitutive expression of pathogenesis-related (PR) genes does not generate a HR response when challenged with *Pseudomonas syringae*, and keeps bacteria growth-rate lower compared to WT[(1)]. *cngc2* showed exclusive hypersensitivity to Ca^{2+} compared to WT[(2)]. This mutant over-expresses a kinase while the expression of auxin-related genes is down-regulated[(3)]. When challenged with bacterial LPS, the generation of NO is lower than in WT[(3)] when inoculated with *P. carotovorum,* external Ca^{2+} is required for initiating cell death in *cngc2*[(4)]	Northern blot analysis revealed expression in the inflorescence stem and leaf while no transcript was found in roots[(5)]	[(1)](Clough et al., 2000) [(2)](Chan et al., 2003) [(3)](Chan et al., 2008) [(4)](Ahn, 2007) [(5)](Kohler et al., 2001)
AtCNGC3	*cngc3* mutants had lower germination rates when grown on 100–140 mM NaCl. No significant reduction of *cngc3* germination rates were noticeable when grown on isomolar concentrations of sorbitol and other salts. When grown on high levels of KCl or NaCl, *cngc3* plants had no difference in Na^+ content, but had 50% lower K^+ than WT[(1)].	Seeds prior to germination, seedlings, root, root hairs, shoot and leaves. Determined using promoter: GUS fusion.	[(1)](Gobert et al., 2006)

(continued)

Table 1 (continued)

Gene name	Knock-out, knock-down, and over expression phenotypes	Expression	reference
AtCNGC4, DND2	*cngc4* is a lesion-mimic mutant. It contains high level of SA, constitutive expression of pathogenesis-related (PR) genes, does not generate a HR response when challenged with *Pseudomonas Syringae* and keeps bacteria growth rate lower compared to WT[1, 2].	GUS expression was detected in Arabidopsis leaves 8–24 hours post inoculation with an avirulent *Xanthomonas*[1]	[1](Balague et al., 2003) [2](Jurkowski et al., 2004) [3](Rostoks et al., 2006)
HvNEC1	Fast neutron-induced mutation in HvNEC1, the barley homolog of the AtCNGC4, results in spontaneous lesion formation and constitutive expression of PR1 and β, 1-3 glucanase genes.[3] (mutants caused by EMS also have the same phenotype)	The expression of *NEC1* was detected in 10- to 20-day-old leaf tissue[3]	
AtCNGC10	AtCNGC10 antisense construct led to 40% lower K^+ levels, high starch accumulation, bigger starch granules, a more frequent appearance of peroxisomes, reduced root growth and elongation, early flowering, and delayed bending in response to gravistimulation[1] *35S::AtCNGC10* construct was able to partially complement the *Atakt1* mutant in K^+ uptake[1]	*CNGC10* was preferably expressed in roots when compared to leaves. Expression was analyzed at the RNA and protein levels [2]	[1](Borsics et al., 2007) [2](Christopher et al., 2007)
AtCNGC11/12	A 3 kb deletion generated a chimeric CNGC from *AtCNGC11* and *AtCNGC12* resulting in a mutant showing altered defense responses, stunted growth, curly leaves and constitutive expression of PR genes.		(Yoshioka et al., 2006)
AtCNGC18	*cngc18* plants are male sterile. *CNGC18* knock-out resulted in aberrant germinating pollen tubes that failed to enter transmission tracks, resulting in no fertilization[1, 2]. Transient over-expression of *AtCNGC18* under the *LAT52* promoter caused shorter and wider germination tubes resulting in depolarized tip growth in tobacco pollen grains.[1]	Protein was detected in PM of pollen using YFP [1] and GFP [2]. GUS staining was visible in pollen grains throughout floral stages 12-14 [2]	[1](Chang et al., 2007) [2](Frietsch et al., 2007)

3 CNGCs, Where Ca^{2+}, Hormones, Reactive Oxygen Species, and Nitric Oxide Meet

All known pathogenesis-related AtCNGC mutants accumulate high levels of SA. This includes the mutants *cngc2* and *cngc4* and the chimeric *cngc11/12*. While CNGC2 and CNGC4 are grouped together in group IVB, CNGC11 and CNGC12 are from group I. Knockout lines of these last two genes do not have a pronounced defense phenotype, leading to the conclusion that the *cngc11/12* chimera is the probable reason for the constitutive expression of pathogenesis-related genes in this mutant (Yoshioka et al. 2006). When challenged with the avirulent bacterium *Pseudomonas syringae*, *cngc2* mutants maintain lower bacteria growth rates thanks to the elevated expression of pathogenesis-related genes. Unlike the classic "gene for gene" resistance where the plant initiates a local programmed cell death at the infection site (also known as the hypersensitive response) and thus limits bacteria propagation, *cngc2* does not exhibit hypersensitive response and yet is more resistant to a wide range of pathogens when compared with wild-type plants (Yu et al. 1998). When *cngc2* was inoculated with *Pectobacterium carotovorum* (previously known as *Erwinia carotovora*), the bacterial agent that causes soft rot disease on Korean cabbage (*Brassica rapa L.* ssp. *pekinensis*), it managed to keep bacterial growth rates lower than in wild-type plants, with no hypersensitive response (Ahn 2007). From this observation and the fact that CNGC2 was shown to transport ions across the plasma membrane (Leng et al. 1999), Ahn (2007) suggests that ion transport provided by CNGC2 is crucial for the progression of soft rot disease in *Arabidopsis*. While hypersensitive response and reactive oxygen species (ROS) such as superoxide and H_2O_2 were detected in the wild type after inoculation with *P. carotovorum*, *cngc2* did not accumulate any of these. Wild-type plants that were treated with catalase, ascorbic acid, and $LaCl_3$ when inoculated with *P. carotovorum* displayed reduced hypersensitive response and lower bacterial growth rates. This treatment did not affect *cngc2* resistance. It was concluded that a functional CNGC2, Ca^{2+} transport across the PM, and Ca^{2+} elevation in the cytosol together with ROS accumulation are necessary to trigger hypersensitive response and susceptibility to soft rot disease in *Arabidopsis*. To assess the role of Ca^{2+} in these processes, wild-type and mutant plants were treated with $CaCl_2$ or ethylene glycol bis(2-aminoethyl ether) *N,N,N′,N′*-tetraacetic acid (EGTA) after inoculation with *P. carotovorum*. While $CaCl_2$ hampered the resistance of *cngc2* mutants against the pathogen, EGTA conferred resistance to wild-type plants. With use of W-7 and chloropromazine, known CaM antagonists, it was shown that CaM is essential for the "defense no death" phenotype since resistance was broken once *cngc2* plants had been treated with the previously mentioned CaM antagonists. Another explanation for the ability of CaM antagonists to break the resistance of *cngc2* to *P. carotovorum* might be the fact that these antagonists cause cytosolic Ca^{2+} bursts (Kaplan et al. 2006). It is therefore possible that these bursts trigger downstream effectors that eventually cause cell death.

The involvement of Ca^{2+} and ROS (Grant et al. 2000) as well as that of SA in response to pathogen attack are well documented (Chen et al. 1993; Draper 1997). Since the *cngc2* mutant combines all of these (elevated levels of SA, lack of ROS upon infection, and Ca^{2+} conductance ability) it is a natural candidate for study in order to understand the connection between all these factors.

As mentioned earlier, *cngc2* microarray data revealed downregulation of auxin-responsive genes in this mutant. The CAX1 transporter, which also can cause auxin-related defects, was downregulated in this mutant as well (Chan et al. 2008). Unpublished work conducted in our laboratory based on the *cngc4* mutant phenotype suggests a role for this gene in auxin function as well. Upon reaching the bolting stage, *cngc4* plants grow several stems, while wild-type plants at this stage grow only one main stem. This apical dominance disruption led us to assume that auxin is involved in some way. Classic plant hormone interaction experiments conducted in the 1930s, showed that application of auxin can imitate the action of a terminal bud and this led to the conclusion that auxin has a pivotal role in controlling apical dominance (Thimann and Skoog 1933). To assess the role of auxin in the "bushy" *cngc4* phenotype, plants were subjected to one of three treatments: decapitation, decapitation and auxin application upon the stump, or no treatment at all (Fig. 3). Plant stems were counted 25–30 days after decapitation and the results were analyzed statistically. These experiments were repeated three times and the results were consistent. Statistical analysis showed a significant difference between the number of flowering branches of wild-type plants (2.6 branches) compared with *cngc4* mutants (5.647 branches) when they were not decapitated. A significant difference was also found between decapitated plants without auxin treatment (wild type 3.6 branches, *cngc4* 5.125 branches), while no significant difference was found between plants that were decapitated and treated with auxin (wild type 2.384 branches, *cngc4* 2.424 branches). For all experiments 25–35 plants were used and the experiments were conducted three times.

The fact that *cngc4* mutants have phenotypes associated with defective auxin responses and that exogenous auxin can restore the apical dominance phenotype, as shown in (Fig. 3), may imply that CNGC4 somehow acts in auxin synthesis or auxin-conjugate hydrolysis. Among all documented AtCNGCs, only *cngc2* and *cngc4* show pleiotropic effects (excluding the CNGC11/12 chimera). Interestingly, among the 20 members of the *Arabidopsis* CNGC family, AtCNGC2 and AtCNGC4 are the closest in amino acid sequence, and are distinct from all other CNGCs (Kaplan et al. 2007). Since auxin-signaling mutants also show similar pleiotropic effects (Lincoln et al. 1990; Leyser et al. 1993; Gray and Estelle 2000), these observations strengthen the hypothesis that CNGC4 function may be coupled to auxin, and accounts for the many aberrant phenotypes exhibited by this mutant. Considering all the above-mentioned data, it is tempting to speculate about some kind of SA–auxin antagonistic effect. Nevertheless, it is known that certain elements within defense-related promoters are activated by SA as well as by auxin (Jupin and Chua 1996), and it was shown that SA, auxin, and H_2O_2 can induce the expression of the pathogenesis-related GST6 gene in *Arabidopsis* (Chen and Singh 1999).

Fig. 3 *Atcngc4* mutants exhibit a "bushy" phenotype but positively respond to exogenous auxin. Panels *a–c* are for wild-type plants; **a** undecapitated, untreated plants, **b** decapitated plants, **c** a decapitated plant treated with 0.5% auxin lanolin paste. Notice how auxin treatment reduces branching in wild-type plants when compared with decapitated, untreated wild-type plants. Panels *d–f* are for *Atcngc4* mutants. **d** undecapitated, untreated plants, **e** decapitated plants, **f** decapitated plants treated with 0.5% auxin lanolin paste. In this mutant's case, massive branching is a clear phenotype suggesting loss of apical dominance. While decapitation does not increase the number of branches, auxin treatment reduces branching to wild-type levels

Another molecule involved in SA signaling, ROS accumulation, hypersensitive response, and cell death is nitric oxide (NO). The involvement of NO in these processes is well documented (Dangl 1998). Interestingly, cyclic nucleotides are known to be secondary messengers in a NO signaling pathway. It was also found that pathogenesis-related gene activation by NO is mediated by SA. While NO treatment caused SA elevation, using transgenic NahG plants that cannot accumulate SA, pathogenesis-related gene induction was abolished (Durner et al. 1998). A more specific mode of action for NO was suggested by Zhang and colleagues (2003). They found that NO acts to extend hypersensitive response signaling from cell to cell and is not responsible directly for triggering the hypersensitive response.

Ali et al. (2007) recently showed a more specific connection between NO, cyclic nucleotides, Ca^{2+}, and Ca^{2+}/CaM. Sodium nitroprusside (SNP) is a NO donor that has been reported to be able to reverse phenotypes of the NO-associated protein

(AtNO1- previously known as AtNOS1) loss of function mutant (Guo et al. 2003). Ali et al. used SNP on *cngc2* mutants inoculated with two strains of avirulent pathogens, *Pseudomonas syringae* pv. *syringae* and *Pseudomonas syringae* pv. *tomato* DC3000, containing the avirulence gene *avrRpt2*. Treatment with SNP followed by inoculation with the pathogens mentioned resulted in hypersensitive response symptoms in *cngc2* mutants, while inoculation with no SNP treatment or SNP treatment alone had no significant effect on hypersensitive response in *cngc2* or wild-type plants. NO generation by a pathogenic stimulus was monitored using a NO-specific fluorescent dye (DAF-2DA). Guard cells isolated from the wild type and *cngc2* were treated with lipopolysaccharide (LPS), a known elicitor of the plant immune system, and NO generation was evaluated under the microscope. LPS triggered NO production in wild-type cells, while *cngc2* cells showed decreased NO production.

To be able to link Ca^{2+} and Ca^{2+}/CaM to LPS-derived NO production, Ali et al used the Ca^{2+} chelator EGTA and the Ca^{2+}-channel blockers Gd^{3+} and the CaM antagonist W7 and its inactive structural analog W5. EGTA- and Gd^{3+}-treated wild-type or *cngc2* cells showed no detectable NO production. W7 treatment abolished NO production similarly to the Ca^{2+} inhibitors. The NOS inhibitor L-NAME, its inactive isomer D-NAME, and TEA, (NAME = N^G-nitro-L-Arg-methyl ester, L and D are the chemical conformations of this material, TEA = tetraethylammonium) a K^+-channel blocker, were tested as well. TEA and D-NAME had no effect on NO production, while L-NAME was able to eliminate NO production. Using patch clamp recordings, Ali et al. (2007) were able to prove that a functional CNGC2 from wild-type guard cells can conduct an inward Ca^{2+} current in a cyclic AMP dependent manner, whereas this current cannot be detected in protoplasts from *cngc2* guard cells.

A model was proposed in which, in response to a pathogen challenge, upon elicitor binding to an unknown receptor, cyclic nucleotides are produced and upregulated in the cytosol by a nucleotide triphosphate cyclase or another unknown mechanism triggered by the receptor. These cyclic nucleotides then bind to the CNGC2 CNBD, opening the channel and thus allowing Ca^{2+} to enter the cell. The spike in intracellular Ca^{2+} results in Ca^{2+}/CaM complexes that, among other things, activate the NO synthase. The generation of NO invokes a hypersensitive response and cell death in a Ca^{2+}/CaM-dependent manner. In the absence of a functional CNGC2, the specific Ca^{2+} transient elevation is missing, resulting in the absence of Ca^{2+}/CaM complexes. Consequently, NO signaling is not initiated, leading to the absence of hypersensitive response and cell death in the affected cells. Combining the conclusions of Ali et al. (2007) and Ahn (2007) reinforces the necessity of Ca^{2+} and Ca^{2+}/CaM in the hypersensitive response.

4 Conclusions and Future Perspectives

In this review we discussed the diverse roles of plant CNGCs. The AtCNGC protein family consists of 20 members with a variable degree of homology between themselves. Even so, when knocked out, most AtCNGCs display a distinct phenotype, and compensation by the other family members is not visible. This is an intriguing phenomenon since functional redundancy in such a family is expected. AtCNGCs are expressed throughout plant development and fulfill a wide range of activities. It seems that phylogenetically close family members carry out the same or similar cellular functions. From group I, CNGC1, CNGC3, and CNGC10 are important in metal uptake from the environment into plant cells. Group II and group III contain CNGCs that are pollen-specific channels. From group IVB, CNGC2 and CNGC4 take part in a Ca^{2+}-dependent signal transduction cascade involved in plant defense against biotic (a variety of bacteria) and abiotic (calcified environment for *cngc2*; Cheng et al. 2003) stresses and heat stress for *cngc4* (Sherman and Fromm, unpublished data). These data and known channel structures from mammalian systems may lead one to ask whether the plant CNGCs work as heterodimers, which possibly consist of family members from the same phylogenetic group. The fact that the *dnd1/dnd1*, *dnd2/dnd2* double mutant displayed an extreme *dnd* phenotype might strengthen this hypothesis (Jurkowski et al. 2004). Another question arising from these data is what differentiates the channels in their functions? For example, if the pollen-specific Ca^{2+} transporter CNGC18 and the root-specific K^+ transporter CNGC10 share 46% identity and 64% similarity, what are the functional residues underlying their specificities? Group IVA remains a mystery, as no papers have been published about members of this group. Interestingly, recent experimental data suggest that some plant CNGCs are localized to internal membranes rather than the PM. The only CNGC predicted to be localized to the chloroplast membrane is CNGC20 (cell eFP Browser prediction, data not shown); however, this needs to be confirmed experimentally.

In summary, in spite of plant CNGCs having been studied for many years, the information on their functions is at best fragmented and completely lacking for some. The analysis of CNGCs requires a massive effort and combination of different approaches, including electrophysiology, genetics, cell and molecular biology, and physiology. Nevertheless, the diverse functions and expression profiles revealed so far attest for the important roles CNGCs play in plant development and in interactions with their biotic and abiotic environment.

References

Ahn I-P (2007) Disturbance of the Ca^{2+}/calmodulin-dependent signalling pathway is responsible for the resistance of Arabidopsis dnd1 against *Pectobacterium carotovorum* infection. Mol Plant Pathol 8:747–759

Ali R, Ma W, Lemtiri-Chlieh F, Tsaltas D, Leng Q, von Bodman S, Berkowitz GA (2007) Death don't have no mercy and neither does calcium: Arabidopsis CYCLIC NUCLEOTIDE GATED CHANNEL2 and innate immunity. Plant Cell 19:1081–1095

Arazi T, Sunkar R, Kaplan B, Fromm H (1999) A tobacco plasma membrane calmodulin-binding transporter confers Ni^{2+} tolerance and Pb^{2+} hypersensitivity in transgenic plants. Plant J 20:171–182

Arazi T, Kaplan B, Fromm H (2000) A high-affinity calmodulin-binding site in a tobacco plasma-membrane channel protein coincides with a characteristic element of cyclic nucleotide-binding domains. Plant Mol Biol 42:591–601

Balague C, Lin B, Alcon C, Flottes G, Malmstrom S, Kohler C, Neuhaus G, Pelletier G, Gaymard F, Roby D (2003) HLM1, an essential signaling component in the hypersensitive response, is a member of the cyclic nucleotide-gated channel ion channel family. Plant Cell 15:365–379

Bock KW, Honys D, Ward JM, Padmanaban S, Nawrocki EP, Hirschi KD, Twell D, Sze H (2006) Integrating membrane transport with male gametophyte development and function through transcriptomics. Plant Physiol 140:1151–1168

Borsics T, Webb D, Andeme-Ondzighi C, Staehelin LA, Christopher DA (2007) The cyclic nucleotide-gated calmodulin-binding channel AtCNGC10 localizes to the plasma membrane and influences numerous growth responses and starch accumulation in *Arabidopsis thaliana*. Planta 225:563–573

Brenner S, Johnson M, Bridgham J, Golda G, Lloyd DH, Johnson D, Luo S, McCurdy S, Foy M, Ewan M, Roth R, George D, Eletr S, Albrecht G, Vermaas E, Williams SR, Moon K, Burcham T, Pallas M, DuBridge RB, Kirchner J, Fearon K, Mao J, Corcoran K (2000) Gene expression analysis by massively parallel signature sequencing (MPSS) on microbead arrays. Nat Biotechnol 18:630–634

Chan CW, Schorrak LM, Smith RK Jr, Bent AF, Sussman MR (2003) A cyclic nucleotide-gated ion channel, CNGC2, is crucial for plant development and adaptation to calcium stress. Plant Physiol 132:728–731

Chan CW, Wohlbach DJ, Rodesch MJ, Sussman MR (2008) Transcriptional changes in response to growth of Arabidopsis in high external calcium. FEBS Lett 582:967–976

Chang F, Yan A, Zhao L, Wu W, Yang Z (2007) A putative calcium-permeable cyclic nucleotide-gated channel, CNGC18, regulates polarized pollen tube growth. J Integr Plant Biol 49:1261–1270

Chen W, Singh KB (1999) The auxin, hydrogen peroxide and salicylic acid induced expression of the Arabidopsis GST6 promoter is mediated in part by an ocs element. Plant J 19:667–677

Chen Z, Silva H, Klessig DF (1993) Active oxygen species in the induction of plant systemic acquired resistance by salicylic acid. Science 262:1883–1886

Cheng NH, Pittman JK, Barkla BJ, Shigaki T, Hirschi KD (2003) The Arabidopsis cax1 mutant exhibits impaired ion homeostasis, development, and hormonal responses and reveals interplay among vacuolar transporters. Plant Cell 15:347–364

Christopher DA, Borsics T, Yuen CY, Ullmer W, Andeme-Ondzighi C, Andres MA, Kang BH, Staehelin LA (2007) The cyclic nucleotide gated cation channel AtCNGC10 traffics from the ER via Golgi vesicles to the plasma membrane of Arabidopsis root and leaf cells. BMC Plant Biol 7:48

Clough SJ, Fengler KA, Yu IC, Lippok B, Smith RK Jr, Bent AF (2000) The Arabidopsis dnd1 "defense, no death" gene encodes a mutated cyclic nucleotide-gated ion channel. Proc Natl Acad Sci U S A 97:9323–9328

Dangl J (1998) Innate immunity. Plants just say NO to pathogens. Nature 394:525–527

Draper J (1997) Salicylate, superoxide synthesis and cell suicide in plant defence. Trends Plant Sci 2:162–165

Durner J, Wendehenne D, Klessig DF (1998) Defense gene induction in tobacco by nitric oxide, cyclic GMP, and cyclic ADP-ribose. Proc Natl Acad Sci U S A 95:10328–10333

Fesenko EE, Kolesnikov SS, Lyubarsky AL (1985) Induction by cyclic GMP of cationic conductance in plasma membrane of retinal rod outer segment. Nature 313:310–313

Frietsch S, Wang YF, Sladek C, Poulsen LR, Romanowsky SM, Schroeder JI, Harper JF (2007) A cyclic nucleotide-gated channel is essential for polarized tip growth of pollen. Proc Natl Acad Sci U S A 104:14531–14536

Gobert A, Park G, Amtmann A, Sanders D, Maathuis FJ (2006) *Arabidopsis thaliana* cyclic nucleotide gated channel 3 forms a non-selective ion transporter involved in germination and cation transport. J Exp Bot 57:791–800

Grant M, Brown I, Adams S, Knight M, Ainslie A, Mansfield J (2000) The RPM1 plant disease resistance gene facilitates a rapid and sustained increase in cytosolic calcium that is necessary for the oxidative burst and hypersensitive cell death. Plant J 23:441–450

Gray WM, Estelle I (2000) Function of the ubiquitin-proteasome pathway in auxin response. Trends Biochem Sci 25:133–138

Guo FQ, Okamoto M, Crawford NM (2003) Identification of a plant nitric oxide synthase gene involved in hormonal signaling. Science 302:100–103

Heazlewood JL, Verboom RE, Tonti-Filippini J, Small I, Millar AH (2007) SUBA: the Arabidopsis subcellular database. Nucl Acids Res 35:D213–218

Hua BG, Mercier RW, Leng Q, Berkowitz GA (2003) Plants do it differently. A new basis for potassium/sodium selectivity in the pore of an ion channel. Plant Physiol 132:1353–1361

Jupin I, Chua NH (1996) Activation of the CaMV as-1 *cis*-element by salicylic acid: differential DNA-binding of a factor related to TGA1a. EMBO J 15:5679–5689

Jurkowski GI, Smith RK Jr, Yu IC, Ham JH, Sharma SB, Klessig DF, Fengler KA, Bent AF (2004) Arabidopsis DND2, a second cyclic nucleotide-gated ion channel gene for which mutation causes the "defense, no death" phenotype. Mol Plant Microbe Interact 17:511–520

Kaplan B, Sherman T, Fromm H (2007) Cyclic nucleotide-gated channels in plants. FEBS Lett 581:2237–2246

Kaplan B, Davydov O, Knight H, Galon Y, Knight MR, Fluhr R, Fromm H (2006) Rapid transcriptome changes induced by cytosolic Ca^{2+} transients reveal ABRE-related sequences as Ca^{2+}-responsive *cis* elements in Arabidopsis. Plant Cell 18:2733–2748

Kohler C, Merkle T, Neuhaus G (1999) Characterisation of a novel gene family of putative cyclic nucleotide- and calmodulin-regulated ion channels in *Arabidopsis thaliana*. Plant J 18:97–104

Kohler C, Merkle T, Roby D, Neuhaus G (2001) Developmentally regulated expression of a cyclic nucleotide-gated ion channel from Arabidopsis indicates its involvement in programmed cell death. Planta 213:327–332

Leng Q, Mercier RW, Yao W, Berkowitz GA (1999) Cloning and first functional characterization of a plant cyclic nucleotide-gated cation channel. Plant Physiol 121:753–761

Leyser HM, Lincoln CA, Timpte C, Lammer D, Turner J, Estelle M (1993) Arabidopsis auxin-resistance gene AXR1 encodes a protein related to ubiquitin-activating enzyme E1. Nature 364:161–164

Lincoln C, Britton JH, Estelle M (1990) Growth and development of the axr1 mutants of Arabidopsis. Plant Cell 2:1071–1080

Ma W, Ali R, Berkowitz GA (2006) Characterization of plant phenotypes associated with loss-of-function of AtCNGC1, a plant cyclic nucleotide gated cation channel. Plant Physiol Biochem 44:494–505

Maathuis FJ, Filatov V, Herzyk P, Krijger GC, Axelsen KB, Chen S, Green BJ, Li Y, Madagan KL, Sanchez-Fernandez R, Forde BG, Palmgren MG, Rea PA, Williams LE, Sanders D, Amtmann A (2003) Transcriptome analysis of root transporters reveals participation of multiple gene families in the response to cation stress. Plant J 35:675–692

Maser P, Thomine S, Schroeder JI, Ward JM, Hirschi K, Sze H, Talke IN, Amtmann A, Maathuis FJ, Sanders D, Harper JF, Tchieu J, Gribskov M, Persans MW, Salt DE, Kim SA, Guerinot ML (2001) Phylogenetic relationships within cation transporter families of Arabidopsis. Plant Physiol 126:1646–1667

Nakamura T, Gold GH (1987) A cyclic nucleotide-gated conductance in olfactory receptor cilia. Nature 325:442–444

Schuurink RC, Shartzer SF, Fath A, Jones RL (1998) Characterization of a calmodulin-binding transporter from the plasma membrane of barley aleurone. Proc Natl Acad Sci U S A 95:1944–1949

Smyth DR, Bowman JL, Meyerowitz EM (1990) Early flower development in Arabidopsis. Plant Cell 2:755–767

Sunkar R, Kaplan B, Bouche N, Arazi T, Dolev D, Talke IN, Maathuis FJ, Sanders D, Bouchez D, Fromm H (2000) Expression of a truncated tobacco NtCBP4 channel in transgenic plants and disruption of the homologous Arabidopsis CNGC1 gene confer Pb^{2+} tolerance. Plant J 24:533–542

Talke IN, Blaudez D, Maathuis FJ, Sanders D (2003) CNGCs: prime targets of plant cyclic nucleotide signalling? Trends Plant Sci 8:286–293

Thimann KV, Skoog F (1933) Studies on the growth hormone of plants. III. The inhibiting action of growth substance on plant development. Proc Nat Acad Sci U S A 19:712–716

Winter D, Vinegar B, Nahal H, Ammar R, Wilson GV, Provart NJ (2007) An "electronic fluorescent pictograph" browser for exploring and analyzing large-scale biological data sets. PLoS ONE 2:e718

Yoshioka K, Moeder W, Kang HG, Kachroo P, Masmoudi K, Berkowitz G, Klessig DF (2006) The chimeric Arabidopsis CYCLIC NUCLEOTIDE-GATED ION CHANNEL11/12 activates multiple pathogen resistance responses. Plant Cell 18:747–763

Yu IC, Parker J, Bent AF (1998) Gene-for-gene disease resistance without the hypersensitive response in Arabidopsis *dnd1* mutant. Proc Natl Acad Sci U S A 95:7819–7824

Zhang C, Czymmek KJ, Shapiro AD (2003) Nitric oxide does not trigger early programmed cell death events but may contribute to cell-to-cell signaling governing progression of the Arabidopsis hypersensitive response. Mol Plant Microbe Interact 16:962–972

Zimmermann P, Hirsch-Hoffmann M, Hennig L, Gruissem W (2004) GENEVESTIGATOR. Arabidopsis microarray database and analysis toolbox. Plant Physiol 136:2621–2632

Signaling in Vesicle Traffic: Protein-Lipid Interface in Regulation of Plant Endomembrane Dynamics

V. Žárský and M. Potocký

Abstract The plant cell is characterized by a complex set of endomembrane compartments constituting the exocytic and endocytic trafficking pathways. Both the function and communication between these compartments are regulated by distinct families of proteins that direct membrane fission, targeting, and fusion. Many of these events are regulated by the cooperative actions of monomeric ARF, Rab, and Rho GTPases with their corresponding effector proteins and membrane (phospho) lipids. Glycerophospholipids, namely phosphoinositides, play a multitude of roles in membrane traffic, by mediating protein recruitment to membranes, by defining organelle identity, or by directly affecting membrane dynamics. Here we summarize recent advances in plant vesicular trafficking analysis with special emphasis on the interplay between membrane phospholipids and proteins in the creation and maintenance of specific endomembrane compartments.

1 Introduction

Cell division and cell expansion are the two basic processes of plant morphogenesis and development in the absence of cell migration. These processes rely on vectorial vesicle trafficking which is largely interconnected with the dynamics of the cytoskeleton. Small GTPases of the Ras superfamily and their effectors are important regulators orchestrating these processes also in plants. During the past few years, it has become clear that phospholipids have essential roles in the spatiotemporal regulation of plant membrane trafficking, in addition to their established roles in signal

V. Žárský (✉) and M. Potocký
Institute of Experimental Botany, Academy of Sciences of the Czech Republic, Rozvojová 263, 165 02 Prague 6, Czech Republic
e-mail: zarsky@ueb.cas.cz

V. Žárský
Department of Plant Physiology, Faculty of Science, Charles University, Viničná 5, 128 44 Prague 2, Czech Republic

F. Baluška and S. Mancuso (eds.), *Signaling in Plants*, Signaling and Communication in Plants, 107
DOI: 10.1007/978-3-540-89228-1_6, © Springer-Verlag Berlin Heidelberg 2009

transduction. They can act as constitutive signals that aid in defining organelle identity, and are short-lived recruiters and regulators of cytoskeletal and membrane dynamics. Recent studies have provided important clues as to how regulated activation of phospholipid-metabolizing enzymes and recruitment of their binding proteins might cooperate in targeting distinct pools of phospholipids to different cell physiological functions. That is why we focus in this short overview mainly on small GTPases together with phospholipid-modifying proteins, recognizing that a plethora of other proteins and protein complexes are important in complex realm of plant endomembrane system dynamics.

The field of signal transduction in vesicle trafficking in plants is so vast that, despite fragmentary knowledge of it, we have to divide it into two interrelated areas – "distal" mechanisms of signaling and integration of vesicle traffic on the physiological level and "proximal" cellular mechanisms of signaling in vesicle trafficking, which will be focus of this chapter. We summarize the main concepts and cover recent progress especially in respect to the interplay between membrane lipids and proteins in the creation and maintenance of specific endomembrane compartments and functional subdomains. We will start with an overview of the dynamics and participation of small GTPases (especially ADP-ribosylation factor, ARF; Rab; and ROP1) in vesicle formation, their transport, and their targeting and look how their regulators and effectors (in a broad sense including coat proteins and soluble *N*-ethylmaleimide sensitive factor attachment protein receptors, SNAREs) network with the membrane-modifying activities, especially lipid kinases, phosphatases, and phospholipases (especially phospholipase C, PLC, and phospholipase D, PLD). The interface between membrane proteins (especially GTPases) and membrane lipids (mainly glycerophospholipids) in plant vesicle trafficking is a major focus of this chapter, which is organized according to proteinaceous regulators, but as a conclusion we will see (based mostly on examples and knowledge from major non-plant-model cell types) that there is lot of reciprocity in the regulatory circuits of secretory pathways - regulators are often themselves regulated by products of "downstream" processes. Membrane proteins regulate membrane lipids and membrane lipids regulate membrane proteins.

In the end we will show that a big task for future research is to uncover how distal mechanisms of signal transduction in response to changes in internal or external plant environments collaborate with the proximal mechanisms regulating turnover of endomembranes on the cellular level. (For a recent general overview of the eukaryotic secretory pathway see Cai et al. 2007 and for a general overview of the secretory pathway in plants see Jürgens 2004.)

In this volume major events where physiological signals feed directly into the secretory pathway (tropic responses) are well covered in specific chapters both for gravitropism (see "Signaling in Plant Gravitropism" by Patrick H. Masson et al., this volume) and for phototropism (see "Signaling in Phototropism" by R. Brandon Celaya et al., this volume) along with the special chapter on auxin (see "Auxin and the Communication Between Plant Cells" by Peter Nick, this volume). Also signaling in pathogen response is covered (see "Vesicle Trafficking in Plant Pathogen Defence" by Hans Thordal-Christensen, this volume) and actin dynamics as a crucial part of vesicle traffic in plant cells is covered by a specific chapter (see "Signaling to the Actin Cytoskeleton During Cell Morphogenesis and Patterning" by Alison Sinclair et al., in this volume). That is also why we concentrate in this chapter on the "proximal" signals directly involved in endomembrane dynamics and from the more physiological perspective we will occasionally cover only selected examples. Owing to the scarcity of molecular data on mechanisms of plant vesicle traffic regulation, we have to refer to the mechanisms described in fungi and animals as they have turned out to be instructive in many areas of plant cell secretory pathway analysis.

2 Ras-Related GTP Binding Proteins Function in the Eukaryotic Cell Morphogenesis Driven by Vectorial Membrane Containers/Vesicle Traffic

Ras-related GTPases function as a molecular switch. Binding of GTP to small GTPase proteins imposes conformational changes on those proteins – they are active in the GTP-bound conformation and inactive in the GDP-bound conformation. Low intrinsic GTPase activity of these proteins results in a rather long life time for their active GTP-bound conformation. Only upon the intervention of a GTPase activating protein (GAP) is the γ-phosphate of the GTP molecule hydrolyzed, turning the GTPase off into the inactive, GDP-bound state.

Most small GTPases are modified posttranslationally at their C terminus by addition of hydrophobic prenyl (farnesylation or geranylgeranylation) or acyl (palmitoylation) moieties, so that they function as peripheral membrane proteins that cycle between membranes and cytoplasm. GDP dissociation inhibitors (GDIs) regulate this cycling by extracting GDP-bound GTPases from the target membranes for possible recycling through GDP/GTP exchange. This reactivation is catalyzed by guanine nucleotide exchange factors (GEFs; for a general review of plant small GTPases see Molendijk et al. 2004). The specificity of localization to particular membranes, and the ability for local activation/deactivation destine small GTPases to function as *organizers of specific membrane domains*. They participate in the initiation and maintenance of targeted processes of cell morphogenesis and signaling within the eukaryotic cells (Zerial and McBride 2001).

Ras-related GTPases (sometimes called "GTP-binding proteins") are divided into several well-defined families or subfamilies; for the purpose of vesicle traffic regulation, small GTPases of the ARF, Rab, and ROP (Rho of plants) families are of main interest. Collectively they regulate secretory pathway, dynamics of cytoskeleton, cell wall synthesis, activity of membrane-bound enzymes and transporters (e.g., NADPH oxidases and ion channels), signal transduction pathways, and metabolic pathways.

2.1 *ARF GTPases, Vesicle Coating, and Endosome Recycling*

ARF GTPases are well-known organizers of vesicle formation domains (export sites) and "nucleators" of vesicle coat formation (for a recent review on plant ARFs see Gebbie et al. 2005; Matheson et al. 2008). ARFs are essential not only for vesicle coating but also for uncoating in all eukaryotic cells. There are also subfamilies of ARF-like proteins (ARLs) which play more diverse roles in different organisms and cell types. In animal cell biology ARF GTPases were among the first where regulatory links to membrane lipid metabolism were discovered – ARF1 directly interacts with PLD and stimulates its activity (for recent update see Perez-Mansilla et al. 2006). ARF proteins were also directly implicated in the regulation of phosphoinositide signaling: ARF6 colocalizes and directly activates phosphatidylinositol 4-phosphate 5-kinase (PtdIns4P-5K) at the plasma membrane (Honda et al. 1999), and ARF6 and phosphatidylinositol 4,5-biphosphate [PtdIns(4,5)P2] colocalize on the plasma membrane and on endosomal structures. ARF6-organized PtdIns(4,5)P2 turnover at the plasma membrane is apparently involved in regulated secretion (Brown et al. 2001).

An important pharmacological tool used to study the function of ARF GTPases is fungal toxin brefeldin A (BFA), which specifically inhibits activation of ARF by BFA-sensitive GEFs (note that in plants there are also BFA-insensitive GEFs – see further), leading in plants partly to Golgi apparatus (GA)–endoplasmic reticulum (ER) coalescence and in high BFA concentrations it induces "BFA compartments," which consist of a mixture of endomembranes including endosomal compartments (Ritzenthaler et al. 2002).

The *Arabidopsis* genome potentially encodes 19 proteins with sequence similarities to ARFs and ARLs. Nine proteins, six of them highly similar, are possible ARFs. The functions of the six highly similar, putative ARFs in whole plant development were probed by suppressing their expression with antisense. Antisense plants were severely stunted because the cell production rate and the final cell size were both reduced (Gebbie et al. 2005). Changed time-to-flowering, apical dominance, and fertility may reflect alterations to hormonal and other signaling pathways with which ARFs may interact. No gross changes in targeting or compartmentalization were seen in antisense plants containing green fluorescent protein (GFP) targeted to the ER and the GA (Gebbie et al. 2005). These effects are less drastic than the effects on endomembranes and wall composition that are seen in short-term experiments with BFA and with dominant negative (DN) ARF mutants as both these treatments affect several ARF-dependent processes at once. Large redundancy is expected in plant ARF functions.

Recent evidence indicates that ARF1 carries out multiple roles in plant cells that may be independent of the established effector complex COPI (Pimpl et al. 2003). Matheson et al. (2007) followed the dynamics of a novel putative effector, the plant golgin GRIP-related ARF binding domain containing *Arabidopsis* protein 1 (GDAP1) in vivo. They showed that ARF1 functions to recruit GDAP1 to membranes. In addition, although ARF1 appears to be central to the recruitment of both COPI components and the golgin, the authors observed different subcellular distribution of these ARF1 effectors – GDAP1 and COPI are both distributed on Golgi membranes; however, GDAP1 is also found on ARF1-labeled structures that lack coatomer, suggesting that the membrane environment, rather than ARF1 alone, influences the differential recruitment of ARF1 effectors. It seems that modulation of the cellular functions of ARF1 in plant cells encompasses not only the intrinsic activities of the effectors, but also differential recruitment onto membranes that is spatially regulated (Matheson et al. 2007).

The same laboratory of Federica Brandizzi studied recently two members of the *Arabidopsis* ARF GTPase family, ARF1 and ARFB, which are membrane-anchored through the same N-terminal myristoyl group but to different target membranes – ARF1 is targeted to the Golgi and post-Golgi structures, whereas ARFB accumulates at the plasma membrane (Matheson et al. 2008). While the subcellular localization of ARFB appears to depend on multiple domains including the C-terminal half of the GTPase, the correct targeting of ARF1 is dependent on two domains: an N-terminal ARF1 domain that is necessary for the targeting of the GTPase to membranes and a core domain carrying a conserved MxxE motif that influences the relative distribution of ARF1 between the Golgi and post-Golgi compartments. The N-terminal ARF1 domain alone was insufficient to maintain an interaction with membranes and correct targeting is a protein-specific property that depends on the status of the GTP switch. Finally, an ARF1–ARFB chimera containing only the first 18 amino acids from ARF1 was shown to compete with ARF1 membrane binding loci. Although this chimera exhibited GTPase activity in vitro, it was unable to recruit coatomer, a known ARF1 effector, onto Golgi membranes. These results suggest that the targeting of plant ARF GTPases to the correct membranes may not only depend on interactions with effectors, but also relies on distinct protein domains and further binding partners on the Golgi surface, including possibly membrane lipid environment.

The best characterized plant secretory pathway regulator is GEF for ARFs and was discovered as *Arabidopsis* embryo mutant *gnom/EMB30* by forward genetic screens (Shevell et al. 1994; Busch et al. 1996). It soon became obvious that it is involved in the PIN auxin efflux carrier recycling and its localization helped to identify the first plant endosome (Geldner et al. 2003; for details see "Signaling in Plant Gravitropism" by Patrick H. Masson et al., in this volume).

Circumstantial evidence suggests that intracellular membrane trafficking pathways diversified independently in the plant kingdom, but documented examples are rare. ARF-GEFs are essential for vesicular trafficking in all eukaryotic kingdoms, but of the eight ARF-GEF families, only the ancestral BIG and GBF types are found in plants. Whereas fungal and animal GBF proteins perform conserved functions at the GA, the *Arabidopsis thaliana* GBF protein GNOM is thought to act in only the process of recycling from endosomes. Teh and Moore (2007) and Richter et al. (2007) showed that the related *Arabidopsis* GBF protein GNOM-LIKE1 (GNL1) has an ancestral function at the GA, but is also required for selective internalization from the plasma membrane in the presence of BFA. They identified *gnl1* mutants that accumulated biosynthetic and recycling ER markers in enlarged internal compartments. Notably, in the absence of functional GNL1, Golgi stacks were rendered sensitive to the ARF-GEF inhibitor BFA, which caused them to fuse with the ER. Furthermore, in BFA-treated *gnl1* roots, the internalization of a polar plasma-membrane marker, the auxin efflux carrier PIN2, was selectively inhibited. Thus, GNL1 is a BFA-resistant GBF protein that functions with a BFA-sensitive ARF-GEF both at the GA and in selective endocytosis, but not in recycling from endosomes. It seems that the evolution of endocytic trafficking in plants was accompanied by neofunctionalization within the GBF family, whereas in other kingdoms it occurred independently by elaboration of additional ARF-GEF families (Teh and Moore 2007).

Jensen et al. (2000) predicted *Arabidopsis* proteins which share an 80-residue zinc finger domain known from ARF GAPs. One of these is a 37-kDa protein, designated ZAC, which has a novel domain structure in which the N-terminal ARF GAP domain and a C-terminal C2 domain are separated by a region without homology to other known proteins. ZAC protein was immunodetected mainly in association with membranes and fractionated with Golgi and plasma membrane marker proteins. ZAC membrane association was confirmed in assays by a fusion between ZAC and the GFP and prompted an analysis of the in vitro phospholipid-binding ability of ZAC. Phospholipid dot-blot and liposome-binding assays indicated that fusion proteins containing the ZAC-C2 domain bind anionic phospholipids nonspecifically, with some variance in Ca^{2+} and salt dependence. Similar assays demonstrated specific affinity of the ZAC N-terminal region (residues 1–174) for phosphatidylinositol 3-phosphate (PtdIns3P). Binding was dependent in part on an intact zinc finger motif, but proteins containing only the zinc finger domain (residues 1–105) did not bind PtdIns3P. Recombinant ZAC possessed GTPase-activating activity on *Arabidopsis* ARF proteins. These data identify a novel PtdIns3P-binding protein region and thereby provide evidence that this phosphoinositide is recognized as a signal in plants.

All the above-mentioned observations indicate that also plant ARFs will be interconnected with the membrane lipid environment and new observations in this area are to be expected.

2.2 *Rab GTPases at the Crossroads of Vesicle Dynamics Regulation*

Rab GTPases are recognized regulators right in the heart of vesicle/membrane trafficking pathways as they are involved in the definition of membrane subdomains (Zerial and McBride 2001; see further) and all steps of vesicle formation, transport on the cytoskeleton (also via direct interaction with the cytoskeletal motors), and docking prior the SNARE-dependent fusion with the target membrane (for reviews of the function of non-plant Rab GTPases see Deneka et al. 2003; Novick et al. 2006; Grosshans et al. 2006). Rab GTPases turned out to be crucial regulators of the endomembrane system also from the evolutionary perspective (Gurkan et al. 2007). Expansion of Rab GTPases in eukaryotic evolution seems to be a major driving force in endomembrane specialization. Rab-regulated interaction networks seem to coordinate many different activities of the formation and fusion of membrane containers through their capacity to control the consecutive assembly of protein complexes that mediate endomembrane structure and dynamics. Specific Rab isoforms differ greatly in respect to effector interactions and specialized functions. An activated Rab GTPase functions as an organizer of a specific membrane domain by recruiting different effectors contributing to vesicle formation and cargo loading to the native vesicle. By direct or indirect (via adaptor protein) interaction with specific actin or microtubular motors, Rab GTPases regulate vectorial transport of vesicles along the cytoskeletal elements. Vesicle tethering at the target membrane is an important mechanism of specificity of interaction control before SNARE complex formation. As a vesicle tether, long coiled-coil proteins and/or protein complexes are responsible for vesicle docking to the target membrane. Multiprotein tethering complexes act either as GEFs (e.g., HOPS) or as effectors (e.g., exocyst) for interacting Rab GTPase. The tethering process seems to be linked, at least in some cases, to the removal of vesicle SNARE and target membrane SNARE autoinhibition and fusogenic SNARE complex activation (Deneka et al. 2003).

In the realm of angiosperm plants, Rab GTPases expanded accordingly and contemporary understanding of their family structure is well analyzed in phylogenetic surveys identifying more than 50 family members (Rutherford and Moore 2002; Vernoud et al. 2003). As the plant Rab nomenclature is not settled, we will use names for Rab subfamilies as proposed by Pereira-Leal et al. (2001) and Vernoud et al. (2003) for *Arabidopsis* Rab GTPases and also sometimes pragmatically as they are used by other authors. Plant Rab involvement in endocytic pathways was nicely summarized by (Nielsen 2006) and an update on Rab and ARF functions in regulation of plant cells was also very recently provided by (Nielsen et al. 2008).

The RabA subfamily related to the Ypt3/Rab11/Rab25 subfamily of Rab GTPases has expanded greatly in *Arabidopsis* and is the largest plant subfamily, with 26 members in six provisional subclasses, RabA1 to RabA6 (Rutherford and Moore 2002; Vernoud et al. 2003). It is supposed to participate mainly in post-Golgi transport and endosome recycling regulation. The analysis of members of this subfamily was the first to point out the richness of different endosomal compartments in plants (Ueda et al. 2004) as two closely related RabA proteins from pea (called Pra2 and Pra3 in the study) turned out to localize to different endomembrane compartments when expressed heterologously in the BY-2 cell line (Inaba et al. 2002). Recently the laboratory of Ian Moore provided evidence that RabA2 and RabA3 together are markers of endosomal/TGN subdomain (convergence of the early endocytic and secretory trafficking pathways in the TGN was proposed by Dettmer et al. 2006), which actively participates in the cell plate formation during plant cytokinesis. Chow et al. (2008) showed that the RabA2 and RabA3 subclasses define a novel post-Golgi membrane domain in *Arabidopsis* root tips. The RabA2/RabA3 compartment was distinct from but often close to Golgi stacks and prevacuolar compartments and partly overlapped the trans-Golgi network (TGN) compartment marked by vacuolar H-ATPase (VHA)-a1 subunit. It was also sensitive to BFA and accumulated FM4-64 before prevacuolar compartments did. Mutations in RabA2a that were predicted to stabilize the GDP- or GTP-bound state shifted the location of the protein to the Golgi or plasma membrane, respectively. In mitosis, t-SNARE KNOLLE accumulated principally in the RabA2/RabA3 compartment. During cytokinesis, RabA2 and RabA3 proteins localized precisely to the growing margins of the cell plate, but VHA-a1, GNOM, and prevacuolar markers were excluded. Inducible expression of dominant-inhibitory mutants of RabA2a resulted in enlarged, polynucleate, meristematic cells with cell wall stubs. The RabA2/RabA3 compartment, therefore, is a TGN compartment that communicates with the plasma membrane and early endosomal system and contributes substantially to the cell plate (Chow et al. 2008). Despite the unique features of plant cytokinesis, membrane traffic to the division plane exhibits surprising molecular similarity across eukaryotic kingdoms in its reliance on Ypt3/Rab11/RabA GTPases.

Detailed analysis of tomato RabA homologue involvement in later stages of the secretory pathway regulated by plasma membrane SNARE proteins SYP121 and SYP122 suggests that this particular RabA (Rab11a) regulates anterograde transport from the TGN to the plasma membrane and points to SYP122, rather than SYP121 as a downstream pathway member. This implies the possibility that SYP121 and SYP122 drive independent secretory events in conjunction with specific Rab GTPases (Rehman et al. 2008).

Pollen tubes along with root hairs are the plant cells with the most effective secretory pathway supporting quick polar, tip-localized cell expansion (for reviews see Samaj et al. 2006; Cole and Fowler 2006). It is therefore not surprising that the first Rab GTPases described in plants turned out to be highly expressed in pollen (RabB in maize; Moore et al. 1997). Also members of the RabA subfamily are prominently localized into the growing pollen tube tip (de Graaf et al. 2005) as well as the root hair tip (Preuss et al. 2004); disturbance of their activity compromises polarized cell growth in both cell types.

Heo et al. (2005) showed that rice Rab GAP has a positive influence on the TGN to the plasma membrane or vacuole transport regulated by RabA/Rab11 homologues, adding at least some data to the largely under-explored plant Rab-GAP proteins.

Recently we have identified in plants one of putative major effectors of both Rab and Rac/Rho GTPases in the final stage of the secretory pathway – tethering complex exocyst. It is an eight-subunit complex participating in vesicle tethering/docking at the target membrane compartment before membrane fusion affective SNARE complex might be formed (Cai et al. 2007). This complex was first described by the laboratory of Peter Novick TerBush et al. 1996 as an effector of yeast Rab8-related GTPase Sec4. It turned out that in animals the Sec15 subunit of exocyst interacts with members of the RabA/Rab11 family (Wu et al. 2005). Over the time it became clear that Rho GTPases in their GTP-charged active conformation also interact with exocyst subunits, making this complex a hot spot of regulation of processes dependent on GTPases (Novick and Guo 2002). While in yeast/animal exocyst active GTPases interact directly with exocyst subunits, (Lavy et al. 2007) showed recently that the *Arabidopsis* Sec3 subunit interacts with active ROP GTPases via adaptor protein ICR1. Mutant phenotypes of *Arabidopsis* exocyst-related mutants (Cole et al. 2005; Synek et al. 2006; Hala et al. 2008) along with the *icr1* mutant all show that exocyst-related processes are at the center of polar cell, meristem, tissue, and plant development and are linked also to polar auxin transport (Synek et al. 2006; Hala et al. 2008).

Recently the Exo70 and Sec3 subunits of exocyst were shown to interact directly with PtdIns(4,5)P2-rich domains at the plasma membrane via a strongly electrostatically charged end (He et al. 2007; Liu et al. 2007; Zhang et al. 2008). This observation provides a mechanism for how Exo70 might be inserted into the membrane before and independently of other exocyst subunits. Our comparative analysis of *Arabidopsis* EXO70 paralogues suggests that most of them have a putative lipid/PtdIns(4,5)P2 interacting domain on the C-terminus conserved.

RabB in angiosperms are related to Rab2 GTPases supposed to be involved in the ER to GA and intra-GA regulation in Opisthokonts. In tobacco pollen tubes, GFP-Rab2/RabB associates efficiently with Golgi bodies. Expression of DN-Rab2s substantially reduced the delivery of GFP-labeled Golgi-located cargo proteins to their target, while cytoplasmic/ER signals were enhanced, suggesting a role for NtRab/RabB in the early secretory pathway (Cheung et al. 2002).

RabD related to human Rab1 was shown to participate, as predicted from the homology, in the ER to GA transport (Batoko et al. 2000). In this study secretory GFP (secGFP) and transient expressions in *Agrobacterium*-infiltrated leaves was systematically used for the first time to monitor the function of Rab GTPase in the plant secretory pathway. In this report, anecdotal observation of cessation of actin-dependent GA movement in cells with disturbed RabD function indicates that also some plant Rab GTPases might be involved in the regulation of cargo movement on the cytoskeleton.

Homologues of human Rab8 form the RabE group in plants and on the basis of homology they are expected to participate in the GA to plasma membrane transport in the final stages of exocytosis. (Zheng et al 2005) showed that the RabE1 DN mutant caused less intracellular secGFP accumulation than the DN mutant of RabD2. Moreover, whereas DN-RabD2 caused secGFP to accumulate exclusively in the ER, RabE1 caused secGFP to accumulate additionally in the GA and a prevacuolar compartment that could be labeled by FM4-64 and yellow fluorescent protein (YFP)-tagged *Arabidopsis* RabF2. These and other observations show that RabE acts downstream of the RabD subclass of GTPases in biosynthetic membrane traffic to the plasma membrane (Zheng et al. 2005).

Rab5 proteins are well-studied regulators of endocytic pathways in animals and fungi (Zerial and McBride 2001). Analysis of AtRabF2b (Ara7) in the same system of tobacco leaf epidermal cells showed that it acts on the vacuolar trafficking pathway (Sohn et al. 2003; Kotzer et al. 2004; Lee et al. 2004; Ueda et al. 2004). As human and yeast RabF homologues (Rab5) are implied mainly in the endosomal transports, this is an example of a plant Rab protein in which functional prediction based solely on homology would be false. It is obvious that in any single case it will be necessary to evaluate experimentally the functions of Rab GTPases as predictions based solely on homology to Opisthokontian proteins are of limited use. RabF1 and RabF2 were observed to localize to different subcompartments of multivesicular bodies/late endosome, supporting the notion of different functional specialization of these closely related proteins (Ueda et al. 2004; Haas et al. 2007).

Rab5/RabF plant homologues are at the moment also best studied in respect to their GEF activator. Only one GEF, named VPS9a, can activate all Rab5 members to GTP-bound forms in vitro in spite of their diverged structures. In the *vps9a-1* mutant, whose GEF activity is completely lost, embryogenesis was arrested at the torpedo stage (Goh et al. 2007). Comparison of two RabFs (ARA5 and ARA6-RabF1) showed that the two types of plant RabF are functionally distinct, despite the regulation by the same activator, VPS9a (Goh et al. 2007). ARA6 was shown to be a very interesting plant-specific member of the RabF subfamily as it completely lacks the classic C′-terminal prenylation consensus sequence. Instead it harbors an extra stretch of amino acid residues containing myristoylation and palmitoylation sites at the N-terminus. ARA6 is tightly associated with membranes in an acylation-dependent manner (Ueda et al. 2001).

Takano et al. (2005) found that *Arabidopsis* Bor1 is a boron exporter for xylem loading and is essential for efficient boron translocation from roots to shoots under boron limitation. BOR1-GFP is localized to the plasma membrane under boron limitation. Shortly after boron application, the protein was observed in a dotlike structures in the cytoplasm before degradation. Colocalization studies of the fusion protein with an endocytic tracer FM4-64 and an endosomal Rab-GTPase RabF/Ara7 suggested that BOR1 is transferred from the plasma membrane via the endosomes to the vacuole for degradation. These results establish that endocytosis and degradation of BOR1 are regulated by boron availability, to avoid accumulation of toxic levels of boron in shoots under high-boron supply, while protecting the shoot from boron deficiency under boron limitation.

Vacuolar Rab7/RabG homologues have not been characterized in plants yet. However, an interesting observation was made in *Arabidopsis* plants lacking the putative effector in a mutant described as *vacuoleless1* (*vcl1*) which blocks vacuole formation and alters the pattern of cell division orientation and cell elongation, resulting in embryo lethality (Rojo et al. 2001). It turned out that the mutant locus encodes the Vps16p homologue of subunit of the HOPS vacuolar vesicle tethering complex. VPS16-containing HOPS complex is functional also in plants and is involved in membrane tethering and fusion both at the prevacuolar compartment as well as at the tonoplast proper (Rojo et al. 2003). Other members of plant Rab GTPase family have been characterized poorly or not at all. As secretory pathway is so crucial to understand plant life, it is a tantalizing task challenging plant cell biologists to fill in the gaps in plant Rab protein studies.

There are examples where Rab GTPases or their regulators/effectors are induced in response to an external signal or stress treatment. Moshkov et al. (2003) determined induction and repression of gene expression of small monomeric GTPases as an integral part of the plant ethylene response. Among the induced genes, Rabs were the most prominent. The gene expression of plant Rab GDIs (Ueda et al. 1996; Zarsky et al. 1997) was shown to be induced by aluminum and pathogen attack stresses in *Arabidopsis* and rice (Kim et al. 1999; Ezaki et al. 2000). Plant Rab GTPases are expected to be involved in osmotic stress response as indicated by Mazel et al. (2004), showing that the overexpression of RabG3e (Rab7 homologue) confers salt and osmotic stress tolerance to transgenic plants.

Kang et al. (2001) reported data suggesting an unprecedented (and unexpected) direct participation of plant Rab GTPase in the regulation of the enzyme involved in brassinosteroid (BR) biosynthesis and in light signal transduction. These surprising data await further analysis and interpretation.

It is to be expected that plant Rab GTPases and their interactors also participate in the general cellular defense pathway as proposed for the secretory pathway by Robatzek (2007) and demonstrated recently for SNARE protein (Kwon et al. 2008, see also Hans Thordal-Christensen, this volume).

Analysis of the barley (*Hordeum vulgare*) HVA22 gene inducible by a variety of stress conditions (dehydration, salinity, extreme temperatures) led to the description of the first plant homologue of Yop1 – yet another Rab binding protein which binds Rabs as a dimer with the Yip1 protein (Brands and Ho 2002). The plant Yop1 homologue seems to be a negative regulator of Rabs (similarly as Rab GDI) in plants. It is also induced by abscisic acid treatment of barley aleurone layers, which blocks GA-stimulated secretion. The yeast Yop1 protein interacts with Sey1, which has a homologue in *Arabidopsis* known as RHD3. The Arabidopsis *ROOT HAIR DEFECTIVE 3* (*RHD3*) gene, whose mutation results in abnormal root hairs with accumulated vesicles, encodes a protein containing an unusual GTP-binding motif (Wang et al. 1997). It remains to be determined whether the interaction between HVA22/Yop1 and RHD3/Sey1 is conserved also in plants.

Recently a mechanism was discovered where Rho GTPase TC10 collaborates with Rab5 in stimulating PtdIns3P production in response to insulin in mammalian cells (Lodhi et al. 2008). As GLUT4 recruitment to the plasma membrane in response to insulin signal is sometimes used as a paradigm for PIN auxin efflux carrier regulation in plants, it might be useful to mention here how this signaling rely works. Insulin promotes the production of PtdIns3P at the plasma membrane by a process dependent on TC10 activation, and Lodhi et al. (2008) showed that insulin-stimulated PtdIns3P production requires the activation of Rab5, a small GTPase that plays a critical role in phosphoinositide synthesis and turnover. This activation occurs at the plasma membrane and is downstream of TC10. TC10 stimulates Rab5 activity via the recruitment of GAPEX-5, a VPS9 domain-containing GEF that forms a complex with TC10. It should be expected that analogous relationships will also operate in the GTPases/membrane lipid interface in plant cells.

Unexpected involvement of the phosphoinositide 3-kinase (PtdIns3K) subunit in Rab activity regulation was observed recently in mammals. The p85α subunit of PtdIns3K was shown to have GAP activity toward Rab5 and Rab4 GTPases. Disrupted Rab GAP function of the p85 subunit of PtdIns3K results in cell transformation (Chamberlain et al. 2008). The essential role of p85 in the regulation of Rab function suggests a novel role for p85 in controlling receptor signaling and trafficking through its effects on Rab GTPases. Again, analogous regulatory short cuts between GTPases (especially those involved in the endocytosis) and phospholipid signaling proteins have to be expected also in plant cells.

The first example of GTPase involved in phosospholipid signaling in plants was described by Kost et al. (1999) in pollen tubes. ROP GTPase active at the very growing pollen tube tip interacts with PtdIns4P-5K and positively regulates its PtdIns(4,5)P2 producing activity contributing to positive feedback mechanisms operating at the tip to sustain highly polarized and specific distribution of pollen tube constituents. Preuss et al. (2006) discovered that, the RabA4b

GTPase labels a novel, TGN compartment displaying a developmentally regulated polar distribution in growing *Arabidopsis thaliana* root hair cells. GTP-bound RabA4b selectively recruits the plant phosphatidylinositol 4-hydroxy kinase β1 (PtdIns-4Kβ1), but not members of other PtdIns-4K families. PtdIns-4Kβ1 colocalizes with RabA4b on tip-localized membranes in growing root hairs, and mutant plants in which both the PtdIns-4Kβ1 and -4Kβ1 genes are disrupted display aberrant root hair morphologies. PtdIns-4Kβ2 interacts with RabA4b through a novel homology domain, specific to eukaryotic type IIIβ PtdIns-4Ks, and PtdIns-4Kβ1also interacts with a Ca^{2+} sensor, AtCBL1, through its N-terminus. It is proposed that RabA4b recruitment of PtdIns-4Kβ1 results in Ca^{2+}-dependent generation of phosphatidylinositol 4-phosphate (PtdIns4P) on this compartment, providing a link between Ca^{2+} and PtdIns(4,5)P2-dependent signals during the polarized secretion of cell wall components in tip-growing root hair cells.

It is interesting that in fact late-acting Rab GTPases may be downstream effectors of early Rab GTPases acting at the outset of secretory pathway. As exemplified by Rab/RabF function in the endocytosis, the Rab machinery can be viewed as a typical modular system, in which specific biochemical interactions between Rab effectors and regulators as well as other compartment-specific proteins create a spatial segregation. By regulating the assembly of a specific membrane domaine, Rab proteins contribute to the compartmental specificity, robustness, and dynamic properties of specific eukaryotic endomembrane compartments (Zerial and McBride 2001).

Peter Novick and Susan Ferro-Novick at Yale University discovered a new type of Rab signaling relay – recruitment of GEF for Rab GTPase active in the next step of vesicle transport by previous Rab GTPase (Wang and Ferro-Novick 2002; Ortiz et al. 2002). It turned out that GEF for the yeast intra-Golgi Rab Ypt32 is the effector of "previous" Rab protein Ypt1 responsible for ER to GA vesicle traffic. Equally Ypt32 recruits the GEF for the "next" yeast exocytic Rab Sec4 known as Sec2 to secretory vesicles. Once on the vesicle, Sec2 activates Sec4, enabling the polarized transport of vesicles to the plasma membrane. How general this mechanism is and if it holds also for plant cells should be clear in a few years.

In endosome maturation, yet another process, related to the Rab cascades mentioned above, was observed and proposed to have more general validity for other endomembrane compartments. This process is called the "Rab conversion mechanism" (Rink et al. 2005; Deretic 2005). Rab5 and Rab7 GTPases are key determinants of early and late endosomes in Opisthokonts, organizing effector proteins into specific membrane subdomains. Rink et al. (2005) found that the level of Rab5 dynamically fluctuates on individual early endosomes, linked by fusion and fission events into a network in time. Within it, degradative cargo concentrates in progressively fewer and larger endosomes that migrate from the cell periphery to the center where Rab5 is rapidly replaced with Rab7. The class C VPS/HOPS complex, an established GEF for Rab7, interacts with Rab5 and is required for Rab5-to-Rab7 conversion. These results reveal unexpected dynamics of Rab domains and suggest Rab conversion is the mechanism of cargo progression between early and late endosomes, with potential implications in other secretory pathway transport steps. Improved coverage of specific endomembrane markers (both for protein as well as for membrane phospholipid composition) for plant cells is boosting greatly understanding of regulatory and kinetic relationships within the plant endomembrane system.

2.3 The ROP GTPase Family and Polarization Processes in the Plant Secretory Pathway

Phylogenetic analysis indicates that the plant ROP genes rapidly evolved prior to the emergence of the embryophyta, to become a group distinct from the Rac genes in other eukaryotes (Brembu et al. 2006). In embryophyta, ROP genes have undergone an expansion through gene duplications (Winge et al. 2000; Christensen et al. 2003; Brembu et al. 2006).

There are two major subfamilies within the ROP family, each with distinct structures, localization mechanisms, and (apparently) functions. Subfamily I (represented in *Arabidopsis* by ROP1-ROP8) contains the conventional C-terminal motif (present in the majority of nonplant Rho GTPases) for prenylation, while members of subfamily II (or type-II) ROPs (represented in *Arabidopsis* by ROP9-ROP11) are not prenylated, but rather palmitoylated at a cysteine-containing motif that appears to be a plant-specific innovation (Ivanchenko et al. 2000; Lavy et al. 2002; Lavy and Yalovsky 2006).

However, the type-I ROP6 GTPase (and possibly other type-I ROPs) is also transiently acylated by palmitic and stearic acid when in the active, GTP-bound conformation (Sorek et al. 2007). This modification stabilizes the membrane localization and induces partitioning into detergent resistant membrane domains (DRMs, also called lipid rafts; Sorek et al. 2007); the type-I ROPs are uncovered also in the tobacco DRM proteome (Morel et al. 2006). The *tip1* mutant in *Arabidopsis*, which acts early in the bulge and root hair formation pathway (Parker et al. 2000), has disrupted *S*-acyltransferase activity (Hemsley et al. 2005) and thus might affect ROP palmitoylation and DRM partitioning. It is possible that activated type-I ROPs are recruited into and help organize DRM domains not only at the tip of elongating root hairs or pollen tubes, but also, e.g., during the cell plate formation at the end of angiosperm cytokinesis.

Recently the C′-terminal polybasic region of human Rac1, which is well conserved also in plant ROPs, was proven to interact directly with the phosphatidic acid (PA) produced by PLD activity – Rac1 mutated in this C′-polybasic region is incapable of translocation to the plasma membrane (Chae et al. 2008). Moreover, these authors also showed that PA is capable of dissociating Rac1-GDI complex functioning in this context as a GDF.

Plant RhoGAPs (in *Arabidopsis*, five genes) in contrast to RhoGAPs in other eukaryotes characterized to date are equipped with a Cdc42/Rac-interactive binding CRIB domain, which not only assists GAP–ROP binding, but is also involved in the

GAP activity itself (Wu et al. 2000). In pollen tubes, the domain of active ROP at the tip is limited by a Rho GAP localized subapical to the tip, where it appears to be regulated by interactions with a 14–3–3 protein (Klahre et al. 2006).

The first ROP GEF was identified as the *spike1* mutant of *Arabidopsis*. It is seedling-lethal and forms simple nonbranched trichomes (Qiu et al. 2002). *SPIKE1* encodes a member of a CZH family of RhoGEFs (Meller et al. 2005).

A breakthrough in the understanding of ROP signaling regulation was the discovery of new class of Biconta-specific GEFs (encoded by 14 genes in *Arabidopsis*) that activate ROP (Berken et al. 2005; Gu et al. 2006). These plant-specific ROP nucleotide exchanger (PRONE) family proteins were actually first described in tomato, based on recovery of one member (kinase partner protein, KPP) as an interactor with the cytoplasmic domain of the pollen-specific receptor-like kinases (RLKs) LePRK1 and LePRK2 (Kaothien et al. 2005). Overexpression in pollen tubes of either KPP (Kaothien et al. 2005) or one of several *Arabidopsis* PRONE genes (Gu et al. 2006) causes depolarized growth. A most exciting possibility, implied by the KPP–PRK interaction, is that this interaction regulates GEF activity, perhaps by releasing it from the autoinhibitory domain at the PRONE family C-terminus (Gu et al. 2006). Furthermore, given that ROP has been recovered in a complex with another RLK (CLV1; Trotochaud et al. 1999), ROP GEF–RLK interactions could be a general mechanism to activate ROP signaling in multiple physiological and cellular contexts in plants.

ROP might stimulate PLC and PLD activities via PtdIns(4,5)P2 synthesis by direct activation of PtdIns4P-5K (Kost et al. 1999); PtdIns4P-5K may be stimulated also by PA (reviewed in Oude Weernink et al. 2007). It is proposed that this reciprocal stimulation of PLD and PtdIns4P-5K is able to generate rapid feed-forward loops for localized and abrupt generation of PA and PtdIns(4,5)P2, which may then govern the recruitment and activation of proteins to a membrane domain to execute specific tasks (reviewed in Oude Weernink et al. 2007). The active de novo formation of the cell wall (as at the tip of growing root hairs and pollen tubes, or cell plate) could represent such domains, as the plasma membrane in this region is PtdIns(4,5)P2-enriched (Vincent et al. 2005; Kost et al. 1999). These processes in plant cells are able to build a specific plasma membrane domain based on reciprocal activation of PtdIns4P-5K, PLD/PA, and ROP GTPase activities.

We have already referred to the exocyst tethering complex in plants and mentioned that it interacts with active ROP GTPase via ICR1 protein which was recently implied in the endosome dynamics (Robert et al. 2008). In fact exocyst involvement is implied also in vesicle recycling via recycling endosome (Jafar-Nejad et al. 2005) and the membrane recycling process was shown to be a very effective vehicle for cell polarization.

Valdez-Taubas and Pelham (2003) showed that localized exocytosis, coupled with endocytosis and recycling, can be a mechanism to maintain dynamic polarization of membrane proteins, if diffusion to an equilibrium distribution in the membrane is slow. Such domains of localized exocytosis/rapid recycling may be found in plant cells during cytokinesis, as well as at the tips of root hairs and pollen tubes (Baluska et al. 2005; Ovecka et al. 2005; Dhonukshe et al. 2006).

3 Endosomes Function in Signal Transduction Via Internalized Activated Receptors: Signaling Endosomes in Plants

The concept of a signaling endosome was originally developed in the field of mammalian signal transduction from cell surface receptors as soon as it became clear that endocytosis is not a mere step to switch off and degrade the activated receptor. On the contrary, in many cases internalization of the receptor–ligand complex by endocytosis (receptor-mediated endocytosis) and coalescence of the endocytic vesicle harboring activated receptor with the endosome is absolutely required to relay the signal further (Vieira et al. 1996; Wunderlich et al. 2001).

The first hint that a similar mechanism also acts in signal transduction in plants was provided by studies of brassinoid receptor in the laboratory of Sacco de Vries. In *Arabidopsis* brassinosteroid (BR), perception is mediated by two leucine-rich-repeat (LRR)RLKs, BRASSINOSTEROID INSENSITIVE1 (BRI1) and BRI1-ASSOCIATED RECEPTOR KINASE1 (BAK1). Coexpression of BRI1 and AtBAK1 results in a change of the steady-state distribution of both receptors because of accelerated endocytosis. Endocytic vesicles contain either BRI1 or AtBAK1 or both of them (Russinova et al. 2004). The final proof that the BRI1-dependent signal transduction pathway uses a signaling endosome was provided by Geldner et al. (2007) – increasing endosomal localization of BRI1 enhances activation of the pathway and upregulation of brassinoid-specific gene expression. Use of endosomes as signaling compartments is possibly a general phenomenon in eukaryotes.

BRI1 and BAK1 are members of the biggest angiosperm receptor kinase family (more than 200 different proteins in *Arabidopsis*), which is characterized by extracellular LRR motifs and the intracellular serine/threonine kinase domain. A member of this receptor family was also among the first shown to be involved in the interaction between plant cells and pathogens playing a key role in plant immunity – genetic analysis proved that the receptor recognizing bacterial flagellin (flg22) is encoded by the *Arabidopsis* locus LRR-RLK FLAGELLIN SENSING2 (FLS2; Gomez-Gomez and Boller 2000). It is involved in perception of so-called microbe-associated molecular patterns. Transgenic lines that express a functional FLS2-GFP fusion driven by its native promoter revealed localization of the nonactivated receptor at the plasma membrane. Stimulation with the flagellin epitope flg22 induces its transfer into intracellular mobile vesicles/endosome, followed by degradation. FLS2 internalization depends on cytoskeleton and proteasome functions, and receptor activation (Robatzek et al. 2006). Internalization and FLS2-endosome formation was blocked by wortmannin (PtdIns3K inhibitor –PtdIns3P is an important endosomal membrane phospholipid) and not by BFA. Signaling of activated FLS2 was shown to be dependent on BAK1 LRR-receptor kinase without a link to the BR signaling pathway (Chinchilla et al. 2007). The importance of endosome signaling in plant cells is despite contemporary scarcity of our knowledge tremendous. What von Zastrow and Sorkin (2007) wrote about endocytosis/endosomes and signaling in animal cells is fully valid also for plants: "Endocytosis regulates many cellular signaling processes by controlling the number of functional receptors available at the cell surface. Conversely, some signaling processes regulate the endocytic pathway. Furthermore, various cellular signaling events appear to occur on endosome membranes. The endocytic pathway, by providing a set of dynamic and biochemically specialized endomembrane structures that physically communicate with the plasma membrane, is increasingly viewed as a highly flexible scaffold for mediating precise spatiotemporal control and transport of diverse biological signals. General principles of endosome-based signaling are beginning to emerge but, in many cases, the physiological significance of signaling on the endocytic pathway remains poorly understood".

For a detailed recent update on plant endosomes and endosome signaling, see recent reviews by (Otegui and Spitzer 2008) and (Geldner and Robatzek 2008).

4 Involvement of Plant Phospholipids in the Control of Membrane Dynamics

4.1 Phosphoinositide Kinases and Phosphatases as Regulators of Vesicle Trafficking

Phosphoinositides, phosphorylated derivatives of phosphatidylinositol, are essential in regulating nuclear processes, cytoskeletal dynamics, signaling, and membrane trafficking.

Glycerolipids (poly)phosphoinositides (PPIs) have been implicated in a plethora of processes, including membrane traffic, signal transduction, and cytoskeletal dynamics (recently reviewed by Krauss and Haucke 2007; Di Paolo and De Camilli 2006). A variety of lipid kinases reversibly phosphorylate positions 3, 4, and 5 on the inositol ring, so combinatorial phosphorylation results with only these three modifications in seven phosphoinositide species. Different endomembrane compartments and even subdomains of the same compartments are characterized by a special composition of phospholipids – in fungal and animal cells PtdIns4P is typically overrepresented in Golgi and the TGN, PtdIns(4,5)P2 and phosphatidylinositol 3,4,5-triphosphate are located mainly on the plasma membrane and possibly enriched in raft-like structures. Early endosomes are enriched with PtdIns3P and phosphatidylinositol 3,4-bisphosphate, while phosphatidylinositol 3,5-biphosphate is present predominantly on late compartments of the endosomal pathway. The location of phosphatidylinositol 5-phosphate remains poorly characterized (Di Paolo and De Camilli 2006). Such compartment-specific composition of phosphoinositides serves as a spatial landmark active in the regulation (and being at the same time regulated) by a plethora of peripheral as well as integral membrane proteins (see further).

In plant cells, six out seven signaling PPIs were detected; importantly phosphatidylinositol 3,4,5-triphosphate, a prevalent signal in animal cells, is missing in plants. Most of the signaling plant PPIs were originally implicated in various stress signaling pathways (Meijer and Munnik 2003) and their direct involvement in membrane trafficking is less documented. PtdIns(4,5)P2 is by far the most studied plant PPI and is well documented for actively growing (apical) membrane domains in pollen tubes and root hairs (Kost et al. 1999; Vincent et al. 2005; van Leeuwen et al. 2007; Kusano et al. 2008; Stenzel et al. 2008). Recently, several reports provided genetic evidence for indispensable role of PtdIns(4,5)P2 in cell polarity, strongly hinting at its involvement in vesicular traffic. PtdIns4P-5K isoform 3, a key enzyme producing PtdIns(4,5)P2, which is preferentially expressed in growing root hair cells, localized to the periphery of the apical region of root hairs, possibly associating with the plasma membrane and/or exocytotic vesicles. Loss-of-function mutants in PtdIns4P-5K isoform 3 are compromised in root hair formation, while overexpression led to multiple protruding sites on a single trichoblast, root hair deformation, and general loss of cell polarity. Moreover, transient heterologous expression of full-length PtdIns4P-5 kinase isoform 3 in tobacco pollen tubes increased plasma membrane association of a PtdIns(4,5)P2-specific reporter (PH domain from human PLCδ fused to GFP) in these tip-growing cells. (Kusano et al. 2008; Stenzel et al. 2008).

The spatial restriction and steady-state levels of specific PPIs are controlled primarily by the concerted action of phosphoinositide kinases and phosphatases, whose localization is tightly regulated. PPI distribution on distinct membranes helps ensure vectorial membrane traffic, which is critical to achieve transport of cargo from one compartment to another. PPI hydrolysis is due to both phospholipases and phosphatases, with fundamentally different physiological consequences. Cleavage by phospholipases gives rise to second messengers that propagate and amplify signaling (see further), whereas dephosphorylation controls steady-state levels of PPIs and turns off their signaling. For example, during endocytosis in animal cells, dephosphorylation removes PtdIns(4,5)P2 from internalized membranes, thus restricting the localization of PtdIns(4,5)P2 to the plasma membrane without coupling the endocytic reaction to the generation of signaling metabolites (Cremona et al. 1999). Orthologous PPI phosphatases are present in plant cells and one might speculate that they are also involved in vesicular traffic control. Recently, the root-hair-defective mutant *rhd4* was identified as a Sac1-type phosphatase specific for PtdIns4P (Thole et al. 2008). Apical growth of root hair cells in *Arabidopsis* is improperly controlled in *rhd4* mutant plants, resulting in root hairs that are shorter and randomly form bulges along their length. Interestingly, mammalian SAC1 was documented as critical for Golgi transport. SAC1 accumulates at the Golgi and downregulates anterograde trafficking by depleting Golgi PtdIns4P. When quiescent cells are stimulated by mitogens, SAC1 rapidly shuttles back to the ER, thus releasing the brake on Golgi secretion (Blagoveshchenskaya et al. 2008). Another candidate PPI phosphatase with function in trafficking is FRA3, a member of plant type-II 5PTase family with the highest substrate affinity toward PtdIns(4,5)P2. FRA3 controls actin organization and secondary cell wall synthesis in fiber cells, which may be due to its role in vesicular transport (Zhong et al. 2004).

Recent genetic and pharmacological evidence also suggests the role of plant phosphoinositide 3-kinase (PtdIns-3K) in endocytosis. Leshem et al. (2007) reported that activation of endocytosis as well as reactive oxygen species (ROS) production in endosomes by salt stress is fully PtdIns-3K dependent. After wortmannin treatment (inhibitor of both PtdIns-3K and PtdIns-4K) and more importantly in PtdIns-3K *Arabidopsis* knockout mutants, both endocytosis and ROS production are suppressed. This suppression is partially rescued by the addition of PtdIns3P. The same laboratory reported PtdIns-3K-dependent endocytosis and ROS production in the establishment of legume–rhizobia interaction (Peleg-Grossman et al. 2007). Inhibition of PtdIns-3K suppressed membrane endocytosis, subsequent ROS burst, and resulted in the inhibition of root hair curling and formation of infection threads. A pharmacological study using inhibitors and activators of signaling of lipid and GTPases also points to the link of PtdIns3P production concomitant with GTPase activation in *Vigna unguiculata* root hair activation by Nod factors (Kelly-Skupek and Irving 2006).

4.2 Plant Phospholipases in Endomembrane Dynamics

The implication of PtdIns(4,5)P2 in regulation of the vesicle traffic raised questions regarding the possible involvement of plant PLC and PLD and products of their reactions in these processes. Interestingly, while the role of animal PLDs in vesicular transport is well documented (Roth 2008), direct involvement of mammalian PLCδ (the isoform most similar to plant PLCs) is not clear. Metazoan PLD1 and PLD2 and/or corresponding PA function to induce or facilitate negative curvature of membranes, act as an allosteric regulator of proteins important for membrane traffic (e.g., PtdIns4P-5K), serve as a binding site for proteins (such as the COPI complex, NSF, kinesin) and recruit them to membranes, and can serve as a precursor for production of diacylglycerol (DAG) for any of these processes. In addition, animal PLDs have recently been found to bind to dynamin and accelerate its GTPase activity (Lee et al. 2006). Fractions of both animal PLDs are located at the Golgi and are required for its structural integrity (Freyberg et al. 2003). PLD activity is required for the final stages of exocytosis (Hughes et al. 2004) and for phagocytosis (Corrotte et al. 2006).

In a similar fashion, PLDs may function in vesicular traffic in plants, as suggested by several lines of evidence. Our laboratory and others have found that PLD activity is required for polar expansion of pollen tubes (Potocky et al. 2003; Monteiro et al. 2005) and PLD inhibition decreases the number of secretory vesicles in the pollen tube tip (Potocky, Derksen and Zarsky, unpublished data). PLD inhibition caused rapid changes in the morphology of the GA in BY-2 cells, resulting in extended curved cisternae. Similarly to BFA, PLD inhibition elicited the release of the GTPase ARF1 from Golgi membranes; on the other hand, ER cisternae did not attach laterally to cisternae structures, and ER-Golgi fusion hybrids were not observed (Langhans and Robinson 2007).

Recently mammalian-like PLD has been described to regulate vesicle trafficking and is required for auxin response. *Arabidopsis* PLDζ2 and PA are required for the normal cycling of PIN2-containing vesicles as well as for function in auxin transport and distribution (Li and Xue 2007). A recent observation from the laboratory of Frantisek Baluska and Stefano Mancuso indicates that auxin is also transported by vesicular secretion and that PLDζ2 drives trafficking of auxin for its polar transcellular transport in the transition zone of the root apex (Mancuso et al. 2007). Interestingly, PLDζ2 is also involved in the reaction of plants to inorganic phosphate starvation, specifically in hydrolyzing phosphatidylcholine and phosphatidylethanolamine to produce DAG for digalactosyldiacylglycerol synthesis and free inorganic phosphate (Cruz-Ramirez et al. 2006). PLDζ2 was recently shown to be localized predominantly on tonoplast (Yamaryo et al. 2008), thus raising the question of the exact mechanism of its action in recycling and secretion. Another mammalian-like PLD isoform in *Arabidopsis*, PLDζ1, was localized to the clear zone of vesicles of root hairs and its inducible overexpression caused branching of root hairs, while PLD inhibition with 1-butanol resulted in loss of root hairs (Ohashi et al. 2003).

In contrast to animal cells, where PLCδ is involved in many signaling pathways but its direct implication in membrane transport is not clear, plant PLC specific for PtdIns(4,5)P2 may also fulfill this role, at least in apically growing cells. PLC activity was found to be crucial for polar growth in pollen tubes in *Petunia* and tobacco (Dowd et al. 2006; Helling et al. 2006). PLC is localized laterally at the pollen tube tip plasma membrane in a pattern complementary to the distribution of PtdIns(4,5) P2. Interestingly, the reaction product DAG visualized with the marker Cys1:YFP displayed an intracellular localization similar to that of PtdIns(4,5)P2. Inhibition of PLC activity led to ceased and partially depolarized pollen tube growth, caused PtdIns(4,5)P2 spreading at the apex, and abolished DAG membrane accumulation. Blocking endocytic membrane recycling affected the intracellular distribution of DAG but not of PtdIns(4,5)P2. Although pollen PLC may affect more phenomena in polar growing cells (e.g., organization of actin cytoskeleton; Dowd et al. 2006), it also maintains, together with endocytic membrane recycling, an apical domain enriched in PtdIns(4,5)P2 and DAG required for polar cell growth (Helling et al. 2006).

4.3 Phosphatidylinositol Transfer Proteins and Their Role in the Plant Secretory Pathway

Lipid transport between membranes of eukaryotic organisms represents an essential aspect of organelle biogenesis. Such transport must be strictly selective and directional to ensure a specific lipid composition of individual membranes. At least some of the mechanisms generating and maintaining a nonrandom distribution of lipids in cells are linked to the action of phosphatidylinositol transfer proteins (PITPs). Sec14p, the major yeast PITP, regulates an essential interface between lipid metabolism and protein transport from Golgi membranes to the cell surface (Routt and Bankaitis 2004).

Vincent et al. (2005) demonstrated that AtSfh1p, a member of a large *Arabidopsis* Sec14p-nodulin domain family, is a PITP that regulates a specific stage in root hair development. AtSfh1p localizes along the root hair plasma membrane and is enriched in discrete plasma membrane domains and in the root hair tip cytoplasm, overlapping with the pattern of PtdIns(4,5)P2 in developing root hairs. Null mutant line *sfh1* caused loss of tip-directed PtdIns(4,5)P2, dispersal of secretory vesicles from the tip cytoplasm, loss of the tip F-actin network, and manifest disorganization of the root hair microtubule cytoskeleton., leading to compromised polarized root hair expansion. This raises the intriguing possibility that Sec14p-nodulin domain proteins define a family of polarized membrane growth regulators in plants.

In addition, another plant subfamily of Sec14 proteins also contains the Golgi dynamic (GOLD) motif that is homologous to the luminal portion of the p24 protein superfamily of secretory cargo receptors. As with the Sec14-nodulin proteins, the Sec14-GOLD proteins also hold promise of simultaneously regulating phosphoinsitide synthesis and organization. Peterman et al. (2004) described a Sec14-GOLD protein in *Arabidopsis* termed patellin1. This protein is recruited from the cytoplasm to the expanding and maturing cell plate, as shown by immunolocalization and biochemical fractionation. In vesicle-binding assays, patellin1 bound to specific phosphoinositides with a preference for phosphatidylinositol 5-phosphate, PtdIns(4,5)P2, and PtdIns3P.

5 Conclusions

Membrane lipids and membrane proteins (both peripheral and integral) inhabit the same cellular niches and it is not surprising that they coevolved to a great extent in many specific regulatory steps in eukaryotic cells. They form an intricate network of signals reciprocally regulating the specificity/identity of membrane domains/compartments, activation of vesicle traffic, polarization of the transport of membrane containers and addressing these to the right targets. While in animal and fungal cells there are quite extensive data from this interface (especially for GTPases and phospholipids), in plants this field should be one of the major foci of inquiry for contemporary cell biology.

From the physiological perspective, it is obvious that also within the plant cell secretory pathway there should be an intricate network of signaling involved in the *regulation of the homeostasis* of all membrane trafficking processes. Coordination of early and late events in the secretory pathway is crucial for the effective responses of vesicle traffic to external and internal signals. Here in plant field we are unfortunately left mostly in darkness – this should also be in the future a hot spot for intensive experimental enquiry to achieve insight comparable to that in animals (Sallese et al. 2006).

As we stressed in Sect. 1, input from whole-plant physiological processes into the secretory pathways of plant cells is a largely unexplored area. There are only very few examples of physiological processes where there are some insights into causal links between them, e.g., whole-plant gravistimulation and secretory pathway regulation. It is obvious that networks regulating the secretory pathway will include specific kinase/phosphatase phosphorylation/dephosphorylation processes. These three aspects of the signaling in plant vesicle traffic represent integrating perspectives – work on them will bring us closer to understanding the real dynamics of living plant cells.

Acknowledgements The related work was supported by the Academy of Sciences of the Czech Republic Grant Agency grants IAA6038410 and IAA601110916, MSMT LC06034 REMOROST, KONTAKT MSMT CR – ME841 and MSM0021620858.

References

Baluska F, Liners F, Hlavacka A, Schlicht M, Cutsem PV, McCurdy DW, Menzel D (2005) Cell wall pectins and xyloglucans are internalized into dividing root cells and accumulate within cell plates during cytokinesis. Protoplasma 225:141–155

Batoko H, Zheng HQ, Hawes C, Moore I (2000) A rab1 GTPase is required for transport between the endoplasmic reticulum and Golgi apparatus and for normal Golgi movement in plants. Plant Cell 12:2201–2218

Berken A, Thomas C, Wittinghofer A (2005) A new family of RhoGEFs activates the Rop molecular switch in plants. Nature 436(7054):1176–1180

Blagoveshchenskaya A, Cheong FY, Rohde HM, Glover G, Knodler A, Nicolson T, Boehmelt G, Mayinger P (2008) Integration of Golgi trafficking and growth factor signaling by the lipid phosphatase SAC1. J Cell Biol 180:803–812

Brands A, hua David Ho T (2002) Function of a plant stress-induced gene, HVA22. Synthetic enhancement screen with its yeast homolog reveals its role in vesicular traffic. Plant Physiol 130:1121–1131

Brembu T, Winge P, Bones AM, Yang Z (2006) A RHOse by any other name: a comparative analysis of animal and plant Rho GTPases. Cell Res 16:435–445

Brown FD, Rozelle AL, Yin HL, Balla T, Donaldson JG (2001) Phosphatidylinositol 4,5-bisphosphate and Arf6-regulated membrane traffic. J Cell Biol 154:1007–1017

Busch M, Mayer U, Jürgens G (1996) Molecular analysis of the Arabidopsis pattern formation of gene GNOM: gene structure and intragenic complementation. Mol Gen Genet 250:681–691

Cai H, Reinisch K, Ferro-Novick S (2007) Coats, tethers, Rabs, and SNAREs work together to mediate the intracellular destination of a transport vesicle. Dev Cell 12:671–682

Chae YC, Kim JH, Kim KL, Kim HW, Lee HY, Heo WD, Meyer T, Suh PG, Ryu SH (2008) Phospholipase D activity regulates integrin-mediated cell spreading and migration by inducing GTP-Rac translocation to the plasma membrane. Mol Biol Cell 19:3111–3123

Chamberlain MD, Chan T, Oberg JC, Hawrysh AD, James KM, Saxena A, Xiang J, Anderson DH (2008) Disrupted RabGAP function of the p85 subunit of phosphatidylinositol 3-kinase results in cell transformation. J Biol Chem 283:15861–15868

Cheung AY, Chen CYh, Glaven RH, de Graaf BHJ, Vidali L, Hepler PK, Wu Hm (2002) Rab2 GTPase regulates vesicle trafficking between the endoplasmic reticulum and the Golgi bodies and is important to pollen tube growth. Plant Cell 14:945–962

Chinchilla D, Zipfel C, Robatzek S, Kemmerling B, Nürnberger T, Jones JDG, Felix G, Boller T (2007) A flagellin-induced complex of the receptor FLS2 and BAK1 initiates plant defence. Nature 448:497–500

Chow CM, Neto H, Foucart C, Moore I (2008) Rab-A2 and Rab-A3 GTPases define a trans-golgi endosomal membrane domain in Arabidopsis that contributes substantially to the cell plate. Plant Cell 20:101–123

Christensen TM, Vejlupkova Z, Sharma YK, Arthur KM, Spatafora JW, Albright CA, Meeley RB, Duvick JP, Quatrano RS, Fowler JE (2003) Conserved subgroups and developmental regulation in the monocot rop gene family. Plant Physiol 133:1791–1808

Cole R, Synek L, Zarsky V, Fowler J (2005) SEC8, a subunit of the putative Arabidopsis exocyst complex, facilitates pollen germination and competitive pollen tube growth. Plant Physiol 138:2005–2018

Cole RA, Fowler JE (2006) Polarized growth: maintaining focus on the tip. Curr Opin Plant Biol 9:579–588

Corrotte M, Chasserot-Golaz S, Huang P, Du G, Ktistakis NT, Frohman MA, Vitale N, Bader MF, Grant NJ (2006) Dynamics and function of phospholipase D and phosphatidic acid during phagocytosis. Traffic 7:365–377

Cremona O, Di Paolo G, Wenk M, Luthi A, Kim W, Takei K, Daniell L, Nemoto Y, Shears S, Flavell R, McCormick D, De Camilli P (1999) Essential role of phosphoinositide metabolism in synaptic vesicle recycling. Cell 99:179–188

Cruz-Ramirez A, Oropeza-Aburto A, Razo-Hernandez F, Ramirez-Chavez E, Herrera-Estrella L (2006) Phospholipase DZ2 plays an important role in extraplastidic galactolipid biosynthesis and phosphate recycling in Arabidopsis roots. Proc Natl Acad Sci U S A 103:6765–6770

de Graaf BHJ, Cheung AY, Andreyeva T, Levasseur K, Kieliszewski M, Wu HM (2005) Rab11 GTPase-regulated membrane trafficking is crucial for tip-focused pollen tube growth in Tobacco. Plant Cell 17:2564–2579

Deneka M, Neeft M, van der Sluijs P (2003) Regulation of membrane transport by Rab GTPases. Crit Rev Biochem Mol Biol 38:121–142

Deretic V (2005) Ay, there's the Rab: organelle maturation by Rab conversion. Dev Cell 9:446–448

Dettmer J, Hong-Hermesdorf A, Stierhof YD, Schumacher K (2006) Vacuolar H+-ATPase activity is required for endocytic and secretory trafficking in Arabidopsis. Plant Cell 18:715–730

Dhonukshe P, Baluska F, Schlicht M, Hlavacka A, Samaj J, Friml J, Gadella TWJ (2006) Endocytosis of cell surface material mediates cell plate formation during plant cytokinesis. Dev Cell 10:137–150

Di Paolo G, De Camilli P (2006) Phosphoinositides in cell regulation and membrane dynamics. Nature 443:651–657

Dowd PE, Coursol S, Skirpan AL, hui Kao T, Gilroy S (2006) Petunia phospholipase C1 is involved in pollen tube growth. Plant Cell 18:1438–1453

Ezaki B, Gardner RC, Ezaki Y, Matsumoto H (2000) Expression of aluminum-induced genes in transgenic arabidopsis plants can ameliorate aluminum stress and/or oxidative stress. Plant Physiol 122:657–665

Freyberg Z, Siddhanta A, Shields D (2003) "Slip, sliding away": phospholipase D and the Golgi apparatus. Trends Cell Biol 13:540–546

Gebbie LK, Burn JE, Hocart CH, Williamson RE (2005) Genes encoding ADP-ribosylation factors in *Arabidopsis thaliana* L. Heyn.: genome analysis and antisense suppression. J Exp Bot 56:1079–1091

Geldner N, Robatzek S (2008) Plant receptors go endosomal: a moving view on signal transduction. Plant Physiol 147(4):1565–1574

Geldner N, Anders N, Wolters H, Keicher J, Kornberger W, Muller P, Delbarre A, Ueda T, Nakano A, Jürgens G (2003) The Arabidopsis GNOM ARF-GEF mediates endosomal recycling, auxin transport, and auxin dependent plant growth. Cell 112:219–230

Geldner N, Hyman DL, Wang X, Schumacher K, Chory J (2007) Endosomal signaling of plant steroid receptor kinase BRI1. Genes Dev 21:1598–1602

Goh T, Uchida W, Arakawa S, Ito E, Dainobu T, Ebine K, Takeuchi M, Sato K, Ueda T, Nakano A (2007) VPS9a, the common activator for two distinct types of Rab5 GTPases, is essential for the development of *Arabidopsis thaliana*. Plant Cell 19:3504–3515

Gomez-Gomez L, Boller T (2000) FLS2: an LRR receptor-like kinase involved in the perception of the bacterial elicitor flagellin in Arabidopsis. Mol Cell 5:1003–1011

Grosshans BL, Ortiz D, Novick P (2006) Rabs and their effiectors: achieving specificity in membrane traffic. Proc Natl Acad Sci U S A 103:11821–11827

Gu Y, Li S, Lord EM, Yang Z (2006) Members of a novel class of Arabidopsis Rho guanine nucleotide exchange factors control Rho GTPase-dependent polar growth. Plant Cell 18:366–381

Gurkan C, Koulov AV, Balch WE (2007) An evolutionary perspective on eukaryotic membrane trafficking. Adv Exp Med Biol 607:73–83

Haas TJ, Sliwinski MK, Martinez DE, Preuss M, Ebine K, Ueda T, Nielsen E, Odorizzi G, Otegui MS (2007) The Arabidopsis AAA ATPase SKD1 is involved in multivesicular endosome function and interacts with its positive regulator LYST-INTERACTING PROTEIN5. Plant Cell 19:1295–1312

Hala M, Cole R, Synek L, Drdova E, Pecenkova T, Nordheim A, Lamkemeyer T, Madlung J, Hochholdinger F, Fowler JE, Zarsky V (2008) An exocyst complex functions in plant cell growth in Arabidopsis and tobacco. Plant Cell 20:1330–1345

He B, Xi F, Zhang X, Zhang J, Guo W (2007) EXO70 interacts with phospholipids and mediates the targeting of the exocyst to the plasma membrane. EMBO J 26:4053–4065

Helling D, Possart A, Cottier S, Klahre U, Kost B (2006) Pollen tube tip growth depends on plasma membrane polarization mediated by tobacco PLC3 activity and endocytic membrane recycling. Plant Cell 18:3519–3534

Hemsley PA, Kemp AC, Grierson CS (2005) The TIP GROWTH DEFECTIVE1 S-acyl transferase regulates plant cell growth in Arabidopsis. Plant Cell 17:2554–2563

Heo JB, Rho HS, Kim SW, Hwang SM, Kwon HJ, Nahm MY, Bang WY, Bahk JD (2005) OsGAP1 functions as a positive regulator of OsRab11-mediated TGN to PM or vacuole trafficking. Plant Cell Physiol 46:2005–2018

Honda A, Nogami M, Yokozeki T, Yamazaki M, Nakamura H, Watanabe H, Kawamoto K, Nakayama K, Morris AJ, Frohman MA, Kanaho Y (1999) Phosphatidylinositol 4-phosphate 5-kinase alpha is a downstream effector of the small G protein ARF6 in membrane ruffle formation. Cell 99:521–532

Hughes WE, Elgundi Z, Huang P, Frohman MA, Biden TJ (2004) Phospholipase D1 regulates secretagogue-stimulated insulin release in pancreatic beta-cells. J Biol Chem 279:27534–27541

Inaba T, Nagano Y, Nagasaki T, Sasaki Y (2002) Distinct localization of two closely related Ypt3/Rab11 proteins on the trafficking pathway in higher plants. J Biol Chem 277:9183–9188

Ivanchenko M, Vejlupkova Z, Quatrano RS, Fowler JE (2000) Maize ROP7 GTPase contains a unique, CaaX box-independent plasma membrane targeting signal. Plant J 24:79–90

Jafar-Nejad H, Andrews HK, Acar M, Bayat V, Wirtz-Peitz F, Mehta SQ, Knoblich JA, Bellen HJ (2005) Sec15, a component of the exocyst, promotes notch signaling during the asymmetric division of Drosophila sensory organ precursors. Dev Cell 9:351–363

Jensen RB, Lykke-Andersen K, Frandsen GI, Nielsen HB, Haseloffi J, Jespersen HM, Mundy J, Skriver K (2000) Promiscuous and specific phospholipid binding by domains in ZAC, a membrane-associated Arabidopsis protein with an ARF GAP zinc finger and a C2 domain. Plant Mol Biol 44:799–814

Jürgens G (2004) Membrane trafficking in plants. Annu Rev Cell Dev Biol 20:481–504

Kang JG, Yun J, Kim DH, Chung KS, Fujioka S, Kim JI, Dae HW, Yoshida S, Takatsuto S, Song PS, Park CM (2001) Light and brassinosteroid signals are integrated via a dark-induced small G protein in etiolated seedling growth. Cell 105:625–636

Kaothien P, Ok SH, Shuai B, Wengier D, Cotter R, Kelley D, Kiriakopolos S, Muschietti J, McCormick S (2005) Kinase partner protein interacts with the LePRK1 and LePRK2 receptor kinases and plays a role in polarized pollen tube growth. Plant J 42:492–503

Kelly-Skupek MN, Irving HR (2006) Pharmacological evidence for activation of phospholipid and small GTP binding protein signaling cascades by Nod factors. Plant Physiol Biochem 44:132–142

Kim Y, Kim JE, Lee SD, Lee TG, Kim JH, Park JB, Han JM, Jang SK, Suh PG, Ryu SH (1999) Phospholipase D1 is located and activated by protein kinase C alpha in the plasma membrane in 3Y1 fibroblast cell. Biochim Biophys Acta 1436:319–330

Klahre U, Becker C, Schmitt AC, Kost B (2006) Nt-RhoGDI2 regulates Rac/Rop signaling and polar cell growth in tobacco pollen tubes. Plant J 46:1018–1031

Kost B, Lemichez E, Spielhofer P, Hong Y, Tolias K, Carpenter C, Chua NH (1999) Rac homologues and compartmentalized phosphatidylinositol 4, 5-bisphosphate act in a common pathway to regulate polar pollen tube growth. J Cell Biol 145:317–330

Kotzer AM, Brandizzi F, Neumann U, Paris N, Moore I, Hawes C (2004) AtRabF2b (Ara7) acts on the vacuolar trafficking pathway in tobacco leaf epidermal cells. J Cell Sci 117:6377–6389

Krauss M, Haucke V (2007) Phosphoinositide-metabolizing enzymes at the interface between membrane traffic and cell signaling. EMBO Rep 8:241–246

Kusano H, Testerink C, Vermeer JEM, Tsuge T, Shimada H, Oka A, Munnik T, Aoyama T (2008) The Arabidopsis phosphatidylinositol phosphate 5-kinase PIP5K3 is a key regulator of root hair tip growth. Plant Cell 20:367–380

Kwon C, Neu C, Pajonk S, Yun HS, Lipka U, Humphry M, Bau S, Straus M, Kwaaitaal M, Rampelt H, Kasmi FE, Jürgens G, Parker J, Panstruga R, Lipka V, Schulze-Lefert P (2008) Co-option of a default secretory pathway for plant immune responses. Nature 451:835–840

Langhans M, Robinson DG (2007) 1-Butanol targets the Golgi apparatus in tobacco BY-2 cells, but in a different way to Brefeldin A. J Exp Bot 58:3439–3447

Lavy M, Yalovsky S (2006) Association of Arabidopsis type-II ROPs with the plasma membrane requires a conserved C-terminal sequence motif and a proximal polybasic domain. Plant J 46:934–947

Lavy M, Bracha-Drori K, Sternberg H, Yalovsky S (2002) A cell-specific, prenylation-independent mechanism regulates targeting of type II RACs. Plant Cell 14:2431–2450

Lavy M, Bloch D, Hazak O, Gutman I, Poraty L, Sorek N, Sternberg H, Yalovsky S (2007) A Novel ROP/RAC effiector links cell polarity, rootmeristem maintenance, and vesicle trafficking. Curr Biol 17:947–952

Lee GJ, Sohn EJ, Lee MH, Hwang I (2004) The Arabidopsis rab5 homologs rha1 and ara7 localize to the prevacuolar compartment. Plant Cell Physiol 45:1211–1220

Lee CS, Kim IS, Park JB, Lee MN, Lee HY, Suh PG, Ryu SH (2006) The phox homology domain of phospholipase D activates dynamin GTPase activity and accelerates EGFR endocytosis. Nat Cell Biol 8:477–484

Leshem Y, Seri L, Levine A (2007) Induction of phosphatidylinositol 3-kinase-mediated endocytosis by salt stress leads to intracellular production of reactive oxygen species and salt tolerance. Plant J 51:185–197

Li G, Xue H (2007) Arabidopsis PLDzeta2 regulates vesicle trafficking and is required for auxin response. Plant Cell 19:281–295

Liu J, Zuo X, Yue P, Guo W (2007) Phosphatidylinositol 4,5-bisphosphate mediates the targeting of the exocyst to the plasma membrane for exocytosis in mammalian cells. Mol Biol Cell 18:4483–4492

Lodhi IJ, Bridges D, Chiang SH, Zhang Y, Cheng A, Geletka LM, Weisman LS, Saltiel AR (2008) Insulin stimulates phosphatidylinositol 3-phosphate production via the activation of Rab5. Mol Biol Cell 19:2718–2728

Mancuso S, Marras AM, Mugnai S, Schlicht M, Zarsky V, Li G, Song L, Xue HW, Baluska F (2007) Phospholipase D zeta2 drives vesicular secretion of auxin for its polar cell-cell transport in the transition zone of the root apex. Plant Signal Behav 2:240–244

Matheson LA, Hanton SL, Rossi M, Latijnhouwers M, Stefano G, Renna L, Brandizzi F (2007) Multiple roles of ADP-ribosylation factor 1 in plant cells include spatially regulated recruitment of coatomer and elements of the Golgi matrix. Plant Physiol 143:1615–1627

Matheson LA, Suri SS, Hanton SL, Chatre L, Brandizzi F (2008) Correct targeting of plant ARF GTPases relies on distinct protein domains. Traffic 9:103–120

Mazel A, Leshem Y, Tiwari BS, Levine A (2004) Induction of salt and osmotic stress tolerance by overexpression of an intracellular vesicle trafficking protein AtRab7 (AtRabG3e). Plant Physiol 134:118–128

Meijer HJG, Munnik T (2003) Phospholipid-based signaling in plants. Annu Rev Plant Biol 54:265–306

Meller N, Merlot S, Guda C (2005) CZH proteins: a new family of Rho-GEFs. J Cell Sci 118:4937–4946

Molendijk AJ, Ruperti B, Palme K (2004) Small GTPases in vesicle trafficking. Curr Opin Plant Biol 7:694–700

Monteiro D, Liu Q, Lisboa S, Scherer GEF, Quader H, Malho R (2005) Phosphoinositides and phosphatidic acid regulate pollen tube growth and reorientation through modulation of $[Ca^{2+}]$ c and membrane secretion. J Exp Bot 56:1665–1674

Moore I, Diefenthal T, Zarsky V, Schell J, Palme K (1997) A homolog of the mammalian GTPase Rab2 is present in Arabidopsis and is expressed predominantly in pollen grains and seedlings. Proc Natl Acad Sci U S A 94:762–767

Morel J, Claverol S, Mongrand S, Furt F, Fromentin J, Bessoule JJ, Blein JP, Simon-Plas F (2006) Proteomics of plant detergent-resistant membranes. Mol Cell Proteomics 5:1396–1411

Moshkov IE, Mur LAJ, Novikova GV, Smith AR, Hall MA (2003) Ethylene regulates monomeric GTP-binding protein gene expression and activity in Arabidopsis. Plant Physiol 131:1705–1717

Nielsen E (2006) Rab GTPases in plant endocytosis. In: Samaj J (ed) Plant Endocytosis, Plant Cell Monographs, Vol 1. Springer-Verlag, Heidelberg pp 177–195

Nielsen E, Cheung AY, Ueda T (2008) The regulatory Rab and ARF GTPases for vesicular trafficking. Plant Physiol 147:1516–1526

Novick P, Guo W (2002) Ras family therapy: Rab, Rho and Ral talk to the exocyst. Trends Cell Biol 12:247–249

Novick P, Medkova M, Dong G, Hutagalung A, Reinisch K, Grosshans B (2006) Interactions between Rabs, tethers, SNAREs and their regulators in exocytosis. Biochem Soc Trans 34:683–686

Ohashi Y, Oka A, Rodrigues-Pousada R, Possenti M, Ruberti I, Morelli G, Aoyama T (2003) Modulation of phospholipid signaling by GLABRA2 in root-hair pattern formation. Science 300:1427–1430

Ortiz D, Medkova M, Walch-Solimena C, Novick P (2002) Ypt32 recruits the Sec4p guanine nucleotide exchange factor, Sec2p, to secretory vesicles; evidence for a Rab cascade in yeast. J Cell Biol 157:1005–1015

Otegui MS, Spitzer C (2008) Endosomal functions in plants. Traffic 9:1589–1598

Oude Weernink PA, de Jesus ML, Schmidt M (2007) Phospholipase D signaling: orchestration by PIP2 and small GTPases. Naunyn Schmiedebergs Arch Pharmacol 374:399–411

Ovecka M, Lang I, Baluska F, Ismail A, Illes P, Lichtscheidl IK (2005) Endocytosis and vesicle trafficking during tip growth of root hairs. Protoplasma 226:39–54

Parker JS, Cavell AC, Dolan L, Roberts K, Grierson CS (2000) Genetic interactions during root hair morphogenesis in Arabidopsis. Plant Cell 12:1961–1974

Peleg-Grossman S, Volpin H, Levine A (2007) Root hair curling and Rhizobium infection in *Medicago truncatula* are mediated by phosphatidylinositide-regulated endocytosis and reactive oxygen species. J Exp Bot 58:1637–1649

Pereira-Leal JB, Hume AN, Seabra MC (2001) Prenylation of Rab GTPases: molecular mechanisms and involvement in genetic disease. FEBS Lett 498:197–200

Perez-Mansilla B, Ha VL, Justin N, Wilkins AJ, Carpenter CL, Thomas GMH (2006) The diffierential regulation of phosphatidylinositol 4-phosphate 5-kinases and phospholipase D1 by ADP-ribosylation factors 1 and 6. Biochim Biophys Acta 1761:1429–1442

Peterman TK, Ohol YM, McReynolds LJ, Luna EJ (2004) Patellin1, a novel Sec14-like protein, localizes to the cell plate and binds phosphoinositides. Plant Physiol 136:3080–3094

Pimpl P, Hanton SL, Taylor JP, Pinto-daSilva LL, Denecke J (2003) The GTPase ARF1p controls the sequence-specific vacuolar sorting route to the lytic vacuole. Plant Cell 15:1242–1256

Potocky M, Elias M, Profotova B, Novotna Z, Valentova O, Zarsky V (2003) Phosphatidic acid produced by phospholipase D is required for tobacco pollen tube growth. Planta 217:122–130

Preuss ML, Serna J, Falbel TG, Bednarek SY, Nielsen E (2004) The Arabidopsis Rab GTPase RabA4b localizes to the tips of growing root hair cells. Plant Cell 16:1589–1603

Preuss M, Schmitz A, Thole J, Bonner H, Otegui M, Nielsen E (2006) A role for the RabA4b effiector protein PI-4Kb1 in polarized expansion of root hair cells in *Arabidopsis thaliana*. J Cell Biol 172:991–998

Qiu JL, Jilk R, Marks MD, Szymanski DB (2002) The Arabidopsis SPIKE1 gene is required for normal cell shape control and tissue development. Plant Cell 14:101–118

Rehman RU, Stigliano E, Lycett GW, Sticher L, Sbano F, Faraco M, Dalessandro G, Sansebastiano GPD (2008) Tomato Rab11a characterization evidenced a difference between SYP121-dependent and SYP122-dependent exocytosis. Plant Cell Physiol 49:751–766

Richter S, Geldner N, Schrader J, Wolters H, Stierhof YD, Rios G, Koncz C, Robinson DG, Jürgens G (2007) Functional diversification of closely related ARF-GEFs in protein secretion and recycling. Nature 448:488–492

Rink J, Ghigo E, Kalaidzidis Y, Zerial M (2005) Rab conversion as a mechanism of progression from early to late endosomes. Cell 122:735–749

Ritzenthaler C, Nebenfuhr A, Movafeghi A, Stussi-Garaud C, Behnia L, Pimpl P, Staehelin LA, Robinson DG (2002) Reevaluation of the effects of brefeldin A on plant cells using tobacco

Bright Yellow 2 cells expressing Golgi-targeted green fluorescent protein and COPI antisera. Plant Cell 14:237–261
Robatzek S (2007) Vesicle trafficking in plant immune responses. Cell Microbiol 9:1–8
Robatzek S, Chinchilla D, Boller T (2006) Ligand-induced endocytosis of the pattern recognition receptor FLS2 in Arabidopsis. Genes Dev 20:537–542
Robert S, Chary SN, Drakakaki G, Li S, Yang Z, Raikhel NV, Hicks GR (2008) Endosidin1 defines a compartment involved in endocytosis of the brassinosteroid receptor BRI1 and the auxin transporters PIN2 and AUX1. Proc Natl Acad Sci U S A 105:8464–8469
Rojo E, Gillmor CS, Kovaleva V, Somerville CR, Raikhel NV (2001) VACUOLELESS1 is an essential gene required for vacuole formation and morphogenesis in Arabidopsis. Dev Cell 1:303–310
Rojo E, Zouhar J, Kovaleva V, Hong S, Raikhel NV (2003) The AtC-VPS protein complex is localized to the tonoplast and the prevacuolar compartment in Arabidopsis. Mol Biol Cell 14:361–369
Roth MG (2008) Molecular mechanisms of PLD function in membrane traffic. Traffic 9:1233–1239
Routt SM, Bankaitis VA (2004) Biological functions of phosphatidylinositol transfer proteins. Biochem Cell Biol 82:254–262
Russinova E, Borst JW, Kwaaitaal M, Caño-Delgado A, Yin Y, Chory J, de Vries SC (2004) Heterodimerization and endocytosis of Arabidopsis brassinosteroid receptors BRI1 and AtSERK3 (BAK1). Plant Cell 16:3216–3229
Rutherford S, Moore I (2002) The Arabidopsis Rab GTPase family: another enigma variation. Curr Opin Plant Biol 5:518–528
Sallese M, Pulvirenti T, Luini A (2006) The physiology of membrane transport and endomembrane-based signaling. EMBO J 25:2663–2673
Samaj J, Müuller J, Beck M, Böhm N, Menzel D (2006) Vesicular trafficking, cytoskeleton and signaling in root hairs and pollen tubes. Trends Plant Sci 11:594–600
Shevell DE, Leu WM, Gillmor CS, Xia G, Feldmann KA, Chua NH (1994) EMB30 is essential for normal cell division, cell expansion, and cell adhesion in Arabidopsis and encodes a protein that has similarity to Sec7. Cell 77:1051–1062
Sohn EJ, Kim ES, Zhao M, Kim SJ, Kim H, Kim YW, Lee YJ, Hillmer S, Sohn U, Jiang L, Hwang I (2003) Rha1, an Arabidopsis Rab5 homolog, plays a critical role in the vacuolar trafficking of soluble cargo proteins. Plant Cell 15:1057–1070
Sorek N, Poraty L, Sternberg H, Bar E, Lewinsohn E, Yalovsky S (2007) Activation status-coupled transient S acylation determines membrane partitioning of a plant Rho-related GTPase. Mol Cell Biol 27:2144–2154
Stenzel I, Ischebeck T, Konig S, Holubowska A, Sporysz M, Hause B, Heilmann I (2008) The type B phosphatidylinositol-4-phosphate 5-kinase 3 is essential for root hair formation in *Arabidopsis thaliana*. Plant Cell 20:124–141
Synek L, Schlager N, Elias M, Quentin M, Hauser MT, Zarsky V (2006) AtEXO70A1, a member of a family of putative exocyst subunits specifically expanded in land plants, is important for polar growth and plant development. Plant J 48:54–72
Takano J, Miwa K, Yuan L, von Wiren N, Fujiwara T (2005) Endocytosis and degradation of BOR1, a boron transporter of *Arabidopsis thaliana*, regulated by boron availability. Proc Natl Acad Sci U S A 102:12276–12281
Teh OK, Moore I (2007) An ARF-GEF acting at the Golgi and in selective endocytosis in polarized plant cells. Nature 448:493–496
TerBush DR, Maurice T, Roth D, Novick P (1996) The Exocyst is a multiprotein complex required for exocytosis in Saccharomyces cerevisiae. EMBO J 15:6483–6494
Thole JM, Vermeer JE, Zhang Y, Gadella Jr TW, Nielsen E (2008) ROOT HAIR DEFECTIVE4 encodes a phosphatidylinositol-4-phosphate phosphatase required for proper root hair development in Arabidopsis thaliana. Plant Cell 20:381–395
Trotochaud AE, Hao T, Wu G, Yang Z, Clark SE (1999) The CLAVATA1 receptor-like kinase requires CLAVATA3 for its assembly into a signaling complex that includes KAPP and a Rho-related protein. Plant Cell 11:393–406
Ueda T, Matsuda N, Anai T, Tsukaya H, Uchimiya H, Nakano A (1996) An Arabidopsis gene isolated by a novel method for detecting genetic interaction in yeast encodes the GDP dissociation inhibitor of Ara4 GTPase. Plant Cell 8:2079–2091

Ueda T, Yamaguchi M, Uchimiya H, Nakano A (2001) Ara6, a plant-unique novel type Rab GTPase, functions in the endocytic pathway of *Arabidopsis thaliana*. EMBO J 20:4730–4741
Ueda T, Uemura T, Sato MH, Nakano A (2004) Functional diffierentiation of endosomes in Arabidopsis cells. Plant J 40:783–789
Valdez-Taubas J, Pelham HRB (2003) Slow diffiusion of proteins in the yeast plasma membrane allows polarity to be maintained by endocytic cycling. Curr Biol 13:1636–1640
van Leeuwen W, Vermeer JEM, Gadella TWJ, Munnik T (2007) Visualization of phosphatidylinositol 4,5-bisphosphate in the plasma membrane of suspension-cultured tobacco BY-2 cells and whole Arabidopsis seedlings. Plant J 52:1014–1026
Vernoud V, Horton AC, Yang Z, Nielsen E (2003) Analysis of the small GTPase gene superfamily of Arabidopsis. Plant Physiol 131:1191–1208
Vieira AV, Lamaze C, Schmid SL (1996) Control of EGF receptor signaling by clathrin-mediated endocytosis. Science 274:2086–2089
Vincent P, Chua M, Nogue F, Fairbrother A, Mekeel H, Xu Y, Allen N, Bibikova TN, Gilroy S, Bankaitis VA (2005) A Sec14p-nodulin domain phosphatidylinositol transfer protein polarizes membrane growth of *Arabidopsis thaliana* root hairs. J Cell Biol 168:801–812
von Zastrow M, Sorkin A (2007) Signaling on the endocytic pathway. Curr Opin Cell Biol 19:436–445
Wang H, Lockwood SK, Hoeltzel MF, Schiefelbein JW (1997) The ROOT HAIR DEFECTIVE3 gene encodes an evolutionarily conserved protein with GTP-binding motifs and is required for regulated cell enlargement in Arabidopsis. Genes Dev 11:799–811
Wang W, Ferro-Novick S (2002) A Ypt32p exchange factor is a putative effector of Ypt1p. Mol Biol Cell 13:3336–3343
Winge P, Brembu T, Kristensen R, Bones AM (2000) Genetic structure and evolution of RAC-GTPases in *Arabidopsis thaliana*. Genetics 156:1959–1971
Wu HM, Wong E, Ogdahl J, Cheung AY (2000) A pollen tube growth-promoting arabinogalactan protein from *Nicotiana alata* is similar to the tobacco TTS protein. Plant J 22:165–176
Wu S, Mehta SQ, Pichaud F, Bellen HJ, Quiocho FA (2005) Sec15 interacts with Rab11 via a novel domain and affiects Rab11 localization *in vivo*. Nat Struct Mol Biol 12:879–885
Wunderlich W, Fialka I, Teis D, Alpi A, Pfeifer A, Parton RG, Lottspeich F, Huber LA (2001) A novel 14-kilodalton protein interacts with the mitogen activated protein kinase scaffiold mp1 on a late endosomal/lysosomal compartment. J Cell Biol 152:765–776
Yamaryo Y, Dubots E, Albrieux C, Baldan B, Block MA (2008) Phosphate availability affects the tonoplast localization of PLDzeta2, an *Arabidopsis thaliana* phospholipase D. FEBS Lett 582:685–690
Zarsky V, Cvrckova F, Bischoffi F, Palme K (1997) At-GDI1 from *Arabidopsis thaliana* encodes a rab-specific GDP dissociation inhibitor that complements the sec19 mutation of *Saccharomyces cerevisiae*. FEBS Lett 403:303–308
Zerial M, McBride H (2001) Rab proteins as membrane organizers. Nat Rev Mol Cell Biol 2:107–117
Zhang X, Orlando K, He B, Xi F, Zhang J, Zajac A, Guo W (2008) Membrane association and functional regulation of Sec3 by phospholipids and Cdc42. J Cell Biol 180:145–158
Zhang H, Camacho L, Wee E, Batoko H, Legen J, Leaver CJ, Malho R, Hussey PJ, Moore I (2005) A Rab-E GTPase mutant acts downstream of the Rab-D subclass in biosynthetic membrane traffic to the plasma membrane in Tobacco leaf epidermis. Plant Cell 17:2020–2036
Zhong R, Burk DH, Morrison WHR, Ye ZH (2004) FRAGILE FIBER3, an Arabidopsis gene encoding a type II inositol polyphosphate 5-phosphatase, is required for secondary wall synthesis and actin organization in fiber cells. Plant Cell 16:3242–3259

Signaling to the Actin Cytoskeleton During Cell Morphogenesis and Patterning

Alison Sinclair, Mike Schenkel, and Jaideep Mathur

Abstract The plant is a supracellular organism whose cells are locked in position through shared walls but maintain apoplastic and symplastic connectivity. Their fixed position places each cell into a unique niche within the organism. Consequently, every environmental cue is perceived slightly differently by each cell. The response of each plant cell varies accordingly. Thus, plant growth and development reflect a progression of accommodative arrangements reached between constituent cells. In recent years the actin cytoskeleton, through its direct involvement in subcellular compartmentation, organelle and vesicle trafficking, and structural reinforcement, has emerged as a key player during accommodative growth and development. Here, using salient actin-cytoskeleton-associated cellular phenotypes, we elaborate upon the molecular-cell biological machinery involved in organizing the actin cytoskeleton during cell shape development in plants.

1 Introduction

A series of division, expansion, differentiation, and cell death events are involved in the fabrication of a multicellular plant. These events occur within a genetically defined framework in response to environmental cues and result in the precise patterns that provide each plant species with a unique, recognizable identity. However, if the millions of cells that make up a plant are considered individually, it becomes apparent that each wall-encased plant cell occupies a unique niche with

A. Sinclair
Laboratory of Plant Development & Interactions, Department of Molecular and Cellular Biology, CBS, University of Guelph, 50 Stone Road, Guelph, ON N1G 2W1, Canada

M. Schenkel
Laboratory of Plant Development & Interactions, Department of Molecular & Cellular Biology, CBS, University of Guelph, 50 Stone Road, Guelph, ON N1G 2W1, Canada

J. Mathur (✉)
Department of Molecular and Cellular Biology, 4463 Science Complex, University of Guelph, 488 Gordon Street, Guelph, ON N1G 2W1, Canada
e-mail: jmathur@uoguelph.ca

F. Baluška and S. Mancuso (eds.), *Signaling in Plants*, Signaling and Communication in Plants, 135
DOI: 10.1007/978-3-540-89228-1_7, © Springer-Verlag Berlin Heidelberg 2009

respect to the environment. Owing to its fixed location within the plant, each cell perceives and responds to environmental cues in a manner slightly different from that of its neighbors. Further, each cell has only limited response options for contributing to the genetically determined architectural master plan: it can divide to add more adjustable units (cells); it can expand in a polar fashion to fill out spaces; or it can die and thus vacate space to help reorganize the overall shape. In addition, as components of a supracellular organism, individual cells can differentiate into specialized forms for fulfilling specific functions. The creation of a precise form by numerous plant cells thus reflects an innate ability to accommodate cell structure and function to that of neighboring cells. How is this cooperation and coordination of developmental activities achieved by wall-locked plant cells?

A number of signaling cascades, including Ca^{2+}/H^{+}, reactive oxygen species (ROS), heterotrimeric G-proteins, phospholipases, phosphoinositides, mitogen-activated protein kinases, cyclic AMP, and GTPases, are recognized in plants (Hirt 2000; Hong-Bo et al. 2008; Sumimoto 2008; Shao et al. 2008; Chen 2008;Cheung and Wu 2008; Kost 2008). In recent years the actin cytoskeleton has emerged as a central responder to many of these signaling channels (Brenner et al. 2006; Nibau et al. 2006; Kost 2008). The actin cytoskeleton also constitutes the signaling hub through which numerous subcellular activities are regulated. These include subcellular compartmentation, organelle and vesicle motility, as well as structural reinforcement in co-operation with microtubules. With each new study it becomes increasingly apparent that almost every actin-associated activity has an impact on cell growth and morphogenesis. Interestingly, seemingly different cues appear to produce similar responses from the actin cytoskeleton. Moreover, as we follow molecular cascades elicited in response to diverse signals, we become aware of the considerable gaps in our ability to directly link environmental cues to the far-downstream behavior of the actin cytoskeleton and the subsequent elicitation of a developmental response. While one can easily envision a kind of inverted pyramid of subcellular events whereby diverse cues get parsed into increasingly narrower response categories, the molecular-cell biological mechanisms behind this thinking applied to the actin cytoskeleton are only beginning to be understood.

As detailed and excellent reviews on different aspects of signaling to the actin cytoskeleton continue being published, here we have adopted a broad definition for the term "signaling." The actin cytoskeleton has been placed at a hub: on one hand we explore the putative molecular-cell biological events that organize actin within the plant cell, and on the other we link information on actin organization to the morphogenesis and development of model cell types in different organisms and assorted mutants of *Arabidopsis thaliana*.

2 Actin Organization in Plant Cells

Our understanding of signaling to the actin cytoskeleton first requires a characterization of the typical actin cytoskeleton organization in plants. The actin gene family in plants is diverse and built up of multiple isoforms with unique patterns of regulation (Meagher et al. 2000). The actin cytoskeleton, comprising actin monomers (G-actin) and polymerized filamentous actin (F-actin) can change rapidly through cue-mediated depolymerization and repolymerization cycles. In living cells, F-actin strands twist around each other to produce helical arrangements of microfilaments possessing an approximate diameter of 7 nm. Immunocytochemical methods have informed us a lot about the actin organization in different cell types in plants (Vitha et al. 2000; Staiger and Schliwa 1987). More recent advances in live imaging of F-actin arrays have greatly shaped our present views on the dynamic actin cytoskeleton, notably, the introduction of fluorescent phalloidin into plant cells through micro injection (Schmit and Lambert 1990), microabrasion (Ramachandran et al. 2000), or after enzymatic digestion and freeze shattering (Wasteneys et al. 1997) has been pursued successfully. Synthetic fluorescent fusion proteins created using the filamentous actin binding domain of the Talin gene (Kost et al. 1998) and the actin binding domain 2 of the Fimbrin 1 gene (Sheahan et al. 2004; Wang et al. 2008) are the most used live probes for visualizing F-actin arrays in plants. G-actin pools in plant cells appear predominantly cytosolic and show up as diffuse fluorescence. Three morphologically distinguishable, but readily interchangeable states characterize the polymerized actin. These are seen as very fine meshworks comprising actin filaments, thick F-actin bundles, and loose F-actin strands (Staiger et al. 2000; Mathur et al. 1999; Wang et al. 2008). Since at the light-microscopy level it is not possible to resolve individual filaments of actin, the states probably reflect different levels of actin bundling. Actin patches as described for yeasts (Young et al. 2004) have not been observed in plant cells; however, brightly fluorescent regional aggregates of F-actin can be clearly observed in a variety of plant cells. Whether these represent true actin patches comprising interwoven actin filaments or reflect subcellular compartmentation is still an open question. Actin cages around large organelles have been described (Kandasamy and Meagher 1999). In addition to the above, short-lived arrays of actin filaments that contribute to the preprophase band and to stages of phragmoplast formation have been observed in dividing cells (Kakimoto and Shibaoka 1987; Sano et al. 2005).

Occasionally, 3–5-μm-long, brightly fluorescent bundles of F-actin might be encountered in cells that are experiencing subcellular stress or have been physically injured. All arrays are readily convertible to each other and reflect the dynamic nature of the actin cytoskeleton.

2.1 Different Actin Regulators Organize the Actin Cytoskeleton

A large number of actin genes and actin-interacting proteins have been identified and cloned from plants (Staiger et al. 2000; Meagher and Fechheimer 2003; Ketelaar et al. 2004; Staiger and Blanchoin 2006; Grunt et al. 2008). Since many actin interactors are conserved between eukaryotes, in vitro studies have been very instructive in ascribing roles for different proteins involved in actin cytoskeleton organization (Ayscough 1998). The direct visualization of actin arrays in plant cells through fluorescent imaging (Miller et al. 1996; Kost et al. 1998; Mathur et al. 1999; Blancaflor, 2000; Kovar et al. 2000b; Chen et al. 2002) has reinforced many of the in vitro observations.

As expected, cells expressing fluorescent protein fused to actin monomer interacting proteins such as profilin (Kovar et al. 2000a) exhibit a diffuse fluorescence (Ramachandran et al. 2000; Fu et al. 2002). However, overexpression of fluorescent protein fusions with different actin depolymerizing factors, also actin monomer interacting proteins, labels filamentous actin (Dong et al. 2001; Chen et al. 2002). Labeling experiments have confirmed that the proteins fimbrin (Kovar et al. 2000b) and villin (Klahre et al. 2000; Tominaga et al. 2000) bind to and bundle F-actin at different strengths. The ARP2/3 complex, an efficient enhancer of actin polymerization, is known to initiate side branches on actin filaments, creating a dendritic actin network (Mullins et al. 1998; Amann and Pollard 2001). Though immunological detection of ARP3 (one of the large subunits of the ARP2/3 complex) has been carried out (Van Gestel et al. 2003), and despite the demonstrated presence of its seven subunits (Mathur 2005a; Szymanski 2005), the complex has not been isolated from plants thus far. Inferences on ARP2/3 complex activity in relation to the actin cytoskeleton in plants are thus largely based on observations made in heterologous systems and actin organization observed in arp2/3 mutants.

More than 20 formins are recognized in plants, and as profilin-binding poly(L-proline)-containing proteins these have important roles in cytoskeletal organization (Cheung and Wu 2004; Deeks et al. 2005; Michelot et al. 2005; Ingouff et al. 2005). A major regulator of actin nucleation and polymerization through its control of the pool of unpolymerized ATP-actin is the adenyl cyclase associated protein (AtCAP1; Chaudhry et al. 2007; Deeks et al. 2007). A gelsolin-related actin binding protein PrABP80 has been identified in poppy pollen tubes (Huang et al. 2004). Biochemical assays indicate that the protein can nucleate actin polymerization from monomers, block the assembly of profilin-actin complex onto actin filament ends, enhance profilin-mediated actin depolymerization, and also sever actin filaments (Huang et al. 2004).

Since the number of proteins that can interact and affect actin organization continues to increase, clearly the key to actin's pivotal role as a subcellular responder lies in the way it is organized through the combinatorial and cumulative activities of its various regulatory proteins.

2.2 *Local Ionic Interactions Modulate Actin Regulators*

The plant cell functions by maintaining a basal level of ionic homeostasis. Any extrinsic or intrinsic cues that perturb this homeostasis are dealt with by the release or sequestration of relevant ionic species (Foreman et al. 2003; Shabala et al. 2006). This provides the cell with a fine-tuned mechanism for sensing environmental cues as well as responding to them. Since proteins affecting actin nucleation, polymerization, and subsequent organizational activities respond rapidly to alterations in their ionic environment (Day et al. 2002), they are direct mediators in the stimulus perception-response elicitation machinery. Many actin-regulating proteins are themselves regulated by ionic conditions in their environment (Bretscher and Weber

1980; Matsudaira and Burgess 1982; Harris and Weeds 1983; Kovar et al. 2000a; Huang et al. 2003; Drøbak et al. 2004; Gu et al. 2004; Jones et al. 2007). The best studied ionic pools playing a role in signaling to actin include Ca^{2+} (Hetherington and Brownlee 2004; Bothwell and Ng 2005), NADPH oxidases (Navazio et al. 2000; Kwak et al. 2003; Foreman et al. 2003) and ROS, such as hydrogen peroxide (H_2O_2), superoxide anion (O^{2-}), hydroxyl ions (OH) and nitric oxide (NO) (Miller et al. 2008; Mori and Schroeder 2004). Low, basal levels of cytosolic free calcium ($[Ca^{2+}]_{cyt}$) are sustained by Ca^{2+}-ATPases, and through the activity of a Ca^{2+}/H^+ antiporter that removes Ca^{2+} from the cytosol. Elevation of $[Ca^{2+}]_{cyt}$ during signaling occurs through the opening of different classes of plasma membrane, vacuolar membrane, or endoplasmic reticulum localized Ca^{2+}-permeable ion channels. Whereas Ca^{2+}-and H^+-mediated signaling in plants is usually achieved through storage and release mechanisms, ROS signaling is usually controlled through rapid production and scavenging.

A major approach adopted for linking subcellular ionic distribution to actin organization and cell growth characteristics has been to observe and quantify ionic gradients and site-specific localizations (Felle and Hepler 1997). This approach has been successfully applied to single-celled, tip-growing cells such as pollen tubes (Cárdenas et al. 2008; Kroeger et al. 2008) and root hairs (Carol and Dolan 2006; Monshausen et al. 2007). In *Arabidopsis*, ROS stimulate the entry of Ca^{2+} into the cell through a calcium plasma membrane hyperpolarization-activated Ca^{2+} channel (Foreman et al. 2003). Recently, Takeda et al. (2008) have elucidated an elegant mechanism involving local positive feedback, where ROS derived through an RHD2 NADPH oxidase stimulates a Ca^{2+} influx into the cytoplasm. Though the targeted actin regulators were not identified in this study, the regulatory loop is clearly involved in cell shape determination of root hair cells. Similar interdependent local loops between ROS and Ca^{2+} gradients have been uncovered during polarized growth of *Fucus serratus* zygotes (Coelho et al. 2008).

Details pertaining to integrated calcium-mediated signaling (see "Integrated Calcium Signaling in Plants" by Narendra Tuteja, this volume), the role of apoplastic and cell-wall mediation in signaling events (see "Signaling and Cell Walls" by Ewelina Rodakowska et al., this volume), and signaling during plant-pathogen interactions utilizing localized ROS production (see "Signaling via Plant Peroxidases" by Claude Penel and Christophe Dunand, this volume) are not further elaborated upon here.

2.3 *Rho-like GTPases of Plants: Master Switches Responding to Environmental Cues*

In tracing back many of the ionic alterations brought about in response to environmental cues, a conserved set of proteins constantly comes up. Rho-like GTPases of plants (ROPs) localize to the cell membrane (Bischoff et al. 2000; Molendijk et al. 2001; Cheung et al. 2003) and are involved in multiple signal transduction events in plants. In *Arabidopsis*, 11 ROPs have been described and extensive reviews have been published on signaling mediated by ROPs and their regulators (Zheng and Yang 2000; Gu et al. 2004; Xu and Scheres 2005; Nibau et al. 2006; Kost 2008). Direct links between ROP activity and Ca2+ spiking (Li et al. 1999; Jones et al. 2002), and ROP mediation in localized ROS activation (Jones et al. 2007) have been demonstrated. ROP-mediated signaling is also one of the best elucidated chains of molecular links directed to the actin cytoskeleton in plants (Fu et al. 2005; Fig. 1).

Specific ROPs and their regulators (RICs, ROP-GEFs/ROP-GDI) have an effect on the actin cytoskeleton but these proteins, high up in the response hierarchy of

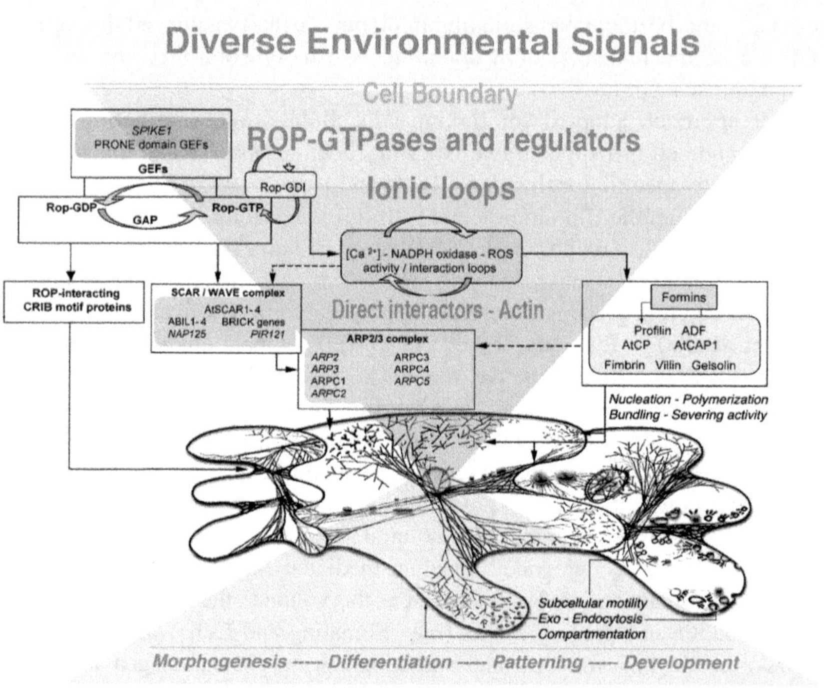

Fig. 1 A generalized depiction of events involved in signaling to the actin cytoskeleton in plants. The leaf epidermal pavement cells drawn here are considered to represent all cell types since the vast majority of actin organizations and actin-mediated processes can be observed in them. Moreover, whereas many different signal cascades might operate in different plant cells, presently pavement cells provide one of the best characterized systems for dissecting signals to the actin cytoskeleton (Fu et al. 2005). An inverted pyramid of events suggests the gradual parsing of signals through the molecular apparatus to produce an impact on the actin cytoskeleton. Diverse environmental cues perceived at the cell boundary (cell wall and plasma membrane) initiate feedback loops of Rho-like GTPases of plants (ROP)-GTPases and ionic activity, which subsequently affect several proteins that bind directly to actin (such as actin monomer interacting, nucleating, bundling, and severing proteins) and bring about changes in its organization. Local actin organization has implications for subcellular motility, exocytosis-endocytosis, cytomembrane organization, and structural reinforcement. The cumulative results of these activities lead to cell differentiation, morphogenesis, and functional diversification during plant development. *Solid lines* and *arrows* depict experimentally demonstrated links between protein complexes, whereas *broken lines* imply putative interactions

the plant cell, do not cause their effects through direct binding to actin. Their effects impact multiple proteins and processes, influencing both the microtubule and the actin cytoskeleton (Fu et al. 2005). From a phenotype-based viewpoint an interference with ROPs and associated machinery does not produce cell forms that can be solely linked to actin defects (e.g., root hair phenotypes depicted in Fig. 2).

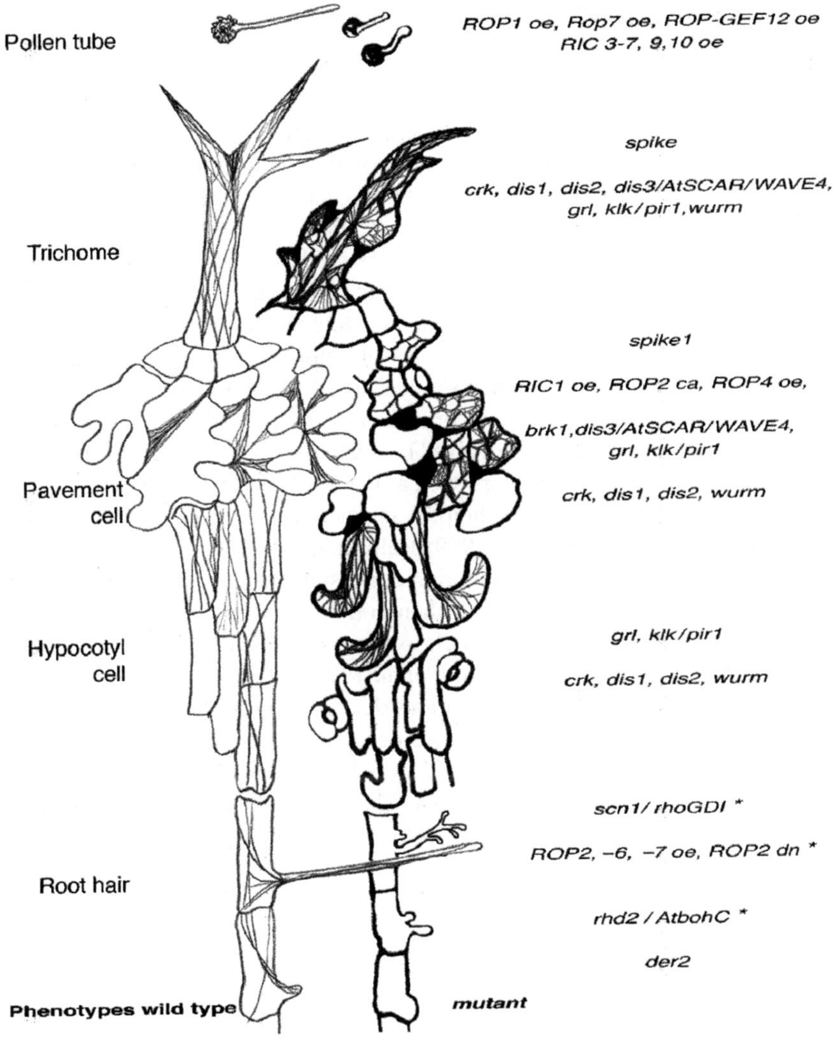

Fig. 2 Salient cellular phenotypes in model cell types of *Arabidopsis thaliana*. The *left side* shows cell morphology displayed by wild-type plants, while the phenotypes drawn in *bold lines* are exhibited by different mutants. In general, cells with an aberrant actin cytoskeleton are smaller in size. Phenotypes displayed by ROPs and their regulators do not strictly follow the pattern obtained by an interference with the actin cytoskeleton only. These phenotypes suggest a strong disturbance in the microtubule cytoskeleton too and could reflect actin-microtubule interactivity during cell shape development

However, neither are the ROP-associated phenotypes indicative of microtubule cytoskeletal defects only. Live imaging of the actin and microtubule cytoskeletons in *Arabidopsis* plants possessing aberrant activity of different ROPs and their regulators (e.g., ROP-GDI) suggests that they exhibit both actin and microtubule defects (Mathur, unpublished data).

3 Developmental Consequences of Different Actin Organizations

Whereas it has been possible to draw correlations between a small region of growth and its ionic environment, such as the one provided by tip-growing cells (Gu et al. 2005; Ovecka et al. 2008; Kroeger et al. 2008), establishing similar correlations has been difficult for diffuse-growing cells that make up the bulk of higher plants. The very nature of diffuse growth involves widespread, possibly transient signaling events that preclude precise localization of ionic concentrations and gradients. Therefore, a reverse approach has been adopted for understanding molecular mechanisms involved in unraveling the signaling to and the roles of actin in diffuse-growing cells. This approach (Mathur et al. 1999; Szymanski et al. 1999; Fu et al. 2002) has largely been based on visualizing defects in actin organization in particular cell types and then tracing them to a molecular lesion involved in creating similar aberrant cellular activity. In this approach, the use of inhibitors such as cytochalasin D, latrunculin, phalloidin, and jasplakinolide has been very helpful in uncovering specific actin-associated cellular phenotypes (Mathur et al. 1999; Baluska et al. 2001; Collings et al. 2006; Foissner and Wasteneys 2007) and hinting at possible processes involved. The actin defective phenotype based approach, when combined with live imaging, has been successful in uncovering new protein classes and complexes involved in mediating signals to the actin cytoskeleton (Mathur et al. 1999, 2003a, b).

The occurrence of the actin organizations discussed above has been demonstrated in different model cell types in plants (representative images in publications: root hair – Ryan et al. 2001; pollen tube – Lovy-Wheeler et al. 2005; hypocotyl cells –Mathur et al. 2003b; pavement cells – Fu et al. 2002, 2005; trichomes – Mathur et al. 1999). Section 3.1 associates actin cytoskeleton organization to a developmental consequence in different cell types.

3.1 Actin Patch Indicative of Cell Polarization

Polar growth is an established facet of plant development. However, the analysis of early events leading to the establishment of polarity in the asymmetric environment inhabited by land plants is a difficult task. Observations on the early development of fucoid zygotes have been informative in highlighting a role for the actin cytoskeleton during polarity establishment in a nearly apolar cell (Kropf et al. 1989; Bisgrove 2007). Upon release from the mother frond, and following their fertilization, spherical fucoid zygotes rely on several vectorial inputs such as blue light, gravity, electrical-ionic impulses, and temperature and osmotic gradients for subsequent development (Brownlee and Bouget 1998). A distinct actin patch has been associated with zygote polarization and creation of the rhizoidal pole (Kropf et al. 1989). A cortical localization of the actin modulator ARP2/3 appears to be intimately involved in actin patch formation (Hable and Kropf 2005). Actin depolymerizing substances disrupt cell polarization and subsequent development (Hable et al. 2003). In later stages of zonation of the zygote, asymmetric localization of ARP2 and actin results in distinct actin arrays (Hable and Kropf 2005).

3.2 Actin Organization Indicative of Focused Growth

Pollen tubes and root hairs have been the most extensively studied tip-growing cells and numerous publications have described the actin cytoskeleton in these cells and reviewed the implications of the polar F-actin observed in them (Geitmann and Emons 2000; Ryan et al. 2001; Ringli et al. 2005; Geitmann 2006). In recent years the moss *Physcomitrella patens* has emerged as an equally informative model system for understanding polarized tip growth.

In *Physcomitrella*, single apical cells of moss filaments undergo polarized tip growth by perceiving and aligning themselves to gravity and light inputs (Cove 2005). Several recent studies point to specialized roles for actin dynamics in modulating these events, prompting the hypothesis that the ARP2/3 complex is pivotal in translating environmental cues by acting as a downstream target for signals regulating polar tip growth (Perroud and Quatrano 2006). Subsequently, there has been a strong focus on evaluating the roles of both the ARP2/3 complex and its upstream regulators Scar/WAVE complexes in polar growth of the apical cell of *P. patens* filament cells. When the arpc4 subunit of the ARP2/3 complex is deleted, the resulting null mutant is viable. However, while tissue morphogenesis in the mutant proceeds normally from filamentous growth to leafy shoot formation, it displays inefficient perception of polarized white light and a marked reduction in apical cell growth (Perroud and Quatrano 2006). Additionally, RNA interference lines knocking down expression of another ARP2/3 member, ARPC1, result in a similar, albeit more severe, phenotype as displayed by the arpc4 mutant. Overexpression of the respective ARP2/3 members (ARPC4, ARPC1) rescues the mutant phenotypes. Moreover, insertion of a yellow fluorescent protein–ARPC4 localizes ARPC4 exclusively to the site of polarized extension in the apical cell tip (Perroud and Quatrano 2006). Moreover, the deletion of BRK1, an upstream regulator of the ARP2/3 complex (Frank and Smith 2002), results in a mutant that though viable and responsive to polarized white light shows significant reduction in apical cell growth (Perroud and Quatrano 2008). BRK1 also localizes to the apical tip. A role for BRK1 in ARP2/3recruitment is supported as the BRK1 mutant of physcomitrella is unable to localize both ARPC4 and AGP (a cell wall proteoglycan) to the apical cell tip (Perroud and Quatrano 2008).

3.3 Actin Organization Indicative of Diffuse Growth

The majority of plant cells grow by diffuse growth. Different model cell types have been informative about the actin-associated molecular–cellular machinery involved in spreading growth occurring over large regions of the cell. These include trichomes (epidermal hair) that grow outward from the aerial epidermis, interconnected jigsaw puzzle shaped pavement cells, and axially elongating epidermal cells of the hypocotyl, cotyledon, and leaf petiole. Each of these epidermal cell types offers a different surface to the environment and their growth responses differ accordingly.

In *Arabidopsis*, trichomes are single-celled and display a rapid expansion phase in the later stage of their development (Mathur 2006b). As the trichome cell grows out of the epidermal plane, unencumbered by growth of neighboring cells, a large portion of it is exposed directly to the environment. In contrast, both pavement and hypocotyl/petiole cells achieve accommodative growth by maintaining connectivity with their neighbors. All three epidermal cell types exhibit long F-actin cables that stretch from end to end, as well as a fine cortical F-actin mesh. In addition, in both trichomes and pavement cells, areas of indention (valleys) display bundled or aggregated F-actin. In general, the presence of bundled F-actin is plant cells has come to be associated with regions of low expansion, whereas the presence of a fine cortical F-actin meshwork suggests regions of increased expansion.

3.3.1 Diffuse Growth in Trichomes

The use of actin inhibitors resulted in random shape alterations in *Arabidopsis* trichomes (Mathur et al. 1999; Szymanski et al. 1999). The drug-induced trichome distortion is phenocopied by mutations in several *Arabidopsis* genes. Molecular characterization of these genes has identified the ARP2/3 complex, an efficient modulator of actin polymerization that initiates the formation of a fine dendritic F-actin mesh (reviewed in Mathur 2005a, b; Szymanski 2005; Hussey et al. 2006), and its upstream regulators, starting from ROPs, ROP-GEFs such as SPIKE1 (Qiu et al. 2002; Basu et al. 2008; Zhang et al. 2008), and including different SCAR/WAVE, HSPC300-like/BRICK proteins (Frank and Smith 2002; Frank et al. 2004; Zhang et al. 2005; Basu et al. 2005; Le et al. 2006; Djakovic et al. 2006), NAP125-like and PIR121-like proteins (reviewed by Szymanski 2005; Smith and Oppenheimer 2005). An elegant interaction map of SCAR genes and their interactions with ARP2/3 proteins obtained through mutant studies, yeast two-hybrid, and bimolecular fluorescence complementation proteins experiments has been presented (Uhrig et al. 2007). In addition to the ROP-SPIKE-SCAR/WAVE-ARP2/3 pathway mentioned above, several other signaling pathways to the actin cytoskeleton might be activated through the involvement of novel ROP-GEFs, termed "plant-specific ROP nucleotide exchanger (PRONE), which are exclusively active towards members of the ROP subfamily (Berken et al 2005).

3.3.2 Interdigitating Growth of Pavement Cells

The final "jigsaw puzzle" shape of epidermal pavement cells arises from multiple local projections (lobes) of a polygonal initial. During expansion, the lobes of one pavement cell fit into the indentions of its neighbors to produce an epidermal surface with an interdigitating pattern. Fu et al. (2005) revealed a ROP-GTPase signaling network underlying the creation of these lobes and indentions by pavement cells. Local activation of a ROP (AtROP2) activates ROP-interactive CRIB-motif-containing protein 4 (RIC4) to enhance actin dynamics and promote localized growth. However, ROP2 activity leads to the inactivation of another target, RIC1, that localizes to cortical microtubules and promotes their ordering into parallel arrays. RIC1-dependent microtubule organization not only inhibits cell outgrowth locally but also suppresses ROP2 activation in the indention zone. Thus, while the ROP2–RIC4 interaction promotes cell outgrowth, the ROP2–RIC1 interaction restrains outgrowth. Such coordinated activity in epidermal cells creates the interdigitations between adjacent pavement cells.

3.3.3 Accommodative Growth of Cylindrical Cells

The growth of cells of the seedling hypocotyl and petioles relies on efficient actin-based communication between cell ends for producing elongation along the longitudinal axis (Baluska et al. 2003). These cells have a diffuse F-actin mesh but display cytoplasmic aggregation at the cell ends. Mutants in different subunits of the ARP2/3 complex, hypocotyl, and petiole cells often display increased F-actin aggregation at their ends and consequently lose cell-cell connectivity (Mathur et al. 2003a, b). The reduced actin polymerization activity in elongating cylindrical cells observed in the arp2/3 mutants has implications for cell-cell communication and polar transport of growth regulators. It thus offers a novel system to understand the initiation and maintenance of multicellularity in higher plants.

4 The Diverse Roles of the Actin Cytoskeleton in Plants

An appraisal of cellular phenotypes associated with different actin cytoskeleton organizations, the nature of actin regulatory proteins that can create the organizations, and the diversity of environmental cues that trigger subcellular activity suggests that actin can be involved in many different kinds of roles inside the plant cell. Some of these roles are discussed in Sect. 4.1.

4.1 Actin as a Barrier

The regional accumulation of actin correlates with increased aggregation and decreased motility of small organelles such as Golgi bodies, peroxisomes, and mitochondria

(Mathur et al. 2003a, b). Upon extrapolation, the observations could suggest that F-actin acts as an intracellular barrier for subcellular trafficking of vesicles. The notion was tested by looking for areas of actin aggregation versus those with fine cortical actin meshwork in different cell types (Mathur et al. 2003a). Regions with actin aggregation invariably displayed lesser growth as compared with regions with a fine meshwork of F-actin (Mathur et al. 2003a; Fu et al. 2005). According to the "actin barrier" concept, a fine F-actin meshwork resulting from increased actin dynamics mediated by the ARP2/3 complex and formins would allow and perhaps even aid the rapid motility of organelles and exocytotic vesicles (Mathur 2004). This leads to a promotion of local growth. In contrast, a dense web comprising F-actin aggregates and bundles in an intracellular locality might hinder organelle motility and vesicular trafficking and could thereby display a lower rate of regional growth. The idea of F-actin behaving as a flexible barrier is appealing when considering the phenotypes described above for pavement and hypocotyl cells impaired in their actin organization. The role of cortical F-actin as a "fence" has also been suggested through three-dimensional electron-tomographic reconstructions of the membrane cytoskeleton in normal rat kidney fibroblast cells (Morone et al. 2006).

4.2 Actin as a Propulsive Force

Asymmetric actin polymerization mediated through the ARP2/3 complex has been implicated in the rocketing motility of enteropathogenic organisms such as *Listeria monocytogenes* and *Shigella* species, as well as vesicles and organelles such as endosomes, phagosomes (Higgs and Pollard 2001; Gouin et al. 2005), and mitochondria (Boldogh et al. 2001). The polymerization of actin is also believed to act as the driving force at the leading edge of motile animal cells (Goley and Welch 2006). In plants, direct evidence for actin polymerization acting as a propulsive force for subcellular structures or as a protrusive force for cell membranes has not been observed. However, given the role of actin in vesicle movement during auxin efflux and translocation (Dhonukshe et al. 2007), the rapid movement of endocytic vesicles (Voigt et al. 2005), and cytoplasmic streaming, the possibility of actin polymerization mediated force generation within the plant cell cannot be totally ruled out (Mathur 2005b).

4.3 Actin as an Interactive Support System

The actin cytoskeleton interacts with microtubules. During root elongation in *Arabidopsis* the hypersensitivity to cytoskeletal antagonists has been used to demonstrate microtubule-microfilament cross-talk (Blancaflor 2000; Collings et al. 2006). One of the molecular mechanisms controlling these interactions has been traced to the antagonistic activities of ROP-RIC proteins (Fu et al. 2005) and elegantly described for the morphogenesis of pavement cells of *Arabidopsis*.

Interestingly, observations in trichome cells suggest that F-actin aggregates coincide with regional clusters of cytoplasmic microtubules (Saedler et al. 2004). The observation suggests that microtubules, with their own set of interactors and regulators, might act to confine F-actin-based regional expansion in a cell (Mathur 2004, 2006a). The role of F-actin in sculpting the cortical microtubule cytoskeleton has also been suggested (Schwab et al. 2003).

4.4 *Actin as a Track*

Observations on different organelles in plant cells have shown that their motility is actin-based (Ovecka et al. 2008) and involves interactions with myosin motors (Samaj et al. 2000; Reisen and Hanson 2007; Sparkes et al. 2008). In this situation, actin filaments act as subcellular tracks for the motor molecules. How cargo specificity and the choice of a specific actin track is decided by the different myosins in plant cells still remains an enigma. There also remains the question of interactivity and switching between myosin motors and microtubule-based motors such as kinesins.

5 Conclusions

While reviews on the minutiae of signaling to the actin cytoskeleton will continue being published, one of the long-term goals of the field is to uncover the mechanisms whereby this global perceiver of, and responder to, myriad environmental cues is able to create local influences within a cell. As described in this chapter, the key to actin's responsiveness lies in its interactivity with numerous regulatory proteins as well as with other components and compartments of cell. Signaling to actin thus becomes an exercise in attempting to understand and sum up everything that goes on inside a living cell. Realistically, we are far from the goal, but given the tremendous scope of actin interactions, the increased availability of novel mutants in model plant species, and the emergence of new tools and techniques for live imaging (Dhanoa et al. 2006), the prospects of unraveling the tangled web around actin looks very promising.

Acknowledgements We gratefully acknowledge funding by the Natural Sciences and Engineering Research Council of Canada (NSERC), Canada Foundation for Innovation (CFI), The Ministry of Research and Innovation (MRI), Ontario, and the Keefer Trust Fund, Guelph.

References

Amann KJ, Pollard TD (2001) The Arp2/3 complex nucleates actin filament branches from the sides of pre-existing filaments. Nat Cell Biol 3:306–310

Ayscough KR (1998) In vivo functions of actin-binding proteins. Curr Opin Cell Biol 10:102–111

Baluska F, Jasik J, Edelmann HG, Salajová T, Volkmann D (2001) Latrunculin B-induced plant dwarfism: Plant cell elongation is F-actin-dependent. Dev Biol 231:113–124

Baluska F, Wojtaszek P, Volkmann D, Barlow P (2003) The architecture of polarized cell growth: the unique status of elongating plant cells. Bioessays 25:569–576

Basu D, Le J, El-Essal Sel-D, Huang S, Zhang C, Mallery EL, Koliantz G, Staiger CJ, Szymanski DB (2005) DISTORTED3/SCAR2 is a putative arabidopsis WAVE complex subunit that activates the Arp2/3 complex and is required for epidermal morphogenesis. Plant Cell 17:502–524

Basu D, Le J, Zakharova T, Mallery EL, Szymanski DB (2008) A SPIKE1 signaling complex controls actin-dependent cell morphogenesis through the heteromeric WAVE and ARP2/3 complexes. Proc Natl Acad Sci U S A 105:4044–4049

Berken A, Thomas C, Wittinghofer A (2005) A new family of RhoGEFs activates the Rop molecular switch in plants. Nature 436:1176–1178

Bischoff F, Vahlkamp L, Molendijk A, Palme K (2000) Localization of AtROP4 and AtROP6 and interaction with the guanine nucleotide dissociation inhibitor AtRhoGDI1 from Arabidopsis. Plant Mol Biol 42:515–530

Bisgrove SR (2007) Cytoskeleton and early development in Fucoid algae. J Integr Plant Biol 49:1192–1198

Blancaflor EB (2000) Cortical actin filaments potentially interact with cortical microtubules in regulating polarity of cell expansion in primary roots of maize (*Zea mays* L.). J Plant Growth Regul 19:406–410

Boldogh IR, Yang HC, Nowakowski WD, Karmon SL, Hays LG, Yates JR3rd, Pon LA (2001) ARP2/3 complex and actin dynamics are required for actin-based mitochondrial motility in yeast. Proc Natl Acad Sci U S A 98:3162–3167

Bothwell JH, Ng CK (2005) The evolution of Ca^{2+} signalling in photosynthetic eukaryotes. New Phytol 166:21–38

Brenner ED, Stahlberg R, Mancuso S, Vivanco J, Baluska F, Van Volkenburgh E (2006) Plant neurobiology: an integrated view of plant signaling. Trends Plant Sci 11:413–419

Bretscher A, Weber K (1980) Villin is a major protein of the microvillus cystoskeleton which binds both G and F actin in a calcium-dependent manner. Cell 20:839–847

Brownlee C, Bouget FY (1998) Polarity determination in Fucus: from zygote to multicellular embryo. Semin Cell Dev Biol 9:179–185

Cárdenas L, Lovy-Wheeler A, Kunkel JG, Hepler PK (2008) Pollen tube growth oscillations and intracellular calcium levels are reversibly modulated by actin polymerization. Plant Physiol 146:1611–1621

Carol RJ, Dolan L (2006) The role of reactive oxygen species in cell growth: lessons from root hairs. J Exp Bot 57:1829–1834

Chaudhry F, Guérin C, von Witsch M, Blanchoin L, Staiger CJ (2007) Identification of Arabidopsis cyclase-associated protein 1 as the first nucleotide exchange factor for plant actin. Mol Biol Cell 18:3002–3014

Chen JG (2008) Heterotrimeric G-proteins in plant development. Front Biosci 13:3321–3333

Chen CY, Wong EI, Vidali L, Estavillo A, Hepler PK, Wu H, Cheung AY (2002) The regulation of actin organization by actin depolymerizing factor in elongating pollen tubes. Plant Cell 14:2175–2190

Cheung AY, Wu HM (2004) Over-expression of an Arabidopsis formin stimulates supernumerary actin cable formation from pollen tube cell membrane. Plant Cell 16:257–269

Cheung AY, Wu HM (2008) Structural and signaling networks for the polar cell growth machinery in pollen tubes. Annu Rev Plant Biol 59:547–572

Cheung AY, Chen CY-, Tao L-, Andreyeva T, Twell D, Wu H- (2003) Regulation of pollen tube growth by Rac-like GTPases. J Exp Bot 54:73–81

Coelho SM, Brownlee C, Bothwell JH (2008) A tip-high, Ca^{2+}-interdependent, reactive oxygen species gradient is associated with polarized growth in *Fucus serratus* zygotes. Planta 227:1037–1046

Collings DA, Lill AW, Himmelspach R, Wasteneys GO (2006) Hypersensitivity to cytoskeletal antagonists demonstrates microtubule-microfilament cross-talk in the control of root elongation in *Arabidopsis thaliana*. New Phytol 170:275–290

Cove D (2005) The moss *Physcomitrella patens*. Annu Rev Genet 39:339–358

Day IS, Reddy VS, Shad Ali G, Reddy AS (2002) Analysis of EF-hand-containing proteins in Arabidopsis. Genome Biol 3(10): 1–24

Deeks MJ, Cvrcková F, Machesky LM, Mikitová V, Ketelaar T, Zársky V, Davies B, Hussey PJ (2005) Arabidopsis group Ie formins localize to specific cell membrane domains, interact with actin-binding proteins and cause defects in cell expansion upon aberrant expression. New Phytol 168:529–540

Deeks MJ, Rodrigues C, Dimmock S, Ketelaar T, Maciver SK, Malhó R, Hussey PJ (2007) Arabidopsis CAP1 – a key regulator of actin organisation and development. J Cell Sci 120:2609–2618

Dhanoa PK, Sinclair AM, Mullen RT, Mathur J (2006) Illuminating subcellular structures and dynamics in plants: a fluorescent protein toolbox. Can J Bot 84:515–522

Dhonukshe P, Aniento F, Hwang I, Robinson DG, Mravec J, Stierhof YD, Friml J (2007) Clathrin-mediated constitutive endocytosis of PIN auxin efflux carriers in Arabidopsis. Curr Biol 17:520–527

Djakovic S, Dyachok J, Burke M, Frank MJ, Smith LG (2006) BRICK1/HSPC300 functions with SCAR and the ARP2/3 complex to regulate epidermal cell shape in Arabidopsis. Development 133:1091–1100

Dong CH, Kost B, Xia G, Chua NH (2001) Molecular identification and characterization of the Arabidopsis AtADF1, AtADFS and AtADF6 genes. Plant Mol Biol 45:517–527

Drøbak BK, Franklin-Tong VE, Staiger CJ (2004) The role of the actin cytoskeleton in plant cell signaling. New Phytol 163:13–30

Felle HH, Hepler PK (1997) The cytosolic Ca^{2+} concentration gradient of *Sinapis alba* root hairs as revealed by Ca^{2+}-selective microelectrode tests and Fura-dextran ratio imaging. Plant Physiol 114:39–45

Foissner I, Wasteneys GO (2007) Wide-ranging effects of eight cytochalasins and latrunculin A and B on intracellular motility and actin filament reorganization in characean internodal cells. Plant Cell Physiol 48:585–597

Foreman J, Demidchik V, Bothwell JH, Mylona P, Miedema H, Torres MA, Linstead P, Costa S, Brownlee C, Jones JD, Davies JM, Dolan L (2003) Reactive oxygen species produced by NADPH oxidase regulate plant cell growth. Nature 422: 442–446

Frank MJ, Smith LG (2002) A small, novel protein highly conserved in plants and animals promotes the polarized growth and division of maize leaf epidermal cells. Curr Biol 12:849–853

Frank M, Egile C, Dyachok J, Djakovic S, Nolasco M, Li R, Smith LG (2004) Activation of Arp2/3 complex-dependent actin polymerization by plant proteins distantly related to Scar/WAVE. Proc Natl Acad Sci U S A 101:16379–16384

Fu Y, Li H, Yang Z (2002) The ROP2 GTPase controls the formation of cortical fine F-actin and the early phase of directional cell expansion during Arabidopsis organogenesis. Plant Cell 14:777–794

Fu Y, Gu Y, Zheng Z, Wasteneys G, Yang Z (2005) Arabidopsis interdigitating cell growth requires two antagonistic pathways with opposing action on cell morphogenesis. Cell 120:687–700

Geitmann A (2006) Plant and fungal cytomechanics: quantifying and modeling cellular architecture. Can J Bot 84:581–593

Geitmann A, Emons AM (2000) The cytoskeleton in plant and fungal cell tip growth. J Microsc 198:218–245

Gouin E, Welch MD, Cossart P (2005) Actin-based motility of intracellular pathogens. Curr Opin Microbiol 8:35–45

Goley ED, Welch MD (2006) The ARP2/3 complex: an actin nucleator comes of age. Nat Rev Mol Cell Biol 7:713–726

Grunt M, Zarsky V, Cvrckova F (2008) Roots of Angiosperm formins: the evolutionary history of plant FH2 domain-containing proteins. BMC Evol Biol 8:115

Gu Y, Wang Z, Yang Z (2004) ROP/RAC GTPase: an old new master regulator for plant signaling. Curr Opin Plant Biol 7:527–536

Gu Y, Fu Y, Dowd P, Li S, Vernoud V, Gilroy S, Yang Z (2005) A Rho family GTPase controls actin dynamics and tip growth via two counteracting downstream pathways in pollen tubes. J Cell Biol. 169:127–138

Hable WE, Kropf DL (2005) The Arp2/3 complex nucleates actin arrays during zygote polarity establishment and growth. Cell Motil Cytoskel 61:9–20

Hable WE, Miller NR, Kropf DL (2003) Polarity establishment requires dynamic actin in fucoid zygotes. Protoplasma 221:193–204

Harris HE, Weeds AG (1983) Plasma actin depolymerizing factor has both calcium-dependent and calcium-independent effects on actin. Biochemistry 22:2728–2741

Hetherington AM, Brownlee C (2004) The generation of Ca^{2+} signals in plants. Annu Rev Plant Biol 55:401–427

Higgs HN, Pollard TD (2001) Regulation of actin filament network formation through ARP2/3 complex: activation by a diverse array of proteins. Annu Rev Biochem 70:649–676

Hirt H (2000) MAP kinases in plant signal transduction. Results Probl Cell Differ 27:1–9

Hong-Bo S, Li-Ye C, Ming-An S (2008) Calcium as a versatile plant signal transducer under soil water stress. Bioessays 30:634–641

Huang S, Blanchoin L, Kovar DR, Staiger CJ (2003) Arabidopsis capping protein (AtCP) is a heterodimer that regulates assembly at the barbed ends of actin filaments. J Biol Chem 278:44832–44842

Huang S, Blanchoin L, Chaudhry F, Franklin-Tong VE, Staiger CJ (2004) A gelsolin-like protein from *Papaver rhoeas* pollen (PrABP80) stimulates calcium-regulated severing and depolymerization of actin filaments. J Biol Chem 279:23364–23375

Hussey PJ, Ketelaar T, Deeks MJ (2006) Control of the actin cytoskeleton in plant cell growth. Annu Rev Plant Biol 57:109–112

Ingouff M, Fitz Gerald JN, Guérin C, Robert H, Sørensen MB, Van Damme D, Geelen D, Blanchoin L, Berger F (2005) Plant formin AtFH5 is an evolutionarily conserved actin nucleator involved in cytokinesis. Nat Cell Biol 7:374–380

Jones MA, Shen JJ, Fu Y, Li H, Yang Z, Grierson CS (2002) The Arabidopsis Rop2 GTPase is a positive regulator of both root hair initiation and tip growth. Plant Cell 14:763–776

Jones MA, Raymond MJ, Yang Z, Smirnoff N (2007) NADPH oxidase-dependent reactive oxygen species formation required for root hair growth depends on ROP GTPase. J Exp Bot 58:1261–1270

Kakimoto T, Shibaoka H (1987) Actin filaments and microtubules in the preprophase band and phragmoplast of tobacco cells. Protoplasma 140:151–156

Kandasamy MK, Meagher RB (1999) Actin-organelle interaction: association with chloroplast in Arabidopsis leaf mesophyll cells. Cell Motil Cytoskel 44:110–118

Ketelaar T, Allwood EG, Anthony R, Voigt B, Menzel D, Hussey PJ (2004) The actin-interacting protein AIP1 is essential for actin organization and plant development. Curr Biol 14:145–149

Klahre U, Friederich E, Kost B, Louvard D, Chua NH (2000) Villin-like actin-binding proteins are expressed ubiquitously in Arabidopsis. Plant Physiol 122:35–48

Kost B (2008) Spatial control of Rho (Rac-Rop) signaling in tip-growing plant cells. Trends Cell Biol 18:119–127

Kost B, Spielhofer P, Chua NH (1998) A GFP-mouse talin fusion protein labels plant actin filaments *in vivo* and visualizes the actin cytoskeleton in growing pollen tubes. Plant J 16:393–401

Kovar DR, Drobak BK, Staiger CJ (2000a) Maize profilin isoforms are functionally distinct. Plant Cell 12:583–598

Kovar DR, Staiger CJ, Weaver EA, McCurdy DW (2000b) AtFim1 is an actin filament crosslinking protein from *Arabidopsis thaliana*. Plant J 24:625–636

Kroeger JH, Geitmann A, Grant M (2008) Model for calcium dependent oscillatory growth in pollen tubes. J Theor Biol 253:363–374

Kropf DL, Berge SK, Quatrano RS (1989) Actin localization during Fucus embryogenesis. Plant Cell 1:191–200

Kwak JM, Mori IC, Pei ZM, Leonhardt N, Torres MA, Dangl JL, Bloom RE, Bodde S, Jones JD, Schroeder JI (2003) NADPH oxidase AtrbohD and AtrbohF genes function in ROS-dependent ABA signaling in Arabidopsis. EMBO J 22:2623–2633

Le J, Mallery EL, Zhang C, Brankle S, Szymanski DB (2006) Arabidopsis BRICK1/HSPC300 is an essential WAVE-complex subunit that selectively stabilizes the Arp2/3 activator SCAR2. Curr Biol 16:895–901

Li H, Lin Y, Heath RM, Zhu MX, Yang Z (1999) Control of pollen tube tip growth by a Rop GTPase-dependent pathway that leads to tip-localized calcium influx. Plant Cell 11:1731–1742
Lovy-Wheeler A, Wilsen KL, Baskin TI, Hepler PK (2005) Enhanced fixation reveals the apical cortical fringe of actin filaments as a consistent feature of the pollen tube. Planta 221:95–104
Mathur J (2004) Cell shape development in plants. Trends Plant Sci 9:583–590
Mathur J (2005a) The ARP2/3 complex: giving plant cells a leading edge. Bioessays 27:377–387
Mathur J (2005b) Conservation of boundary extension mechanisms between plants and animals. J Cell Biol 168:679–682
Mathur J (2006a) Local interactions shape plant cells. Curr Opin Cell Biol 18:40–46
Mathur J (2006b) Trichome cell morphogenesis in Arabidopsis: a continuum of cellular decisions. Can J Bot 84:581–593
Mathur J, Spielhofer P, Kost B, Chua N (1999) The actin cytoskeleton is required to elaborate and maintain spatial patterning during trichome cell morphogenesis in *Arabidopsis thaliana*. Development 126:5559–5568
Mathur J, Mathur N, Kirik V, Kernebeck B, Srinivas BP, Hülskamp M (2003a) Arabidopsis CROOKED encodes for the smallest subunit of the ARP2/3 complex and controls cell shape by region specific fine F-actin formation. Development 130:3137–3146
Mathur J, Mathur N, Kernebeck B, Hülskamp M (2003b) Mutations in actin-related proteins 2 and 3 affect cell shape development in Arabidopsis. Plant Cell 15:1632–1645
Matsudaira PT, Burgess DR (1982) Partial reconstruction of the microvillus core bundle: characterization of villin as a Ca^{2+}-dependent, actin-bundling/depolymerizing protein. J Cell Biol 92:648–656
Meagher RB, Fechheimer M (2003) The Arabidopsis cytoskeletal genome. In: Somerville CR, Meyerowitz EM (eds) The Arabidopsis book. American Society of Plant Biologists, Rockville
Meagher RB, McKinney EC, Kandasamy MK (2000) The significance of diversity in the plant actin gene family. In: Staiger CJ et al (eds) Actin: a dynamic framework for multiple plant cell functions. Kluwer, Dordrecht, pp 3–27
Michelot A, Guérin C, Huang S, Ingouff M, Richard S, Rodiuc N, Staiger CJ, Blanchoin L (2005) The formin homology 1 domain modulates the actin nucleation and bundling activity of Arabidopsis FORMIN1. Plant Cell 17:2296–2313
Miller DD, Lancelle SA, Hepler PK (1996) Actin microfilaments do not form a dense meshwork in *Lilium longiflorum* pollen tube tips. Protoplasma 195:123–132
Miller G, Shulaev V, Mittler R (2008) Reactive oxygen signaling and abiotic stress. Physiol Plant 253:481–489
Molendijk AJ, Bischoff F, Rajendrakumar CS, Friml J, Braun M, Gilroy S, Palme K (2001) *Arabidopsis thaliana* Rop GTPases are localized to tips of root hairs and control polar growth. EMBO J 20:2779–2788
Monshausen GB, Bibikova TN, Messerli MA, Shi C, Gilroy S (2007) Oscillations in extracellular pH and reactive oxygen species modulate tip growth of Arabidopsis root hairs. Proc Natl Acad Sci U S A 104:20996–21001
Mori IC, Schroeder JI (2004) Reactive oxygen species activation of plant Ca^{2+} channels. A signaling mechanism in polar growth, hormone transduction, stress signaling, and hypothetically mechanotransduction. Plant Physiol 135:702–708
Morone N, Fujiwara T, Murase K, Kasai RS, Ike H, Yuasa S, Usukura J, Kusumi A (2006) Three-dimensional reconstruction of the membrane skeleton at the plasma membrane interface by electron tomography. J Cell Biol 174:851–862
Mullins RD, Heuser JA, Pollard TD (1998) The interaction of Arp2/3 complex with actin: Nucleation, high affinity pointed end capping, and formation of branching networks of filaments. Proc Natl Acad Sci U S A 95:6181–6186
Navazio L, Bewell MA, Siddiqua A, Dickinson GD, Galione A, Sanders D (2000) Calcium release from the endoplasmic reticulum of higher plants elicited by the NADP metabolite nicotinic acid adenine dinucleotide phosphate. Proc Natl Acad Sci U S A 97:8693–8698
Nibau C, Wu HM, Cheung AY (2006) RAC/ROP GTPases: 'hubs' for signal integration and diversification in plants. Trends Plant Sci 11:309–315

Ovecka M, Baluska F, Lichtscheidl I (2008) Non-invasive microscopy of tip-growing root hairs as a tool for study of dynamic and cytoskeleton-based vesicle trafficking. Cell Biol Int 32:549–553

Perroud PF, Quatrano RS (2006) The role of ARPC4 in tip growth and alignment of the polar axis in filaments of Physcomitrella patens. Cell Motil Cytoskel 63:162–171

Perroud PF, Quatrano RS (2008) BRICK1 is required for apical cell growth in filaments of the moss *Physcomitrella patens* but not for gametophore morphology. Plant Cell 20:411–422

Qiu JL, Jilk R, Marks MD, Szymanski DB (2002) The Arabidopsis SPIKE1 gene is required for normal cell shape control and tissue development. Plant Cell 14:101–118

Ramachandran S, Christensen HE, Ishimaru Y, Dong CH, Chao-Ming W, Cleary AL, Chua NH (2000) Profilin plays a role in cell elongation, cell shape maintenance, and flowering in Arabidopsis. Plant Physiol 124:1637–1647

Reisen D, Hanson MR (2007) Association of six YFP-myosin XI-tail fusions with mobile plant cell organelles. BMC Plant Biol 7:6

Ringli C, Baumberger N, Keller B (2005) The Arabidopsis root hair mutants der2-der9 are affected at different stages of root hair development. Plant Cell Physiol 46:1046–1053

Ryan E, Steer M, Dolan L (2001) Cell biology and genetics of root hair formation in *Arabidopsis thaliana*. Protoplasma 215:140–149

Saedler R, Mathur N, Srinivas BP, Kernebeck B, Hülskamp M, Mathur J (2004) Actin control over microtubules suggested by DISTORTED2 encoding the Arabidopsis ARPC2 subunit homolog. Plant Cell Physiol 45:813–822

Samaj J, Peters M, Volkmann D, Baluska F (2000) Effects of myosin ATPase inhibitor 2,3-butanedione 2-monoxime on distributions of myosins, F-actin, microtubules, and cortical endoplasmic reticulum in maize root apices. Plant Cell Physiol 41:571–582

Sano T, Higaki T, Oda Y, Hayashi T, Hasezawa S (2005) Appearance of actin microfilament 'twin peaks' in mitosis and their function in cell plate formation, as visualized in tobacco BY-2 cells expressing GFP-fimbrin. Plant J 44:595–605

Schmit AC, Lambert AM (1990) Microinjected fluorescent phalloidin *in vivo* reveals the F-Actin dynamics and assembly in higher plant mitotic cells. Plant Cell 2:129–138

Schwab B, Mathur J, Saedler R, Schwarz H, Frey B, Scheidegger C, Hülskamp M (2003) Regulation of cell expansion by the DISTORTED genes in *Arabidopsis thaliana*: actin controls the spatial organization of microtubules. Mol Genet Genom 269:350–360

Shabala S, Shabala L, Gradmann D, Chen Z, Newman I, Mancuso S (2006) Oscillations in plant membrane transport: model predictions, experimental validation, and physiological implications. J Exp Bot 57:171–184

Shao HB, Chu LY, Shao MA, Jaleel CA, Hong-Mei M (2008) Higher plant antioxidants and redox signaling under environmental stresses. C R Biol 331:433–441

Sheahan MB, Staiger CJ, Rose RJ, McCurdy DW (2004) A green fluorescent protein fusion to actin-binding domain 2 of Arabidopsis fimbrin highlights new features of a dynamic actin cytoskeleton in live plant cells. Plant Physiol 136:3968–3978

Smith LG, Oppenheimer DG (2005) Spatial control of cell expansion by the plant cytoskeleton. Annu Rev Cell Dev Biol 21:271–295

Sparkes IA, Teanby NA, Hawes C (2008) Truncated myosin XI tail fusions inhibit peroxisome, Golgi, and mitochondrial movement in tobacco leaf epidermal cells: a genetic tool for the next generation. J Exp Bot 59:2499–2512

Staiger CJ, Blanchoin L (2006) Actin dynamics: old friends with new stories. Curr Opin Plant Biol 9:554–562

Staiger CJ, Schliwa M (1987) Actin localization and function in higher plants. Protoplasma 141:1–12

Staiger CJ, Baluska F, Volkmann D, Barlow PW (2000) Actin: a dynamic framework for multiple plant cell functions. Springer, Berlin

Sumimoto H (2008) Structure, regulation and evolution of Nox-family NADPH oxidases that produce reactive oxygen species. FEBS J 275:3249–3277

Szymanski DB (2005) Breaking the WAVE complex: the point of Arabidopsis trichomes. Curr Opin Plant Biol 8:103–1012

Szymanski DB, Marks MD, Wick SM (1999) Organized F-actin is essential for normal trichome morphogenesis in Arabidopsis. Plant Cell 11:2331–2347

Takeda S, Gapper C, Kaya H, Bell E, Kuchitsu K, Dolan L (2008) Local positive feedback regulation determines cell shape in root hair cells. Science 319:1241–1244

Tominaga M, Yokota E, Vidali L, Sonobe S, Hepler PK, Shimmen T (2000) The role of plant villin in the organization of the actin cytoskeleton, cytoplasmic streaming and the architecture of the transvacuolar strand in root hair cells of Hydrocharis. Planta 210:836–843

Uhrig JF, Mutondo M, Zimmermann I, Deeks MJ, Machesky LM, Thomas P, Uhrig S, Rambke C, Hussey PJ, Hülskamp M (2007) The role of Arabidopsis SCAR genes in ARP2-ARP3-dependent cell morphogenesis. Development 134:967–977

Van Gestel K, von Witsch M, Samaj J, Baluska F (2003) Immunological evidence for the presence of plant homologues of the actin- related protein Arp3 in tobacco and maize: subcellular localization to actin-enriched pit fields and emerging root hairs. Protoplasma 222:45–52

Vitha S, Baluska F, Braun M, Šamaj J, Volkmann D, Barlow PW (2000) Comparison of cryofixation and aldehyde fixation for plant actin immunocytochemistry: aldehydes do not destroy F-actin. Histochem J 32:457–466

Voigt B, Timmers AC, Samaj J, Hlavacka A, Ueda T, Preuss M, Nielsen E, Mathur J, Emans N, Stenmark H, Nakano A, Baluška F, Menzel D (2005) Actin-based motility of endosomes is linked to the polar tip growth of root hairs. Eur J Cell Biol 84:609–621

Wang YS, Yoo CM, Blancaflor EB (2008) Improved imaging of actin filaments in transgenic Arabidopsis plants expressing a green fluorescent protein fusion to the C- and N-termini of the fimbrin actin-binding domain 2. New Phytol 177:525–536

Wasteneys GO, Willingale-Theune J, Menzel D (1997) Freeze shattering: a simple and effective method for permeabilizing higher plant cell walls. J.Microsc 188:51–61

Xu J, Scheres B (2005) Cell polarity: ROPing the ends together. Curr Opin Plant Biol 8:613–618

Young ME, Cooper JA, Bridgman PC (2004) Yeast actin patches are networks of branched actin filaments. J Cell Biol 166:629–635

Zhang C, Mallery EL, Schlueter J, Huang S, Fan Y, Brankle S, Staiger CJ, Szymanski DB (2008) Arabidopsis SCARs function interchangeably to meet actin-related protein 2/3 activation thresholds during morphogenesis. Plant Cell 20:995–1011

Zhang X, Dyachok J, Krishnakumar S, Smith LG, Oppenheimer DG (2005) IRREGULAR TRICHOME BRANCH1 in Arabidopsis encodes a plant homolog of the actin-related protein2/3 complex activator Scar/WAVE that regulates actin and microtubule organization. Plant Cell 17:2314–2326

Zheng Z, Yang Z (2000) The Rop GTPase switch turns on polar growth in pollen. Trends Plant Sci 5:298–303

Signaling via Plant Peroxidases

Claude Penel and Christophe Dunand

Abstract Plant peroxidases are versatile enzymes. They are known to use hydrogen peroxide (H_2O_2) as an electron acceptor during the oxidation or the cross-linking of various molecules. However, in certain conditions, they can also generate reactive oxygen species (ROS), namely, superoxide ($O_2^{-\cdot}$), H_2O_2, or hydroxyl radical ($\cdot OH$). Most of them are localized in cell walls, where they play a central role in ROS metabolism, since $O_2^{-\cdot}$ and H_2O_2 may be both their substrates and their products, depending on the chemical environment. Their participation in the oxidative burst taking place in the apoplast has been shown in several situations, including defense against pathogens, abiotic stresses, wound healing, and cell death. It is now widely accepted that ROS have important signaling functions. Peroxidases can be considered as being at the crossroads between two different signal transduction systems in plant cells. First, they generate and interconvert the different ROS, thus modulating their signaling function in many physiological processes. Second, they are Ca^{2+}-dependent enzymes, whose activity could be regulated by this second messenger.

1 Introduction

1.1 Reactive Oxygen Species

Molecular oxygen normally acts as a terminal electron acceptor in cellular respiration and is fully reduced; however, in the course of several metabolic processes, it is only partially reduced. This partial reduction is unavoidable. It yields harmful by-products, the reactive oxygen species (ROS), that are scavenged by many enzymatic systems. The three main ROS are superoxide radical ($O_2^{-\cdot}$), hydrogen peroxide (H_2O_2), and hydroxyl radical ($\cdot OH$). Singlet oxygen ($^1O_2$) is another important ROS formed in

C. Penel (✉) and C. Dunand
Laboratoire de Physiologie Végétale, Université de Genève, Quai Ernest-Ansermet 30, 1211, Geneva 4, Switzerland
e-mail: claude.penel@bioveg.unige.ch

F. Baluška and S. Mancuso (eds.), *Signaling in Plants*, Signaling and Communication in Plants,
DOI: 10.1007/978-3-540-89228-1_8,

Table 1 Plant enzymes able to form $O_2^{-\cdot}$, H_2O_2, and $\cdot OH$

Enzymes	Cellular compartment	Substrates	References
	Formation of $O_2^{-\cdot}$		
NADPH oxidases	Plasmalemma/apoplast	$NADPH/O_2$	Doke and Miura (1995)
Class III peroxidases	Apoplast	$Auxin/O_2$	Smith et al. (1988)
	Formation of H_2O_2		
Diamine oxidases	Apoplast	$Spermidine/O_2$	Angelini and Federico (1989)
Oxalate oxidases	Apoplast	$Oxalate/O_2$	Berna and Bernier (1999)
Class III peroxidases	Apoplast	$NAD(P)H/O_2$	Elstner and Heupel (1976)
Glycolate oxidase	Peroxisome	$Glycolate/O_2$	
	Formation of $\cdot OH$		
Class III peroxidases	Apoplast	$NADPH/O_2/H_2O_2$	Liszkay et al. (2003)

chloroplasts. $O_2^{-\cdot}$ and H_2O_2 are produced by the one- and two-electron reduction of O_2, respectively. $\cdot OH$ results from the one-electron reduction of H_2O_2. The two redox couples H_2O_2/H_2O and, above all, $\cdot OH/H_2O$ are characterized by high redox potentials, meaning that they have a strong oxidizing capacity and are very reactive. Consequently, H_2O_2 and $\cdot OH$ have very short life spans. In living tissues, the fate of $O_2^{-\cdot}$ is to be dismutated into H_2O_2 and O_2, either spontaneously or by superoxide dismutase (SOD). Since living organisms have to cope with the presence of these toxic chemical species, they have developed many scavenging mechanisms. An increase in the concentration of ROS is perceived by cells and results in the induction of processes for lowering their concentration. This was most likely a first step in the function of ROS as signaling agents: the perception of their accumulation by cells resulted in the induction of ROS scavenging enzymes (Shao et al. 2008).

However, ROS are not only metabolic by-products. $O_2^{-\cdot}$ and H_2O_2 are purposely formed in plants by many different enzymes. $\cdot OH$ too is produced by a dedicated mechanism. Plant enzymes responsible for ROS generation are listed in Table 1. SOD also plays an important role in ROS metabolism, since it catalyzes the conversion of $O_2^{-\cdot}$ into H_2O_2 (Kliebenstein et al. 1998).

1.2 Plant Peroxidases

Peroxidases are oxidoreductases using H_2O_2 as an electron acceptor. The electrons used for reducing H_2O_2 to water are taken from various donors. Peroxidase classification is based on the identity of the donor. Fifteen different types of peroxidases are listed in the enzyme nomenclature of the International Union of Biochemistry and Molecular Biology (EC 1.11.1.1 to EC 1.11.1.16). These enzymes are necessary for all living organisms to cope with ambient oxygen. At the origin, their role was only to scavenge H_2O_2, a ROS harmful to cells. Plants contain five types of peroxidases: catalases (EC 1.11.1.6), peroxidases (EC 1.11.1.7), glutathione peroxidases (EC 1.11.1.9), ascorbate peroxidases (EC 1.11.1.11), and peroxiredoxins (EC 1.11.1.15).

Catalases are peroxidases that use H_2O_2 as both a hydrogen acceptor and a hydrogen donor, yielding H_2O and O_2 as products. Their role is to scavenge H_2O_2 in peroxisomes, mitochondria, and cytosol (Scandalios 1974). Glutathione and ascorbate peroxidases, as well as peroxiredoxins, also have as a only function the destruction of H_2O_2 in different cell compartments. The last category, peroxidases, also called class III (Welinder 1992), secreted, or guaiacol peroxidases, have more diversified functions. They are hemoproteins, with ferriprotoporphyrin IX as coenzyme, like ascorbate peroxidases (class I), fungal lignin peroxidases (class II), and catalases.

Class III peroxidases (hereafter peroxidases) are glycoproteins synthesized in the endoplasmic reticulum and transported via the Golgi apparatus either towards the apoplast or to vacuoles. The peroxidase proteins are encoded by multigene families. They comprise around 350 amino acids, all have a signal peptide, and some conserved amino acid motifs, especially in the region of interaction with the heme (Passardi et al. 2004). They all contain eight conserved cysteines and one or several glycosylation sites. *Arabidopsis thaliana* has 73 genes encoding a peroxidase (Tognolli et al. 2002) and *Oryza sativa* has 138 (Passardi et al. 2004).

Peroxidases are mainly localized in the cell wall and the apoplast and in vacuoles. They are certainly among the most studied plant proteins. This is probably due to the simplicity of the biochemical assays to reveal their enzymatic activity either in vivo or in vitro. The first studies on plant peroxidases were published during the first years of the twentieth century (Bach and Chodat 1903) and since then interest in these enzymes has never faded. Another aspect of plant peroxidases is their use in many important biotechnological techniques in medicine, biology, and chemistry (Azevedo et al. 2003). Their functions in plants have always been a subject of controversy. It is now known that all the peroxidase-encoding genes are expressed (Valério et al. 2004). One particular gene, *AtPrx42*, is expressed constitutively in all organs throughout the life of the plant. The expression of the other genes is either developmentally controlled or regulated by internal or external factors. It is well known that the peroxidase spectrum obtained by an electrophoretic separation may be modified by all kinds of physical or chemical changes to the plant environment (Gaspar et al. 1982). The isoperoxidase pattern is also extremely sensitive to developmental stages and differentiation (Key and Basile 1987). The same observations are also made at the transcriptional level (Valério et al. 2004). The overall expression pattern of the peroxidase multigene family is therefore extremely complex. One of the central issues concerning peroxidases is to determine whether or not they are interchangeable for achieving a particular function in plants or if each individual enzyme has its own specialized function. At a first glance, it can be hypothesized that since they all have the same coenzyme and the same protein organization, their catalytic characteristics should be very similar. However, there are some experimental indications showing that there are significative catalytic differences among the peroxidases of the same plant. This was shown, for example, with *Arabidopsis* peroxidases. Using 14 recombinant enzymes produced by insect cells infected with baculovirus, Tognolli et al. (2000) showed that their hydrogen donor specificity and their optimum pH were quite different. This study dealt with the peroxidase activity, namely, the oxidation of a donor molecule into a colored product in the presence of H_2O_2.

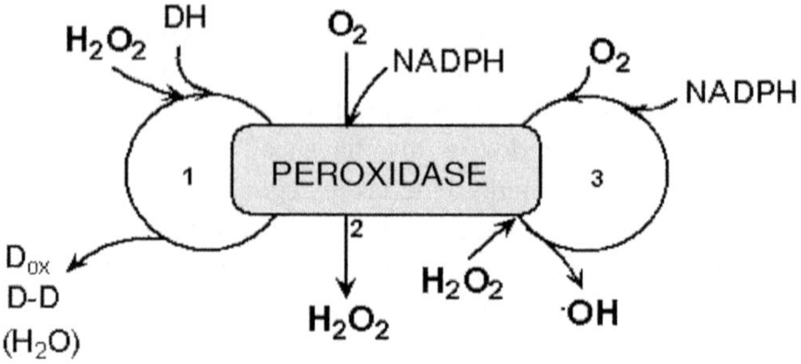

Fig. 1 The three possible catalytic pathways of peroxidases. Only the main substrates and products are shown. *1* peroxidase cycle, *2* formation of H_2O_2 from a reducing molecule and O_2, *3* formation of ·OH. NADPH any suitable reducing molecule, DH electron/hydrogen donor, D_{ox} oxidized DH, D-D polymerized DH

This reaction is due to a multistep mechanism. In the presence of physiological concentrations of H_2O_2, the native peroxidase (oxidation state +3) follows the classic peroxidase cycle in which it binds H_2O_2 and forms compound I (+5). It returns to the native state via compound II (+4) thanks to a hydrogen donor which provides the necessary electrons (Fig. 1, pathway 1). It is generally assumed that it is this activity which is exerted by peroxidases *in planta*. Actually, this is the case when peroxidases oxidatively cross-link cell wall constituents such as pectins, xyloglucans, extensin, and other structural proteins (Fry 1986), polymerize lignin and suberin (Quiroga et al. 2000), or mediate the biochemical modification of secondary metabolites in vacuoles. However, in addition to the peroxidase cycle described above, peroxidases can either react with superoxide or be reduced by molecules such as NADPH or indole-3-acetic acid into ferroperoxidase and bind molecular oxygen as does hemoglobin (Rodriguez-Lopez et al. 1997). In the two cases, the resulting complex, oxyferroperoxidase (+6), may yield ·OH in the presence of H_2O_2 (Fig. 1, pathway 3). ·OH plays a role in cell wall loosening (Liszkay et al. 2003). Oxyferroperoxidase may also be involved in H_2O_2 generation (Fig. 1, pathway 2), leading to oxidative burst (Bolwell et al. 1995). Many works have been devoted to the ability of peroxidases to produce O_2^{-}·, H_2O_2, or ·OH (Halliwell 1977; Pedreño et al. 1995). The main requirement for such a production is the simultaneous presence of O_2 and a suitable reductant promoting the formation of oxyferroperoxidase. This peroxidase-mediated reaction was observed in isolated cell wall (Elstner and Heupel 1976) and at the surface of plant cells (Vianello and Macri 1991). The reducing molecules used to show in vitro the ROS-generating ability of peroxidases are NAD(P)H (Elstner and Heupel 1976; Mäder et al. 1980; Vianello and Macri 1991), cysteine (Pichorner et al. 1992), and indole-3-acetic acid (Smith et al. 1988; Kawano 2003), but there are most likely other, yet unidentified, compounds involved in vivo in the peroxidase-mediated ROS formation (Davies et al. 2006). It appears therefore that plant peroxidases are versatile enzymes, able to react with the main ROS, but also to produce all of them. They are key players in the metabolism of these chemical species in vivo.

2 ROS Signaling

Increasing evidence indicates that ROS have signaling roles in plant. It can be speculated that, at the origin, ROS were only by-products of the aerobic metabolisms that must be efficiently removed to avoid cellular damage. Scavenging enzymes appeared and regulatory mechanisms were selected to relate ROS accumulation to the induction of these enzymes. Progressively, the regulatory mechanisms were used for other purposes and ROS became produced by dedicated enzyme systems. Many recent reviews have described these systems and the physiological processes they control (Xiong et al. 2002; Apel and Hirt 2004; Gechev et al. 2006; Pitzschke et al. 2006; Stone and Yang 2006; Veal et al. 2007; Miller et al. 2008). The increase in ROS concentration, called "oxidative burst" in some cases, triggers many responses (Vranová et al. 2002). At the molecular level, ROS provoke oxidative damages to proteins, nucleic acids, and lipids. Their interaction with proteins having a regulatory role is of particular importance. They mediate redox regulation through the reversible oxidation of essential cysteine residues to form disulfide bonds (Jones et al. 2004). The resulting conformational change activates the protein (Federoff 2006), leading to the modulation of gene expression. It seems that some transcription factors, such as heat shock transcription factors, could act as direct sensors of the presence of ROS (Miller and Mittler 2006). In addition, different ROS have distinct effects on gene transcription (Gadjev et al. 2006). H_2O_2 may also induce mitogen-activated protein kinase cascades and Ca^{2+} fluxes through plasma membrane (Mori and Schroeder 2004; Gechev and Hille 2005). Salicylic acid (Durrant and Dong 2004), nitric oxide (Delledonne et al. 1998), and ethylene (Love et al. 2005) are messengers that interact with ROS in the control of many physiological processes.

ROS, especially H_2O_2, have been shown to play a role as signaling agents in response to various kinds of biotic and abiotic stresses, including pathogen attack (Bestwick et al. 1998; Bolwell et al. 1998; Asselbergh et al. 2007; Choi et al. 2007), nutrient deprivation (Shin and Schachtman 2004), soil and air pollution (Ortega-Villasante et al. 2007), low temperatures (Cheng et al. 2007), and water deficit (Zhu et al. 2007). They play a role in wound healing (Angelini et al. 2008) and in symbiotic interactions (Jamet et al. 2007). They are also involved in programmed cell death (Gechev and Hille 2005; De Pinto et al. 2006), stomatal closure (Mori et al. 2001; Zhang et al. 2001), and elongation growth and development (Joo et al. 2001; Foreman et al. 2003; Gapper and Dolan 2006). They may have a role in the signaling mechanisms of several hormones (Kwak et al. 2006).

3 Peroxidases as ROS Producers During Physiological Processes

Peroxidases are generally viewed as enzymes executing tasks downstream of different signaling pathways. They are considered as simple workers performing tasks controlled by upstream mechanisms. The tasks they have to achieve are supposed to be only related to the modification of cell wall architecture, auxin catabolism, or the chemical transformation of secondary metabolites. However, their involvement in ROS formation and interconversion and their sophisticated redox properties suggest that they could play a more complex role in cell functioning, especially in the apoplast. Some examples of mechanisms in which peroxidases could play a signaling role are presented in the following sections.

3.1 Responses to Stresses

The responsiveness of plant peroxidases to biotic and abiotic stresses has been known for a long time. Numerous reports have described the changes in peroxidase activity following exposition to pathogens, pollutants, or physical stresses (Gaspar et al. 1982). The advent of molecular techniques allowed the analyses to be refined and show that the regulation often occurs at the transcriptional level. On the other hand, increasing evidence showed that all these detrimental events have in common that they trigger an oxidative stress, which is due to an imbalance between generation and elimination of ROS (Scandalios 2002). This oxidative stress induces the activation or downregulation of genes encoding ROS-scavenging enzymes, but also transcription factors and other proteins. It may also lead to cell death. The role of ROS as a secondary messenger between stresses and the responses of the plant cell is now obvious.

As indicated above, the fact that peroxidases are able to produce $O_2^{-\cdot}$ and H_2O_2 in the presence of O_2 and a suitable reductant has been known for many decades. The demonstration of the involvement of these enzymes in the generation of ROS in vivo during oxidative burst is more recent. Bolwell et al. (1995) observed that cells of French bean exposed to an elicitor from *Colletotrichum lindemuthianum* exhibit a rapid increase of oxygen uptake rapidly followed by H_2O_2 accumulation. An essential factor in this H_2O_2 production appears to be the alkalinization of the apoplast. A cell-wall-bound peroxidase isolated from French bean cells seemed to be responsible for the production of H_2O_2. This hypothesis was confirmed by further works using French bean cells (Bolwell et al. 1998) and *Arabidopsis* cells (Davies et al. 2006). While apoplastic alkalinization is sufficient in French bean to observe peroxidase-dependent generation of H_2O_2, a secretory component is required in *Arabidopsis*. In both cases, the identity of the strong reductant remains elusive. The involvement of peroxidases in ROS formation has been suspected in many other experimental systems: hypersensitive reaction of lettuce cells to *Pseudomonas syringae* (Bestwick et al. 1997), application of nonpurified cellulase on grape leaves (Papadakis and Roubelakis-Angelakis 1999), incompatible interaction of tobacco suspension cells inoculated with *Phytophthora nicotianae* (Able et al. 2000), defense of cotton against bacterial blight (Delannoy et al. 2003), infection of tomato leaves by *Botrytis cinerea* (Patykowski and Urbanek 2003), and many others. In all these cases, peroxidases are responsible for the accumulation of H_2O_2, alone or in cooperation with plasmalemma NADPH oxidases (Papadakis and

Roubelakis-Angelakis 1999). The H_2O_2 generated by peroxidases may also induce its further accumulation by activating NADPH oxidase and lead to the hypersensitive response and cell death (Davies et al. 2006). The formation of H_2O_2 by peroxidases in response to pathogen invasion was recently confirmed in a work on pepper leaves infected with *Xanthomonas campestris* (Choi et al. 2007). The expression of a particular peroxidase gene, *CaPO2*, is induced by the bacteria. *CaPO2*-silenced plants are highly susceptible to *X. campestris* infection, do not accumulate H_2O_2, and do not exhibit hypersensitive cell death. In contrast, the overexpression of *CaPO2* in *Arabidopsis* confers enhanced disease resistance, together with H_2O_2 accumulation, pathogenesis-related protein synthesis, and cell death.

There are several pieces of evidences indicating that specific peroxidase genes are upregulated when plant cells are challenged with pathogens or elicitors and that the encoded peroxidases are involved in the production of ROS, mainly H_2O_2. This occurs in the apoplast, but apoplastic H_2O_2 may easily migrate into the cell, through aquaporins, which may help the signaling function of H_2O_2 by facilitating its diffusion across the plasmalemma (Dynowski et al. 2008). Apoplastic H_2O_2 produced by peroxidase exerts its signaling function by activating gene expression and eventually triggering hypersensitive reaction and cell death. It can also be used locally by other peroxidases to stiffen cell wall by cross-linking polymers and structural proteins (Bruce and West 1989; Moerschbacher et al. 1990; Brisson et al. 1994; Ribeiro et al. 2006; Asselbergh et al. 2007). The formation of cell wall appositions (papillae) is another defense strategy used by plants which is dependent on the presence of H_2O_2. These appositions are structures formed with callose, pectin, lignin, and proteins between the plasma membrane and the cell wall allowing the penetration of bacteria or fungi to be blocked. In French bean mesophyll infected with *Xanthomonas campestris*, a particular peroxidase is associated with these structures and could be responsible for the formation of H_2O_2 and/or for the polymerization of the material forming papillae (Brown et al. 1998). These appositions are also associated with H_2O_2 accumulation in other experimental systems (Hückelhoven et al. 1999; An et al. 2006).

Physical and chemical factors are the source of various stresses. Salt, drought, heat, cold, and chemical pollution are accompanied by the formation of ROS which induce a secondary, oxidative, stress and may act as secondary messengers involved in the response of the plant (Wang et al. 2003). ROS formation is often attributed to proteins named "respiratory burst oxidase homolog" (Torres and Dangl 2005), but, as in the case of biotic stresses, peroxidases can also be a H_2O_2 or an $O_2^{-\cdot}$ producer. In this context, numerous reports have been published on the impact of various abiotic stresses on the level of peroxidase activity and on the number of peroxidase isoforms (Penel et al. 1992). The expression of peroxidase-encoding genes is also modified in a complex way (Kim et al. 2007). Peroxidases belong to the proteins that allow plants to withstand adverse conditions. The expression of these proteins is upregulated by specific transcription factors (Hu et al. 2008), leading to an enhanced resistance to stresses. In the maize root apical zone, six proteins are upregulated following water stress. These proteins, oxalate oxidase, SOD, and two peroxidases, are all involved in ROS production (Zhu et al. 2007).

Peroxidase induction also leads to lignification (Lee et al. 2007). Kim et al. (2008) have shown that the overexpression of a particular peroxidase gene results in increased H_2O_2 production and enhances stress tolerance. This peroxidase gene, *swpa4*, is naturally upregulated in sweet potato in response to several different environmental stresses. Transgenic tobacco expressing *swpa4* is more tolerant to many biotic or abiotic stresses and contains more H_2O_2 and a higher level of pathogenesis-related proteins. These observations provide another example of the implication of a peroxidases in a signal transduction pathway.

3.2 Growth

Overexpressing a peroxidase gene can induce a stimulation of growth. This is the case when a horseradish peroxidase is expressed in hybrid aspen (Kawaoka et al. 2003) or when an *Arabidopsis* peroxidase is overexpressed (Passardi et al. 2006). ROS and growth are closely related. In roots, the distribution of H_2O_2 and $O_2^{-\cdot}$ is different according to the zone considered (meristematic, elongating cells, differentiating cells) and this could be related to the function of ROS (Dunand et al. 2007). H_2O_2 is found in the apoplast of differentiating root cells and, to a lesser extent, in the elongating cells (Córdoba-Pedregosa et al. 2003) and $O_2^{-\cdot}$ is found mainly in the elongating cells (Dunand et al. 2007). This shows that ROS accumulate in the apoplast not only when cells are challenged with external stresses, but also in their regular functioning. Decreasing ROS concentration in the apoplast reduces tissue elongation (Rodríguez et al. 2004). The presence of H_2O_2 and $O_2^{-\cdot}$ in the apoplast most likely promotes cell elongation through the peroxidase-mediated production of $^{\cdot}OH$, which cleaves various cell wall polymers (Liszkay et al. 2003). The positive effect of peroxidases on growth observed in some transgenic plants could be explained by their role in generating $^{\cdot}OH$, by a contribution to the accumulation of H_2O_2, or both.

3.3 Stomatal Closure

Stomatal closure induced by abscisic acid is dependent on the formation H_2O_2 which induces Ca^{2+}-channel opening, leading to an increase of cytosolic Ca^{2+} concentration (Pei et al. 2000). The role of H_2O_2 as an intermediate in this mechanism has been confirmed and its generation was ascribed to plasma membrane NADPH oxidase (Wang and Song 2008). However, it has been shown that the uptake of O_2 by guard cell protoplasts may be due to peroxidases in the presence of a reductant such as NAD(P)H (Pantoja and Willmer 1988). A direct involvement of peroxidases was suggested by Mori et al. (2001). They showed that salicylic acid induces stomatal closure and the formation of $O_2^{-\cdot}$ and that these two related effects are suppressed by the peroxidase inhibitor salicylhydroxamic acid. In this case too, peroxidase could be responsible for H_2O_2 generation in parallel with NADPH oxidase.

3.4 Nodulation and Mycorrhization

ROS are involved in the establishment of both root nodules (Santos et al. 2001; D'Haeze et al. 2003) and mycorrhiza (Baptista et al. 2007). In the latter case, three peaks of H_2O_2 production were detected in the roots of *Castanea sativa* following contact with *Pisolithus tinctorius* mycelia. The first one coincides with SOD induction and catalase downregulation. The enzyme responsible for H_2O_2 formation has not been identified, but since there are three distinct peaks, different enzyme systems could be involved. It should be noted that a thorough study of gene expression in rice root during mycorrhization showed that the greatest upregulation concerned a peroxidase-encoding gene (Guimil et al. 2005). H_2O_2 is necessary for nodulation. *Medicago sativa* overexpressing a catalase-encoding gene does not form normal infection threads when infected with *Sinorhizobium meliloti* (Jamet et al. 2007). The involvement of a peroxidase was shown in pea nodulation with *Rhizobium leguminosarum*. Inhibition of peroxidases with salicylhydroxamic acid prevents the insolubilization of a matrix glycoprotein associated with the lumen of the infection threads (Wisniewski et al. 2000). The H_2O_2 necessary for this reaction seems to be produced by a diamine oxidase. In the case of *Sesbania rostrata* nodulation, a peroxidase gene is transiently upregulated during nodulation process (Den Herder et al. 2007). This induction is dependent on a bacterial nodulation factor and is modulated by H_2O_2. As a root nodule extensin colocalizes with the peroxidase, it is assumed that the former is cross-linked by the latter.

3.5 Other Processes

Angiosperm stigmas exhibit a strong peroxidase activity when they are ready to receive pollen (Dupuis and Dumas 1990). ROS, especially H_2O_2, are also massively generated in stigma papillae and their levels decrease when pollen grains adhere (McInnis et al. 2006a). The concomitant presence of H_2O_2 and peroxidases in stigma cells may be explained in two ways. Either peroxidases are involved in the generation of H_2O_2, or they use it to perform an unknown task (McInnis et al. 2006b).

Wound healing is another physiological process known to be associated with peroxidases and ROS. Peroxidases are responsible for the synthesis of lignin and suberin necessary for healing. This synthesis is H_2O_2-dependent. NADPH oxidase (Razem and Bernards 2003) and polyamine oxidase (Angelini et al. 2008) have been proposed as sources of H_2O_2 in two different experimental systems. There is no indication of the involvement of peroxidase in this process.

Finally, seed germination is usually accelerated by the presence of exogenously supplied H_2O_2 (Ogawa and Iwabuchi 2001). Peroxidase activity increases in seeds at the beginning of germination and this increase is promoted by the addition of H_2O_2 to the germination medium (Sarath et al. 2007). It is not known whether endogenous H_2O_2 is also involved in the germination process. In *Arabidopsis* seeds, there is an accumulation of $O_2^{-\cdot}$ and peroxidase at the micropylar endosperm, where the embryo radicle emerges. $O_2^{-\cdot}$ does not appear in seeds treated with the peroxidase inhibitor salicylhydroxamic acid. It is also absent in seeds of a transgenic *Arabidopsis* that does not produce a peroxidase normally present in seeds. This mutant exhibits a delay in endosperm rupture (Dunand and Penel, unpublished data).

4 Conclusion

As already noted, a signaling function may be attributed to class III peroxidases because of their ability to produce or, eventually, destroy ROS in response to various external or internal stimuli. In fact, the putative H_2O_2-scavenging role of apoplastic peroxidases has not been demonstrated. It would require the existence of a biochemical cycle regenerating an electron donor necessary for reducing H_2O_2, as is the case of ascorbate for ascorbate peroxidase (Asada 1997). It has also been suggested that extracellular peroxidases could work as catalases if H_2O_2 is present in high concentration (Mika et al. 2004), but this catalase-like activity seems to be weak (Hernández-Ruiz et al. 2001). In constrast, ROS generation by peroxidases is well established. It requires the presence of a reductant with suitable characteristics. Dihydroxyfumaric acid, indole-3-acetic acid, NADD(P)H, and sulfhydryl-containing molecules have been shown to meet these characteristics and allow peroxidases to produce H_2O_2 in vitro. The issue of the identity of the natural reductant(s) allowing peroxidases to generate $O_2^{-\cdot}$ or H_2O_2 in the apoplast is central to understanding the role of these enzymes in oxidative burst. The reducing molecule(s) should be secreted to come in contact with the peroxidases already present in the cell wall. According to the work of Davies et al. (2006) on the production of ROS by *Arabidopsis* cell suspension cultures in response to an elicitor from *Fusarium oxysporum*, this secretory component is sensitive to monensin and *N*-ethylmaleimide and insensitive to brefeldin A, suggesting the involvement of post-Golgi vesicles.

Peroxidases possess complex biochemical properties. In vivo, their action may be regulated at several levels. First, a plant contains a large range of different peroxidase proteins. These enzymes probably have different catalytic properties owing to differences in the amino acid sequences interacting with the prosthetic group and the substrate access channel (Veitch 2004). It seems that only some peroxidases are able to generate ROS (Blee et al. 2001; Kim et al. 2008). In addition to the functional variability, each peroxidase-encoding gene has its own regulatory mechanism of expression, owing to a great heterogeneity of the *cis*-acting sequences (Valério et al. 2004). The presence of a particular apoplastic enzyme at the proper site and the right time to perform its function depends on four processes: gene transcription, protein synthesis, secretion, and targeting to the site of action. Examples of control of peroxidase gene expression were given in the preceding sections. The spatial control of peroxidase secretion has been observed, for example, the accumulation of peroxidases at the site of microbe invasion (Bestwick et al. 1997), but the mechanism has not yet been elucidated. It is known that some peroxidases exhibit a strong affinity for the Ca^{2+}-mediated conformation of pectin (Carpin et al. 2001).

This could provide a mechanism to attract peroxidases to particular sites of the cell wall, where they could produce ROS (Delannoy et al. 2003). In addition, the chemical environment prevailing in the apoplast is likely to orientate peroxidases towards ROS production or H_2O_2-mediated oxidation or polymerization.

Four apoplastic enzymes can generate H_2O_2 (Table 1), making it difficult to uncover the identity of the enzyme(s) responsible for ROS generation in a given physiological process. Inhibitors such as sodium azide and salicylhydroxamic acid for peroxidases and diphenyleneiodonium for NADPH oxidases are commonly used to identify the ROS-producing enzme. Unfortunately these inhibitors have side effects (Frahry and Schopfer 1998; Ros Barceló 1998) and do not give unquestionable results. The powerful tools offered by reverse genetics have now been used to unambiguously identify peroxidases as ROS-producing enzymes in different systems (Choi et al. 2007; Kim et al. 2008).

Besides their role as ROS-producing enzymes, peroxidases are also related to the cellular signaling network through Ca^{2+} ions, because they are Ca^{2+}-containing proteins (Shiro et al. 1986), whose activity may be enhanced by this cation (Converso and Fernández 1996). In addition, it was recently shown that an extracellular peroxidase isolated from the *Euphorbia* latex has an affinity for calmodulin and is strongly activated by the simultaneous presence of calmodulin and Ca^{2+} (Mura et al. 2005). Calmodulin, a Ca^{2+}-dependent regulatory protein found mainly in cytoplasm, is also

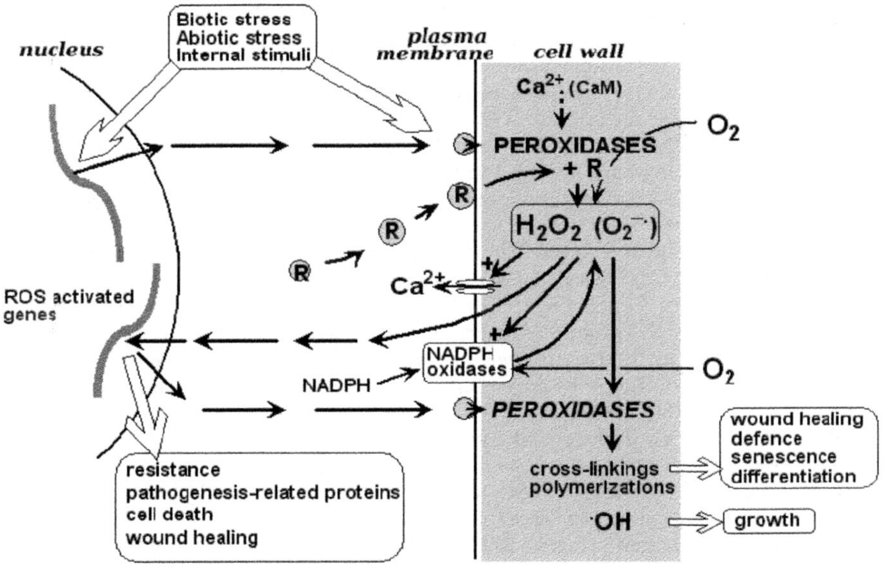

Fig. 2 Possible involvement of apoplastic peroxidases in the reactive oxygen species and Ca^{2+} signaling network. Apoplastic Ca^{2+} concentration could be necessary for peroxidase activation. In the presence of O_2 and a suitable reductant (*R*), peroxidases generate H_2O_2 (and possibly O_2^-). This H_2O_2 can diffuse into the cell and modulate gene expression, open Ca^{2+} channels, activate NADPH oxidases, or be used by other peroxidases for polymerization or cross-linking reactions. ·OH can also be produced from H_2O_2 and reduced peroxidases

present in the apoplast, where it exerts important functions (Chen et al. 2008). In *Vicia faba* guard cells, extracellular calmodulin stimulates the formation of apoplastic H_2O_2 and the increase of cytosolic Ca^{2+} concentration, leading to stomatal closure (Chen et al. 2008). It is also known that H_2O_2 induces Ca^{2+}-channel aperture and the accumulation of Ca^{2+} in the cell. More generally, ROS and Ca^{2+} are often involved as messengers in the same physiological process (Rentel and Knight 2004; Hu et al. 2007). Brought together, these observations allow us to hypothesize that, in some cases, apoplastic peroxidases activated by Ca^{2+}/calmodulin could produce H_2O_2, triggering a cascade of events in the symplast. Figure 2 recapitulates the current knowledge concerning the possible participation of apoplastic peroxidases in ROS and Ca^{2+} signaling networks. Future work will allow us to determine their exact role.

Acknowledgement Funding from the Swiss National Science Foundation (3100A0-109325) is gratefully acknowledged.

References

Able AJ, Guest DI, Sutherland MW (2000) Hydrogen peroxide yields during the incompatible interaction of tobacco suspension cells inoculated with *Phytphthora nicotianae*. Plant Physiol 124:899–910

An Q, Ehlers K, Kogel KH, van Bel AJ, Hückelhoven R (2006) Multivesicular compartments proliferate in susceptible and resistant MLA 12-barley leaves in response to infection by the biotrophic powdery mildew fungus. New Phytol 172:563–576

Angelini R, Federico R (1989) Histochemical evidence of polyamine oxidation and hydrogen peroxide generation in the cell wall. J Plant Physiol 135:212–217

Angelini R, Tisi A, Rea G, Chen MM, Botta M, Federico R, Cona A (2008) Involvement of polyamine oxidase in wound healing. Plant Physiol 146:162–177

Apel K, Hirt H (2004) Reactive oxygen species: metabolism, oxidative stress, and signal transduction. Annu Rev Plant Biol 55:373–399

Asada K (1997) The role of ascorbate peroxidase and monodehydroascorbate reductase in H_2O_2 scavenging in plants. In: Scandalios JG (ed) Oxidative stress and the molecular biology of antioxidants defenses. Cold Spring Harbor Laboratory Press, Cold Spring Harbor, pp 715–735

Asselbergh B, Curvers K, França SC, Audenaert C, Vuylsteke M, Van Breusegem F, Höfte M (2007) Resistance to *Botrytis cinerea* in *sitiens*, an abscisic acid-deficient tomato mutant, involves timely production of hydrogen peroxide and cell wall modifications in the epidermis. Plant Physiol 144:1863–1877

Azevedo AM, Martins VC, Prazeres DM, Vojinovi V, Cabral JM, Fonseca LP (2003) Horseradish peroxidase: a valuable tool in biotechnology. Biotechnol Annu Rev 9:199–247

Bach AN, Chodat R (1903) Untersuchungen über die Rolle der Peroxyde in der Chemie der lebenden Zelle. IV: Über Peroxydase. Ber Dtsch Chem Ges 36:600–605

Baptista P, Martins A, Pasi MS, Tavares RM, Lino-Neto T (2007) Involvement of reactive oxygen species during early stages of ectomycorrhiza establishment between *Castanea sativa* and *Pisolithus tinctorius*. Mycorrhiza 17:185–193

Berna A, Bernier F (1999) Regulation by biotic and abiotic stress of a wheat germin gene encoding oxalate oxidase, a H_2O_2-producing enzyme. Plant Mol Biol 39:539–549

Bestwick CS, Brown IR, Bennett MHR, Mansfield JW (1997) Localization of hydrogen peroxide accumulation during the hypersensitive reaction of lettuce cells to *Pseudomonas syringae* pv *phaseolicola*. Plant Cell 9:209–221

Bestwick CS, Brown IR, Mansfield JW (1998) Localized changes in peroxidase activity accompany hydrogen peroxide generation during the development of a nonhost hypersensitive reaction in lettuce. Plant Physiol 118:1067–1078

Blee KA, Jupe SC, Richard G, Zimmerlin A, Davies DR, Bolwell GP (2001) Molecular identification and expression of the peroxidase responsible for the oxidative burst in French bean (*Phaseolus vulgaris* L.) and related members of the gene family. Plant Mol Biol 47:607–620

Bolwell GP, Butt VS, Davies DR, Zimmerlin A (1995) The origin of the oxidative burst in plants. Free Radic Res 23:517–532

Bolwell GP, Davies DR, Gerrish C, Auh CK, Murphy TM (1998) Comparative biochemistry of the oxidative burst produced by rose and French bean cells reveals two distinct mechanisms. Plant Physiol 116:1379–1385

Brisson LF, Tenhaken R, Lamb C (1994) Function of oxidative cross-linking of cell wall structural proteins in plant disease resistance. Plant Cell 6:1703–1712

Brown I, Trethowan J, Kerry M, Mansfield J, Bolwell GP (1998) Localization of components of the oxidative cross-linking of glycoproteins and of callose synthesis in papillae formed during the interaction between non-pathogenic strain of *Xanthomonas campestris* and French bean mesophyll cells. Plant J 15:333–343

Bruce RJ, West CA (1989) Elicitation of lignin biosynthesis and isoperoxidase activity by pectic fragments in suspension cultures of castor bean. Plant Physiol 91:889–897

Carpin S, Crèvecoeur M, De Meyer M, Simon P, Penel C (2001) Identification of a Ca^{2+}-pectate binding site on an apoplastic peroxidase. Plant Cell 13:511–520

Chen Y-L, Huang R, Xiao Y-M, Lü P, Chen J, Wang X-C (2008) Extracellular calmodulin-induced stomatal closure is mediated by heterotrimeric G protein and H_2O_2. Plant Physiol 136:4096–4103

Cheng C, Yun K-Y, Ressom HW, Mohanty B, Bajic VB, Jia Y, Yun SJ, De Los Reyes BG (2007) An early response regulatory cluster induced by low temperature and hydrogen peroxide in seedlings of chilling-tolerant japonica rice. BMC Genomics 8:175–192

Choi HW, Kim YJ, Lee SC, Hong JK, Hwang BK (2007) Hydrogen peroxide generation by the pepper extracellular peroxidase CaPO2 activates local and systemic cell death and defense response to bacterial pathogens. Plant Physiol 145:890–904

Converso DA, Fernández ME (1996) Ca^{2+} activation of wheat peroxidase: a possible physiological mechanism of control. Arch Biochem Biophys 333:59–65

Córdoba-Pedregosa MC, Córdoba F, Villalba JM, González-Reyes JA (2003) Zonal changes in ascorbate and hydrogen peroxide contents, peroxidase, and ascorbate-related enzyme activities in onion roots. Plant Physiol 131:697–706

Davies DR, Bindschedler LV, Strickland TS, Bolwell GP (2006) Production of reactive species in *Arabidopsis thaliana* cell suspension cultures in response to an elicitor from *Fusarium oxysporum*: implications for basal resistance. J Exp Bot 57:1817–1827

Delannoy E, Jalloul A, Assigbetsé K, Marmey P, Geiger JP, Lherminier J, Daniel JF, Martinez C, Nicole M (2003) Activity of class III peroxidases in the defense of cotton to bacterial blight. Mol Plant Microbe Interact 16:1030–1038

Delledonne M, Xia Y, Dixon RA, Lamb CJ (1998) Nitric oxide functions a signal in plant disease resistance. Nature 394:585–588

Den Herder J, Lievens S, Rombauts S, Holsters M, Goormachtig S (2007) A symbiotic plant peroxidase involved in bacterial invasion of the tropical legume *Sesbania rostrata*. Plant Physiol 144:717–727

De Pinto MC, Paradiso A, Leonetti P, De Gara L (2006) Hydrogen peroxide, nitric oxide and cytosolic ascorbate peroxidase at the crossroad between defence and cell death. Plant J 48:784–795

D'Haeze W, De Rycke R, Mathis R, Goormachtig S, Pagnotta S, Verplancke C, Capoen W, Holsters M (2003) Reactive oxygen species and ethylene play a positive role in lateral root base nodulation of a semiaquatic legume. Proc Natl Acad Sci U S A 100:11789–11794

Doke N, Miura Y (1995) In vitro activation of NADPH-dependent O2 . generating system in a plasma membrane-rich fraction of potato tuber tissues by treatment with an elicitor from *Phytophthora infestans* or with digitonin. Physiol Mol Plant Pathol 46:17–28

Dunand C, Crèvecoeur M, Penel C (2007) Distribution of superoxide and hydrogen peroxide in *Arabidopsis* root and their influence on root development: possible interaction with peroxidases. New Phytol 174:332–341
Dupuis I, Dumas C (1990) Biochemical markers of female receptivity in maize (*Zea mays* L.) assessed using *in vitro* fertilization. Plant Sci 70:11–20
Durrant WE, Dong X (2004) Systemic acquired resistance. Annu Rev Phytopathol 42:185–209
Dynowski M, Schaaf G, Loque D, Moran O, Ludewig U (2008) Plant plasma membrane water channels conduct the signaling molecule H_2O_2. Biochem J. doi:10.1042/BJ20080287
Elstner EF, Heupel AL (1976) Formation of hydrogen peroxide by isolated cell walls from horseradish (*Armoracia lapathifolia* Gilib.). Planta 130:175–180
Federoff N (2006) Redox regulatory mechanisms in cellular stress responses. Ann Bot 98:289–300
Foreman J, Demidchik V, Bothwell JHF, Mylona P, Mledema H, Torres MA, Linstead P, Costa S, Brownlee C, Jones JDG, Davies JM, Dolan L (2003) Reactive oxygen species produced by NADPH oxidase regulate plant cell growth. Nature 422:442–446
Frahry G, Schopfer P (1998) Inhibition of O_2-reducing activity of horseradish peroxidase by diphenyleneiodonium. Phytochemistry 48:223–227
Fry SC (1986) Cross-linking of matrix polymers in the growing cell walls of angiosperms. Annu Rev Plant Physiol 37:165–186
Gadjev I, Vanderauwera S, Gechev TS, Laloi C, Minkov IN, Shulaev V, Apel K, Inzé D, Mittler R, Van Breusegem F (2006) Transcriptomic footprints disclose specificity of reactive oxygen species signaling in Arabidopsis. Plant Physiol 141:436–445
Gapper C, Dolan L (2006) Control of plant development by reactive oxygen species. Plant Physiol 141:341–345
Gaspar T, Penel C, Thorpe T, Greppin H (1982) Peroxidase 1970–1980. A survey of their biochemical and physiological roles in higher plants. University of Geneva, Geneva
Gechev TS, Hille J (2005) Hydrogen peroxide as a signal controlling plant programmed cell death. J Cell Biol 168:17–20
Gechev TS, Van Breusegem F, Stone JM, Denev I, Laloi C (2006) Reactive oxygen species as signals that modulate plant stress response and programmed cell death. Bioessays 28:1091–1101
Guimil S, Chang H-S, Zhu T, Sesma A, Roux C, Ioannidis V, Oakeley EJ, Docquier M, Descombes P, Briggs SP, Paszkowski U (2005) Comparative transcriptomics of rice reveals an ancient pattern of response to microbial colonization. Proc Natl Acad Sci U S A 102:8066–8070
Halliwell B (1977) Generation of hydrogen peroxide, superoxide and hydroxyl radicals during the oxidation of dihydroxyfumaric acid by peroxidase. Biochem J 163:441–448
Hernández-Ruiz J, Arnao MB, Hiner ANP, García-Cánovas F, Acosta M (2001) Catalase-like activity of horseradish peroxidase: relationship to enzyme inactivation by H_2O_2. Biochem J 354:107–114
Hu X, Jiang M, Zhang A, Lin F, Tan M (2007) Calcium-calmodulin is required for abscisic acid-induced antioxidant defense and functions both upstream and downstream of H_2O_2 production in leaves of maize (*Zea mays*) plants. New Phytol 173:27–38
Hu H, You J, Fang Y, Zhu X, Qi Z, Xiong L (2008) Characterization of transcription factor gene *SNAC2* conferring cold and salt tolerance in rice. Plant Mol Biol. doi:10.1007/s11103–008–9309–5
Hückelhoven R, Fodor J, Preis C, Kogel K-H (1999) Hypersensitive cell death and papilla formation in barley attacked by the powdery mildew fungus are associated with hydrogen peroxide but not with salicylic acid accumulation. Plant Physiol 119:1251–1260
Jamet A, Mandon K, Pippo A, Hérouart D (2007) H_2O_2 is required for optimal establishment of the *Medicago sativa/Sinorhizobium meliloti* symbiosis. J Bacteriol 187:8741–8745
Jones DP, Go Y-M, Andersen CL, Ziegler TR, Kinkade JM, Kirlin WG (2004) Cysteine/cystine couple is a newly recognized node in the circuitry for biologic redox signaling and control. FASEB J. doi:10.1096/fj.03–0971fje
Joo JH, Bae YS, Lee JS (2001) Role of auxin-induced reactive oxygen species in root gravitropism. Plant Physiol 126:1055–1060
Kawano T (2003) Roles of the reactive oxygen species-generating peroxidase reactions in plant defense and growth induction. Plant Cell Rep 21:829–837

Kawaoka A, Matsunaga E, Endo S, Kondo S, Yoshida K, Shinmyo A, Ebinuma H (2003) Ectopic expression of a horseradish enhances growth rate and increases oxidative stress resistance in hybrid aspen. Plant Physiol 132:1177–1185

Key LE, Basile DV (1987) Specific peroxidase isoenzymes are correlated with organogenesis. Plant Physiol 84:99–105

Kim Y-H, Lim S, Han S-H, Lee J-C, Song W-K, Bang J-W, Kwon S-Y, Lee H-S, Kwak S-S (2007) Differential expression of 10 sweetpotato peroxidases in response to sulfur dioxide, ozone, and ultraviolet radiation. Plant Physiol Biochem 45:908–914

Kim Y-H, Kim CY, Song WK, Park D-S, Kwon S-Y, Lee H-S, Bang J-W, Kwak S-S (2008) Overexpression of sweetpotato *swpa4* peroxidase results in increased hydrogen peroxide production and enhances stress tolerance in tobacco. Planta. doi:10.1007/s00425–007–0663–3

Kliebenstein DJ, Monde R-A, Last RL (1998) Superoxide dismutase in Arabidopsis: an eclectic enzyme family with disparate regulation and protein localization. Plant Physiol 118:637–650

Kwak JM, Nguyen V, Schroeder JI (2006) The role of reactive oxygen species in hormonal responses. Plant Physiol 141:323–329

Lee B-R, Kim K-Y, Jung W-J, Avice J-C, Ourry A, Kim T-H (2007) Peroxidases and lignification in relation to the intensity of water-deficit stress in white cover (*Trifolium repens* L.). J Exp Bot 58:1271–1279

Liszkay A, Kenk B, Schopfer P (2003) Evidence for the involvement of cell wall peroxidase in the generation of hydroxyl radicals mediated extension growth. Planta 217:658–667

Love AJ, Yun B-W, Lava LV, Loake GJ, Milner JL (2005) *Cauliflower mosaic virus*, a compatible pathogen of Arabidopsis, engages three distinct defense-signaling pathways and activates rapid systemic generation of reactive oxygen species. Plant Physiol 139:935–948

Mäder M, Ungemach J, Schloss P (1980) The role of peroxidase isoenzyme groups of *Nicotiana tabacum* in hydrogen peroxide formation. Planta 147:467–470

McInnis SM, Desikan R, Hancock JT, Hiscock SJ (2006a) Production of reactive oxygen species and reactive nitrogen species by angiosperm stigmas and pollen: potential signalling crosstalk? New Phytol 172:221–228

McInnis SM, Emery DC, Porter R, Desikan R, Hancock JT, Hiscock SJ (2006b) The role of stigma peroxidases in flowering plants: insights from further characterization of a stigma-specific peroxidase (SSP) from *Senecio squalidus* (Asteraceae). J Exp Bot 57:1835–1846

Mika A, Minibayeva F, Beckett R, Lüthje S (2004) Possible function of extracellular peroxidases in stress-induced generation and detoxification of active oxygen species. Phytochem Rev 3:173–193

Miller G, Mittler R (2006) Could heat shock transcription factors function as hydrogen peroxide sensors in plants? Ann Bot 98:279–288

Miller G, Shulaev V, Mittler R (2008) Reactive oxygen signaling and abiotic stress. Physiol Plant. doi:10.1111/j.1399–3054.2008.01090.x

Moerschbacher BM, Noll U, Gorrichon L, Reisener H-J (1990) Specific inhibition of lignification breaks hypersensitive resistance of wheat to stem rust. Plant Physiol 93:465–470

Mori IC, Schroeder JI (2004) Reactive oxygen species activation of plant Ca^{2+} channels. A signaling mechanism in polar growth, hormone transduction, stress signaling, and hypothetically mechanotransduction. Plant Physiol 135:702–708

Mori IC, Pinontaon R, Kawano T, Muto S (2001) Involvement of superoxide generation in salicylic acid-induced stomatal closure in *Vicia faba*. Plant Cell Physiol 42:1383–1388

Mura A, Medda R, Longu S, Floris G, Rinaldi AC, Padiglia A (2005) A Ca^{2+}/calmodulin binding peroxidase from *Euphorbia* latex: novel aspects of calcium-hydrogen peroxide cross-talk in the regulation of plant defenses. Biochemistry 44:14120–14130

Ogawa K, Iwabuchi M (2001) A mechanism for promoting the germination of *Zinnia elegans* seeds by hydrogen peroxide. Plant Cell Physiol 42:286–291

Ortega-Villasante C,Hernández LE, Rellán-Álvarez R, Del Campo FF, Carpena-Ruiz RO (2007) Rapid alteration of cellular redox homeostasis upon exposure to cadmium and mercury in alfalfa seedlings. New Phytol 176:96–107

Pantoja O, Willmer CM (1988) Redox activity and peroxidase activity associated with the plasma membrane of guard-cell protoplasts. Planta 174:44–50

Papadakis AK, Roubelakis-Angelakis KA (1999) The generation of active oxygen species differs in tobacco and grapevine mesophyll protoplasts. Plant Physiol 121:197–205
Passardi F, Longet D, Penel C, Dunand C (2004) The class III peroxidase multigenic family in rice and its evolution in land plants. Phytochemistry 65:1879–1893
Passardi F, Tognolli M, De Meyer M, Penel C, Dunand C (2006) Two cell wall peroxidase from *Arabidopsis* influence root elongation. Planta 223:965–974
Patykowski J, Urbanek H (2003) Activity of enzymes related to H_2O_2 generation and metabolism in leaf apoplastic fraction of tomato leaves infected with *Botrytis cinerea*. J Phytopathol 151:153–161
Pedreño MA, Ferrer MA, Gaspar T, Muñoz R, Ros Barceló A (1995) The polyfunctionality of cell wall peroxidases avoids the necessity of an independent H_2O_2-generating system for phenolic coupling in the cell wall. Plant Perox Newsl 5:3–8
Pei Z-M, Murata Y, Benning G, Thomine S, Klusener B, Allen GJ, Grill E, Schroeder JI (2000) Calcium channels activated by hydrogen mediate abscisic acid signaling in guard cells. Nature 406:731–734
Penel C, Gaspar T, Greppin H (1992) Plant peroxidases 1980–1990. Topics and detailed literature on molecular, biochemical, and physiological aspects. University of Geneva, Geneva
Pichorner H, Couperus A, Korori SAA, Ebermann R (1992) Plant peroxidase has a thiol oxidase function. Phytochemistry 31:3371–3376
Pitzschke A, Forzani C, Hirt H (2006) Reactive oxygen species signaling in plants. Antiox Redox Signal 8:1757–1764
Quiroga M, Guerrero C, Botella MA, Barceló A, Amaya I, Medina MI, Alonso FJ, de Forchetti SM, Tigier H, Valpuseta V (2000) A tomato peroxidase involved in the synthesis of lignin and suberin. Plant Physiol 122:1119–1127
Razem FA, Bernards MA (2003) Reactive oxygen species production in association with suberization: evidence for an NADPH-dependent oxidase. J Exp Bot 54:935–941
Rentel MC, Knight MR (2004) Oxidative stress-induced calcium signaling in Arabidopsis. Plant Physiol 135:1471–1479
Ribeiro JM, Pereira CS, Soares NC, Vieira AM, Feijo JA, Jackson PA (2006) The contribution of extensin network formation to rapid, hydrogen peroxide-mediated increases in grapevine callus wall resistance to fungal lytic enzymes. J Exp Bot 57:2025–2035
Rodriguez-Lopez JN, Smith AT, Thorneley RNF (1997) Effect of distal cavity mutations on the binding and activation of oxygen by ferrous horseradish peroxidase. J Biol Chem 272:389–395
Rodríguez AA, Córdoba AR, Ortega L, Taleisnik E (2004) Decreased reactive oxygen species concentration in the elongation zone contributes to the reduction in maize leaf grown under salinity. J Exp Bot 55:1383–1390
Ros Barceló A (1998) Use and misuse of peroxidase inhibitors. Trends Plant Sci 3:418
Santos R, Herouart D, Sigaud S, Touati D, Puppo A (2001) Oxidative burst in alfalfa-*Sinorhizobium meliloti* symbiotic interaction. Mol Plant-Microbe Interact 14:86–89
Sarath G, Hou G, Baird LM, Mitchell RB (2007) Reactive oxygen species, ABA and nitric oxide interactions on the germination of warm-season C4-grasses. Planta 226:697–708
Scandalios JG (1974) Subcellular localization of catalase variants coded by two genetic loci during maize development. J Hered 65:28–32
Scandalios JG (2002) Oxidative stress responses - what have genome-scale studies taught us? Gen Biol 3:1019.1–1019–6
Shao H-B, Chu L-Y, Lu Z-H, Kang C-M (2008) Primary antioxidant free radical scavenging and redox signaling in higher plant cells. Int J Biol Sci 4:8–14
Shin R, Schachtman DP (2004) Hydrogen peroxide mediates plant root cell response to nutrient deprivation. Proc Natl Acad Sci U S A 101:8827–8832
Shiro Y, Kurono M, Morishima I (1986) Presence of endogenous calcium ion and its functional and structural regulation in horseradish peroxidase. J Biol Chem 261:9382–9390
Smith AM, Morrison WL, Milham PJ (1988) Oxidation of indole-3-acetic acid by peroxidase: involvement of reduced peroxidase and compound III with superoxide as a product. Biochemistry 21:4414–4419

Stone JR, Yang S (2006) Hydrogen peroxide: a signaling messenger. Antioxid Redox Signal 8:243–270
Tognolli M, Overney S, Penel C, Greppin H, Simon P (2000) A genetic and enzymatic survey of *Arabidopsis thaliana* peroxidases. Plant Perox Newsl 14:3–12
Tognolli M, Penel C, Greppin H, Simon P (2002) Analysis and expression of the class III peroxidase large gene family in *Arabidopsis thaliana*. Gene 288:129–138
Torres MA, Dangl JL (2005) Functions of the respiratory burst oxidase in biotic interactions, abiotic stress and development. Curr Opin Plant Biol 8:397–403
Valério L, De Meyer M, Penel C, Dunand C (2004) Expression analysis of the *Arabidopsis* peroxidase multigene family. Phytochemistry 65:1331–1342
Veal EA, Day AM, Morgan BA (2007) Hydrogen peroxide sensing and signaling. Mol Cell 26:1–14
Veitch NC (2004) Structural determinants of plant peroxidase function. Phytochem Rev 3:3–18
Vianello A, Macri F (1991) Generation of superoxide and hydrogen peroxide at the surface of plant cells. J Bioenerg Biomembr 23:409–423
Vranová E, Inzé D, Van Breusegem F (2002) Signal transduction during oxidative stress. J Exp Bot 53:1227–1236
Wang P, Song C-P (2008) Guard-cell signalling for hydrogen peroxide and abscisic acid. New Phytol 178:703–718
Wang W, Vinocur B, Altman A (2003) Plant response to drought, salinity and extreme temperatures: towards genetic engineering for stress tolerance. Planta 218:1–14
Welinder KG (1992) Superfamily of plant, fungal and bacterial peroxidases. Curr Opin Struct Biol 2:388–393
Wisniewski J-P, Rathbun EA, Knox JP, Brewin NJ (2000) Involvement of diamine oxidase and peroxidase in insolubilization of the extracellular matrix: implications for pea nodule initiation by *Rhizobium leguminosarum*. Mol Plant Microbe Interact 13:413–420
Xiong L, Schumaker KS, Zhu J-K (2002) Cell signaling during cold, drought, and salt stress. Plant Cell 14:S165–S183
Zhang X, Zhang L, Dong F, Gao J, Galbraith DW, Song CP (2001). Hydrogen peroxide is involved in abscisic acid-induced stomatal closure in *Vicia faba*. Plant Physiol 126:1438–1448
Zhu J, Alvarez S, Marsh EL, LeNoble ME, Cho I-J, Sivaguru M, Chen S, Nguyen HT, Wu Y, Schachtman P, Sharp RE (2007) Cell wall proteome in the maize primary root elongation zone. II. Region-specific changes in water soluble and lightly ionically bound proteins under water deficit. Plant Physiol 145:1533–1548

Signaling and Cell Walls

Ewelina Rodakowska, Marta Derba-Maceluch, Anna Kasprowicz, Paweł Zawadzki, Agnieszka Szuba, Daniel Kierzkowski, and Przemysław Wojtaszek

Abstract Cell walls are dynamic entities providing the link between cellular and organismal features of the plant. Being external to protoplasts, they provide not only the transportation pathways for signaling molecules, but also constitute a rich source of such molecules. Moreover, their spatial placement also makes them a specific "zone of the first contact" with the surrounding environment. Here, we provide a brief overview of the functions the walls play in plant signaling. We describe the physicochemical properties of the walls and discuss their impact on the type of signaling molecules transported via the apoplast, and the types of extracellular domains of receptor molecules. The role of the walls in the formation and maintenance of the structural and functional continuum between cell walls, the plasma membrane, and the cytoskeleton is then considered, especially with respect to mechanosensing, transduction of mechanical signals, and monitoring of cell wall integrity. Finally, a range of signaling molecules and their activities are reviewed.

1 Introduction

The contemporary view of the plant cell wall is not only that of a structural element and a physical barrier between the cell and its immediate surroundings, but also that of a physiologically active and dynamic cellular compartment. This responsive and adaptive subcellular region constitutes a pathway for and a source of signaling molecules, enabling communication with its own protoplast, surrounding environment, and other cells. Cell walls not only form a part of a cell wall–plasma

E. Rodakowska, M. Derba-Maceluch, A. Kasprowicz, P. Zawadzki, D. Kierzkowski, and P. Wojtaszek (✉)
Department of Molecular and Cellular Biology, Faculty of Biology, Adam Mickiewicz University, Umultowska 89, 61–614 Poznań, Poland
e-mail: przemow@amu.edu.pl

A. Szuba and P. Wojtaszek
Institute of Bioorganic Chemistry, Polish Academy of Sciences, Noskowskiego 12/14, 61-704 Poznań, Poland

F. Baluška and S. Mancuso (eds.), *Signaling in Plants*, Signaling and Communication in Plants,
DOI: 10.1007/978-3-540-89228-1_9,

membrane–cytoskeleton (WMC) continuum, integrating the cells structurally and functionally, but also form part of a supracellular structure – the apoplast (Wojtaszek 2000). Through provision of mechanical strength, walls define cell shapes and thus plant overall morphology. On the other hand, walls have to be plastic and dynamic enough to enable smooth execution of developmental programs and responses to changes in the environment. Although changes in cell wall composition are usually tightly controlled by the protoplast, they also have some reciprocal functions as the state of wall chemical and mechanical properties constitutes a source of information affecting cell fate. Thus, it seems that plant cell walls also have a role in integration of and participating in complex signaling pathways. Owing to the space limitations of this volume, only some aspects will be covered here. The reader is directed to several excellent reviews which have been published in recent years (Kohorn 2000; Brownlee 2002, Pilling and Höftc 2003; Fleming 2005).

2 Chemical and Structural Features of the Walls

Although the apoplast represents approximately 5% of the total plant volume, the properties of the constituting walls affect strongly its transport abilities, limiting the range of molecules able to move with respect to their size, charge, and relative hydrophilicity (Brett and Waldron 1996). Moreover, local and domain-like differentiation of the chemical composition of the walls seems to be crucial for the dynamic generation of specific signaling molecules at the right place and at the right time.

2.1 *Cell Walls as Molecular Sieves and pH-Stats*

The structure of the walls is currently presented as a set of several intertwining networks forming a composite material with remarkable mechanical and physiological properties. For the most part plant walls are built from polysaccharides, supplemented with proteins and phenolic compounds, mainly lignins. Typical walls of a higher-plant species are built from three major classes of polysaccharides: cellulose, hemicellulose, and pectins (Popper 2008). It is believed that the pectic network is the major determinant of the pore size, creating effectively a kind of molecular sieve. This is mainly due to the formation of pectic Ca^{2+} bridges and rhamnogalacturonan II–boron bridges. These pores have been estimated to be on the order of 3–5 nm in diameter (Fleischer et al. 1999), which for proteins translates into the 10–50-kDa limit. Thus, ions, sucrose, peptides, and small organic compounds, such as phytohormones and amino acids, can diffuse through the wall. Additionally, it is assumed that owing to the presence of a high proportion of uronic acids in pectins, the walls bear a net negative charge. The pH of the walls is usually in the range 4.5–6.5 and depends on the species (Canny 1995). The net charge invokes electrostatic interactions with mobile molecules, affecting also the speed of their movement.

2.2 Water Content and Distribution

Although in general cell walls are reasonably well hydrated, with water content reaching 70%, the distribution of water molecules is not even throughout them. The cellulose–hemicellulose network forms a relatively "water-free" area, enabling the creation of strong hydrophobic interactions within and between the components of the network leading to, e.g., the paracrystalline structure of cellulose microfibrils. Water is distracted from this network by the pectic matrix, which, in effect, leads to the formation of a mosaic pattern of water distribution with "water-rich" pectic channels.

2.3 Apoplastic Domains of Receptor Molecules

Physical or chemical stimuli, either crossing the walls or transported through them, finally arrive at the protoplast surface. Here, they are either directly transmitted, e.g., via the WMC continuum, or perceived by receptors that transduce those signals to the cellular interior. Plant receptor-like kinases (RLKs) play fundamental roles in this process. RLKs form a large gene family in *Arabidopsis thaliana* with more than 610 members, among which 417 are characterized by the full receptor configuration, i.e., the presence of a ligand-binding ectodomain, a transmembrane domain, and a cytoplasmic kinase domain. The kinase domain usually phosphorylates serine and threonine residues, although some RLKs can also phosphorylate tyrosine residues. The extracellular domains of RLKs are quite divergent and can be classified into more than 21 structural classes. This diversity suggests the involvement of RLKs in the perception of a wide range of signals. However, as yet the biological function is known for less than half of the RLK classes (Shiu and Bleecker 2001).

2.3.1 Leucine-Rich Repeat

In the *Arabidopsis* genome, leucine-rich repeat (LRR) RLKs are coded by approximately 216 genes, constituting the largest group within RLK family. LRRs mostly mediate protein–protein interaction and are also involved in ligand binding. Sequence analysis suggested the existence of at least seven different domain organizations with variation in the number and the structure of the repeats. It is believed that a repetitive structure may be beneficial in processes in which the rapid generation of new variants is required, such as in plant disease resistance (Kobe and Kajava 2001).

2.3.2 Carbohydrate-Binding Motifs

Domains specialized for carbohydrate binding (Lysin motif, L-lectin, C-lectin, or PAN-AP subfamily of APPLE domain) could be found in many RLKs and together they represent the second largest class of external receptor RLK domains. The ability to bind wall components makes this group an ideal candidate for monitoring cell wall integrity and plant defense responses. For example, L-lectins are involved in recognition of pathogen glycans (Senchou et al. 2004), while the LysM receptor recognizes rhizobial Nod factors (Madsen et al. 2003). Domains such as thaumatin and chitinase-related domains are implicated in binding of the components of fungal cell walls.

2.3.3 Wall-Binding Motifs

Several RLK domains contain motifs which suggest their putative involvement in creating physical linkages between the protoplast and the cell wall. These include wall-associated kinases (WAKs; Kohorn 2001), containing epidermal growth factor repeats, and proline-rich extensin-like receptor kinases (Haffani et al. 2006). The *Arabidopsis* WAKs are encoded by five genes, while the proline-rich extensin-like receptor kinases are encoded by 15 genes. Biochemical and immunological analyses showed that external WAK domains bind directly to pectins (Decreux and Messiaen 2005; Decreaux et al. 2006). It is suggested that this physical linkage is necessary for the WAK function as a regulator of cell expansion (Reise et al. 2003).

2.4 Plasmodesmata as Wall Communication Channels

Plasmodesmata cross the walls and connect cytoplasms of neighboring cells, allowing for cell-to-cell transport of various solutes, among them proteins and RNA (Heinlein and Epel 2004). Such trafficking plays a critical role in intercellular signaling, enabling coordination of differentiation and development of cells located in various symplastically isolated domains, e.g., epidermis. Maintenance of symplastic continuity is under strict spatial and temporal control (Rinne and van der Schoot 1998; Gisel et al. 1999). Plasmodesmata are very dynamic structures that transiently oscillate between closed, open, and dilated states (Zambryski and Crawford 2000). Regulation of the plasmodesmata state is tightly interrelated with the functioning of the WMC continuum. Interactions between callose, callose synthases, and plant-specific myosin of class VIII, F-actin, and membranes of the endoplasmic reticulum occur particularly at plasmodesmata and pit fields (Baluška et al. 2003; Golomb et al. 2008). Plasmodesmal transport is affected by the state of actin as stabilization of filaments inhibits the flow rate through plasmodesmata, whereas disruption of filaments increases it (Ding et al. 1996). The composition of cell walls in plasmodesmata-rich areas is highly specific. Owing to an increase of callose and nonesterified pectins content, hemicellulose and cellulose content is reduced (Turner et al. 1994; Roy et al. 1997). Callose can be deposited on both sides of the tunnel, and also inside during dormancy (Radford et al. 1998). There is evidence (Levy et al. 2007) that callose-degrading β-1,3-glucanases are associated with plasmodesmata and can function in their gating.

3 Supracellular Determinants of Cell Wall Signaling to the Protoplast

3.1 Cell Walls and Mechanosensing

Plants have developed adaptive strategies allowing them to survive in harsh environments, maintaining proper osmotic conditions and withstanding gravitational force. Turgidity and growth of plant cells depend on the tight interactions between the protoplast's turgor pressure and the mechanical properties of the wall (Cosgrove 2005). In the terms of mechanics, plant protoplast functions like a hydrostat, exerting a circumferential tensile force and a radial compressive force within the plasma membrane, and pressing against the surrounding cell wall. The wall–protoplast interactions are particularly important for mechanical stabilization of the cells subjected to rapidly changing environmental conditions, especially during the plasmolytic cycle (Lang-Pauluzzi and Gunning 2000). The forces generated at the cell wall–plasma membrane interface are transduced onto the cytoskeleton, making the plant cell a tensegral system (Ingber 2003). The physical forces generated in such a system are sufficient to overcome gravitational force and allow for plant growth and proper water distribution over the whole plant body (Niklas 1992).

In suspension-cultured cells the mechanical stimuli are not transduced to neighboring cells, whereas in plants changes in wall mechanics and/or turgor pressure of individual cells affect the shape of adjacent cells, and, in consequence, their functioning. The plant tissue organization is a result of coordinated alignment of the planes of cell divisions. In contrast, growth of calluses is usually disordered. Application of an external mechanical force could restore coordinated cell divisions within the callus. This suggests that (1) plant protoplasts are able to recognize vectorial information provided by physical forces and (2) tissue organization *in planta* is the result of a proper mechanical environment (Lintilhac and Vesecky 1984; Lynch and Lintilhac 1997). It is assumed that in axial plant organs such "tensile skin" is provided by the outer walls of epidermal cells (Kutschera 2008). These walls are significantly thicker (up to 10 times), and more resistant to mechanical stress than the walls of other cells. In effect they form a rigid cylinder inside which other cells are tightly packed. This creates an additional supracellular pressure allowing the inner cells of the organ to sense and adjust to the immediate mechanical environment as well as to transduce mechanical signals throughout the organ (Kutschera 1995, 2008). Such supracellular features enable plants to function also as very sensitive detectors of mechanical stimuli. Even subtle, short touch can induce not only mechanisms of signal transduction, but also immediate wall remodeling as evidenced by an induced expression of the TCH4 gene coding for xyloglucan endotransglycosidase (Braam and Davies 1990; Braam 2005).

Plant growth and form are optimized to meet the biomechanical properties of cell walls (Green 1999). It is assumed that a protoplast's turgor is a growth propeller, whereas growth directionality is determined by differentiated mechanics of various wall domains surrounding the protoplast, and the coordinated cooperation of the actin and tubulin cytoskeleton (Panteris and Galatis 2005). Localized, mechanically weak wall domains are created through either directional deposition of additional polymers into surrounding wall domains or localized release of loosening proteins, such as expansins. Expansins are highly specific proteins that reversibly break hydrogen bonds between cellulose microfibrils and hemicelluloses (reviewed by McQueen-Mason and Rochange 1999; Sampedro and Cosgrove 2005). Their action provides a driving mechanism for both intrusive growth of differentiating xylem cells (Im et al. 2000), and formation of root hairs (Baluška et al. 2000) or leaf primordia (Fleming et al. 1997).

The biochemical composition of cell walls is an important determinant of their function in mechanosignaling. In *Arabidopsis*, the cell wall polysaccharide composition is a key feature of particular cell types and tissues (Somerville et al. 2004), and depends on the growth and developmental stage (Knox et al. 1990). For example, pectin distribution is not random, as homogalacturonan is preferentially located in the middle lamella, and rhamnogalacturonan II is located close to the plasma membrane (Williams et al. 1996). The pattern of cellular growth can be dependent on changes in wall composition (Suzuki et al. 1990) or spatial wall organization (DeWitt et al. 1999). Changes in mechanical properties of the walls can also affect the functioning of the whole cell, especially the organization of the actin cytoskeleton. Selective disruption of wall polysaccharides or covalent interactions between wall components causes changes in the architecture and dynamics of the actin cytoskeleton, and leads to disturbances of the mechanical stability of the whole cell (Wojtaszek et al. 2007). As such alterations are not isotropic, the plausible explanation points to differential composition and organization of polysaccharides in various wall domains. Another factor to be considered is the formation and maintenance of tightly controlled connections between the cell wall and the protoplast. *Chrysanthemum* suspension cells subjected to an external mechanical force react by elongating perpendicularly to the vector of the force applied, and this process depends on the proper structure of the WMC continuum (Zhou et al. 2007). Another interesting experimental model involves cellular reactions to osmotic conditions as, e.g., hypertonicity introduces tension into the WMC continuum. This immediately induces signaling pathways, and the involvement of mechanosensitive stretch-activated Ca^{2+} ion channels (Hayashi et al. 2006) and phospholipid signaling (Zonia and Munnik 2004) has already been demonstrated.

3.2 Monitoring of Cell Wall Integrity

Maintenance of cell wall integrity or regulation of modifications of the wall's composition or assembly to coordinate growth and development requires the generation and perception of signals providing information on the state of the wall. Studies of transgenic lines or mutants impaired in the synthesis of wall components demonstrated clearly that defects in the wall structure/properties activate specific compensatory reactions of the protoplast. In poplar, repression of lignin biosynthesis led to the accumulation of cellulose (Hu et al. 1999). On the other hand, *Arabidopsis* plants defective in cellulose synthesis, owing to either mutations in cellulose synthase (CesA) genes or chemical inhibitor treatment, display changes in wall polysaccharide composition and/or ectopic lignin deposition (reviewed by Hématy and Höfte 2008). In *Saccharomyces cerevisiae* complex signaling pathways were identified that are involved in the monitoring of cell wall integrity during growth and development, and in responses to wall stressors, such as elevated temperature, osmotic conditions, and pathogen attack. The yeast cell wall integrity pathway comprises a family of protein cell surface sensors (Wsc1–3, Mid2, Mtl1) with periplasmic ectodomain. These detect and transmit signals to a small G-protein, Rho1, and further to effectors, such as the protein kinase C-mediated mitogen-activated protein kinase pathway, responsible for the amplification of the signal from Rho1. The yeast cell wall integrity pathway is linked to other signaling pathways, e.g. regulating the actin cytoskeleton architecture, and in this way correlating cellular processes such as cell wall biosynthesis and cell polarization (reviewed by Gustin et al. 1998; Heinisch et al. 1999; Levin 2005). In contrast, wall integrity sensing mechanisms in plants are poorly understood. A plasma membrane RLK, *THESEUS*, has been identified in *A. thaliana* that could be a player in sensing wall integrity and activation of the signaling pathway only in response to cellulose deficiency. Mutation in the *THE* locus attenuated the compensatory effect in some CesA mutants without changing the cellulose content, while its overexpression induced lignin accumulation and phenotypical changes (Hématy et al. 2007). On the other hand, some data indicate that monitoring of wall integrity might be tightly coupled to osmosensing (Marshall and Dumbroff 1999; Nakagawa and Sakurai 2001).

3.3 WMC Continuum: Potential Linkers

The existence of the WMC continuum in plants, analogous to that found in animals, has been proposed relatively recently (Wyatt and Carpita 1993). WMC functions, and especially interactions between its component macromolecules, have an immediate bearing on the mechanosensing abilities of plant cells. Many potential proteinaceous linkers have been proposed, but as yet direct and fully convincing evidence has been provided for none of them. This topic has been recently covered elsewhere (Wojtaszek 2000; Baluška et al. 2003), allowing us to limit ourselves to several remarks here.

As extracellular domains of WAKs are embedded in a pectin matrix (see Sect. 2.3.3), these proteins are thought to function as signal transmitters from the wall to cytoplasm. Their cytoplasmic active serine/threonine kinase domain physically interacts with KAPP – a type 2C protein phosphorylase (Anderson et al. 2001). Despite this suggested role, there is still lack of direct evidence for their participation in signal transduction. *Arabidopsis* mutants with decreased level of WAKs messenger RNA displayed a dwarf phenotype that could be rescued by treatment with gibberellins, indicating some link with the gibberellin signaling pathway in regulation of cell expansion. Some interesting explanations for dwarfism in WAK mutants come from the analysis of *Arabidopsis wak2* mutants. These plants are also impaired in the activity of vacuolar invertase involved in the regulation of solute concentration in cells. A tempting suggestion is that WAK2 senses the state of cell expansion and modulates accordingly invertase activity to adjust cell turgor (Kohorn et al. 2006).

Many exocellular proteins are anchored to the plasma membrane by the glycosylphosphatidylinositol (GPI) moiety and these could also be strong candidates to transmit signals within the WMC continuum. Most probably the signaling potential of these proteins arises from the possibility of the GPI anchor cleavage that could generate both extracellular and intracellular signal molecules (Borner et al. 2003). Among GPI-anchored proteins, putative signaling function has been already ascribed to arabinogalactan proteins (AGPs). AGPs are involved in cell division, expansion, and differentiation, root development, somatic embryogenesis, pollen tube growth, and programmed cell death (reviewed by Majewska-Sawka and Nothnagel 2000; Seifert and Roberts 2007). Other feasible linkers within the WMC continuum are formins (Deeks et al. 2002) and lectin receptor kinases (Gouget et al. 2006). The association of mechanosensory (stretch-activated) calcium-selective ion channels with AGPs and WAKs (Gens et al. 2000) suggests other possible ways of sensing and transducing mechanical signals within the WMC continuum.

3.4 Interplay Between Chemistry and Mechanics During Plant Wall Invading Pathogen Interactions

Wounding by herbivores and pathogen penetration evokes rapid local defense reactions and longer-term systemic changes. As the walls constitute the physical barrier and the zone of first contact with pathogens, the invader recognition at cell walls and the activation of wall-located defense mechanisms are crucial for plant survival. In the early response, mechanical resistance to penetration is a key mechanism preventing further invasion and spreading of disease. One of the most important elements of penetration resistance is the formation of physical and chemical protective barriers, cell wall appositions (CWA) called "papillae" (Schmelzer 2002). Attempted penetration of the plant wall by a pathogen evokes rapid translocation of subcellular components to the infection site. This is achieved via actin cytoskeleton and endomembrane system reorganization, accumulation of endoplasmic reticulum and Golgi stacks just below the infection site, followed by actin-dependent polarized secretion of wall components (reviewed by Hardham et al. 2007). Interestingly, identical events accompany mechanical stimulation of epidermal cells, suggesting that at the cellular level some basal defense reactions could have evolved from plant mechanosensory responses (Hardham et al. 2008). In addition, extensive synthesis of callose is induced by activation of plasma-membrane-located callose synthase complexes accumulated in the membrane microdomain of the CWA region (Bhat et al. 2005).

For better understanding of the penetration resistance, it is usually assumed that the plasma membrane of invaded cells contains some receptors able to recognize chemical signals, and to activate further signaling pathway(s) inducing CWA formation. It seems that RLKs could play a crucial role here. Those participating in defense responses are collectively called "pattern-recognition receptors," and their ligands, which include, e.g., bacterial flagellin, fungal chitin, and oomycete INF1, are termed "pathogen-associated molecular patterns" (Jones and Dangl 2006). The most studied receptor in *Arabidopsis* is AtFLS2 (FLAGELLINSENSING 2) – a LRR-RLK-recognizing bacterial flg22 (Felix et al. 1999; Chinchilla et al. 2006) Although the role of these proteins in binding of bacterial flagellin is known, other elements of the activated signaling pathway remain elusive. Somatic embryogenesis receptor kinase 3/brassinosteroid-associated kinase 1 is a LRR RLK that forms complexes with flg22-activated AtFLS2 in *Arabidopsis* (Heese et al. 2007). This RLK was previously shown to function in brassinolide signaling through brassinosteroid 1 receptor (Nam and Li 2002). This suggests that immune and hormone signaling pathways are somehow linked, but this intriguing hypothesis has to be proved.

In animal systems, transduction of mechanical stimuli from the extracellular matrix to the cytoskeleton involves integrins. Extracellular ligands that activate integrins possess a characteristic RGD (Arg-Gly-Asp) motif. Although genes coding for integrins were not found in the *Arabidopsis* genome, proteins containing β-integrin-like domains were identified immunologically in many plant species (reviewed by Baluška et al. 2003). The existence of high-affinity binding sites for RGD-containing peptides (Canut et al. 1998), as well as responses to such peptides (Schindler et al. 1989; Barthou et al. 1999), was demonstrated. Exogenous application of RGD-containing peptides evokes the loss of cell wall-plasma membrane connections in plasmolyzed onion cells (Canut et al. 1998) or *Pelvetia* zygotes (Henry et al. 1996). These observations suggest that plants contain RGD-binding proteins which although structurally different could be functionally similar to animal integrins. Their involvement in defense responses was first suggested by the demonstration of interference of RGD peptides with accumulation of phytoalexin in pea epicotyls responding to fungal elicitors (Kiba et al. 1998). Disruption of cell wall-plasma membrane connections during pathogen attack by RGD peptides may lead to reduction of wall-associated reactions and, in consequence, to increased plant susceptibility (Mellersh and Heath 2001). Interestingly, an *Arabidopsis* plasma membrane protein able to bind RGD-containing peptides, such as the protein secreted by the oomycete (IPI-O; Senchou et al. 2004), was identified as a lectin-like receptor kinase (Gouget et al. 2006).

4 Cell Walls as a Source of Signaling Molecules

4.1 Oligosaccharide-Containing Molecules

Mechanical damage or activity of glycohydrolases, either residing in or arriving at plant walls, might yield oligosaccharide fragments. It is commonly accepted that such fragments can have biological activity, evoking various responses of plant cells. Oligosaccharins have been defined as oligosaccharides that at very low concentration affect physiological functioning of plants by acting as signal molecules in the cell wall (Darvill et al. 1992; Aldington and Fry 1993). Although initially identified as elicitors of defense responses, they were later found to influence normal growth and development. The latter issue has recently been thoroughly reviewed elsewhere (Fleming 2005); therefore, here we will concentrate on the role of oligosaccharide fragments in plant interactions with other organisms.

Oligogalacturonides (OGAs) are linear stretches of about two to 20 α-1,4-linked units of galacturonic acid. They originate from the pectin matrix where they are released from homogalacturonan of the primary cell walls, usually by enzymatic degradation by endopolygalacturonases. OGAs act as very powerful elicitors of plant defense reactions to pathogen attack, activating locally wall fortification, generation of reactive oxygen species (ROS), and accumulation of phytoalexins and protease inhibitors (Côté and Hahn 1994; Creelman and Mullet 1997). Although OGAs are not mobile in plants, they also activate more general defense responses (reviewed by Ridley et al. 2001). However, OGA perception and the accompanying signaling pathways are still poorly understood. An interesting feature of OGAs is that degree of polymerization (DP) seems to be crucial for determination of their biological activity. For example, only OGAs with DP = 10–15 induce a transient increase of cytosolic free Ca^{2+} (Mathieu et al. 1998; Moscatiello et al. 2006). Data from in vitro studies indicate plausible involvement of a putative protein kinase cascade signaling network (Farmer et al. 1991). Because of the differentiated, DP-dependent potency of OGAs, it is postulated that plant cells could possess multiple OGA-perceiving receptor-like molecules with different affinities. In addition, the possibility exists that OGAs might by bound by some wall components (Mathieu et al. 1998).

Lipochitooligosaccharides (LCO) are synthesized by soil bacteria collectively known as rhizobia. These highly decorated oligomers of *N*-acetyl-D-glucosamine (Nod factors) were shown to participate in the establishment of symbiotic interactions between legumes and nitrogen-fixing rhizobia (Lerouge et al. 1990). Sensed by plant roots, Nod factors act as activators of morphogenetic events culminating in the formation of symbiotic nodules (Truchet et al. 1991). Nod-induced processes are accompanied by changes in cytoskeleton organization, and plasma membrane and cell wall properties (Timmers et al. 1999). In response to Nod factors, plants activate nodulin genes, among them those coding for proline-rich cell wall proteins and peroxidases (reviewed by D´Haeze and Holsters 2002). The core structure of Nod factors resembles short chitin fragments. Pure chitooligosaccharides can induce various defense responses in plants (Felix et al. 1993; Baier et al. 1999). Among LCO-induced pathogenesis-related (PR) genes, a significant group is chitinases, mostly apoplastic, that hydrolyze β-1,4-*N*-acetyl-D-glucosamine polymers and oligomers. These were shown to directly regulate the lifespan and specific biological activity of Nod factors (Staehelin et al. 1994a, b). Such interplay between LCO-type signals and chitinases seems to be functional also during developmental processes. In the carrot line *temperature sensitive 11* (*ts11*) in which somatic embryogenesis is arrested at nonpermissive temperatures, an addition of Nod factor or supplementation of medium with carrot EP3 chitinase breaks the developmental arrest (De Jong et al. 1992, 1993). Moreover, carrot somatic embryogenesis is regulated by endogenous *N*-acetylglucosamine- or glucosamine-containing signals released by chitinases from oligosaccharide side chains of AGPs (van Hengel et al. 2001).

4.2 Plant Peptides

Owing to physicochemical and structural features of cell walls, it was long thought that in plants only small lipophilic compounds, such as auxins, cytokinins, gibberellins, abscisic acids, ethylene, brasinosteroids, and jasmonates, can function as regulatory molecules. The occurrence of peptide ligands was thought to be limited to animals. However, biochemical, molecular, and functional genomics research in recent years revealed clearly that there are a lot of different peptides in plants with extremely high informative potential, because of their diversity in length, sequence, charge, structure, and modification. This topic has recently been covered by many excellent reviews (Bisseling 1999; Matsubayashi 2003; Xia 2005; Matsubayashi and Sakagami 2006; Farrokhi et al. 2008) and thus we will concentrate on a few selected points.

Several signaling peptides have been quite well characterized, and their involvement in either the regulation of growth and development or plant defense reactions has been clearly demonstrated. However, recent genomic and bioinformatic approaches in *Arabidopsis* indicated that the presently assumed diversity of secreted plant peptides might in fact be much broader (Lease and Walker 2006). This notion is corroborated by the observation that the majority of hormone peptides are released from much bigger precursors via regulatory protein cleavage by respective proteases. This might suggest that the number of peptide hormones in plants might still be underestimated.

Most peptide hormones exert their signaling action extracellularly. This requires firstly the coordinated activity of the whole machinery for protein sorting and transport to allow for protein secretion outside the protoplast. On the other hand, this also requires secretion and precise localization of at least two other components of the signaling system, i.e., an appropriate receptor and a protease that generates peptide signal. These stages of peptide signaling introduce, however, several possibilities of regulatory action, making the system very sensitive and responsive to various additional cues. Two peptide signaling systems will be used to illustrate the issue. It should also be noted, however, that as yet none of the peptide signaling systems characterized is known in its full complexity.

The identity of the cells in the shoot apical meristem (SAM) is determined among other factors by the activity of CLAVATA system. Three *CLV* genes have been characterized. CLV1 belongs to the LRR class of RLKs (Clark et al. 1997). CLV2 is similar to CLV1, but lacks the kinase domain and is thought to form functional heterodimers (Jeong et al. 1999). On the other hand, *CLV3* encodes a 96-amino acid peptide (Fletcher et al. 1999), a member of a large family of CLAVATA3/ ENDOSPERM SURROUNDING REGION (CLE)-related peptides (Jun et al. 2008). Although a functional CLV3 peptide is 78 amino acids long, it is postulated that proteolysis of CLV3 in the apoplast leads to release of an active dodecapeptide containing two hydroxyprolines, derived from a conserved motif characteristic for the whole CLE family (Kondo et al. 2006). The protease processing CLV3 is still not known. CLV3 is secreted to the apoplast by cells of the L1 and L2 layers of the SAM, but acts as a ligand for CLV1/CLV2 complexes located in the L3 layer to activate a signaling cascade leading to cell division inhibition in the SAM (Clark 2001). CLV3 binds directly the CLV1 ectodomain (Ogawa et al. 2008), but recent data indicate also that CLV3 might be sensed independently by another RLK system composed of CLV2 and a receptor kinase CORYNE (Müller et al. 2008).

Peptide hormones are also required for the regulation of stomatal development (Bergmann and Sack 2007). Current model points to an apoplastic serine protease of the subtilase family - STOMATAL DENSITY AND DISTRIBUTION 1 (SDD1; Berger and Altmann 2000) as a local generator of unknown positional signal detected by the LRR-receptor-like protein TOO MANY MOUTHS (TMM; Nadeau and Sack 2002). The recently identified secretory peptide EPIDERMAL PATTERNING FACTOR 1 probably acts upstream of TMM, but independently of SDD1 (Hara et al. 2007). As TMM lacks an intracellular kinase domain, it is suggested that it exerts its activity via interactions with lectin receptor kinases, most probably of the ERECTA family (Shpak et al. 2005). Mutations in genes encoding these proteins result in disturbed number and patterning of stomata.

4.3 Hydrogen Peroxide

The signaling function of ROS can be achieved in two different, but partially overlapping, ways (Van Breusegem et al. 2008). First, it was demonstrated in animal models that ROS could form complexes with respective proteinaceous receptors, predominantly located in the membranes, which leads to the subsequent activation of the signaling pathway (DeYulia et al. 2005). Second, it is assumed that ROS could change the functioning of the cell by chemically modifying biological macromolecules without leading to their destruction (Neill et al. 2002, 2003).

Hydrogen peroxide (H_2O_2) is the most stable ROS. It is electrically neutral, is not a radical, and thus is very mobile and can penetrate cell membranes (Wojtaszek 1997; Neill et al. 2002). Interestingly, recent data suggest the existence of specialized aquaporins channeling H_2O_2 actively through the membranes (Bienert et al. 2007). The signaling power of H_2O_2 has been elegantly demonstrated for plant defense responses against avirulent pathogens where its accumulation at sites distant to attempted infections mediated a reiterative signal network underlying systemic as well as local resistance responses (Alvarez et al. 1998). Although $O_2^{\cdot-}$ is mainly produced in many cases of defense responses, its quick dismutation leads to the accumulation of H_2O_2, which can then diffuse inside the walls and reach neighboring cells (Wojtaszek 1997). Interestingly, however, it is superoxide radical which has been identified as a signaling molecule responsible in *Dictyostelium discoideum* for initiation of the transition from unicellular growth to the formation of multicellular aggregates (Bloomfield and Pears 2003).

A very interesting phenomenon is the existence of various sources of H_2O_2 which are switched on and off depending on the physiological context (Bolwell and Wojtaszek 1997). In the exocellular compartments, ROS are thought to be generated either by membrane NADPH oxidases (Foreman et al. 2003) or by apoplastic enzymes, such as peroxidases (Bindschedler et al. 2006). These can be activated separately or in parallel, depending on the species, and, e.g., its strategy to react to pathogens. It has also been noted that in response to various stresses (biotic vs. abiotic) ROS generation could be either apoplastic or intracellular, e.g., from NADPH oxidase internalized in endosomes (Leshem et al. 2007).

4.4 Nitric Oxide

In contrast to animal cells, the sources of nitric oxide (NO) in plants are very diverse (Wendehenne et al. 2001). As the exocellular compartments are considered, both enzymatic and nonenzymatic mechanisms are operating here, and it is usually thought that plants function in "an open system" with respect to NO (Yamasaki 2005). In the apoplast, a putative nonenzymatic pathway generates NO via reduction of nitrite in an acidic environment (Bethke et al. 2004). NO can also arrive directly at the walls from the soil where it can be produced by soil bacteria (Stöhr and Ullrich 2002). Finally, there are a few membrane enzymes which release NO into the apoplast. NO can be synthesized by nitrite:NO reductase, closely associated with plasma-membrane-bound nitrate reductase, which may also release NO to the apoplast. Together with cytosolic nitrate reductase, nitrite:NO reductase is one of the most important NO sources in plant roots. It should also be remembered that owing to the chemical nature of NO, this small hydrophobic molecule when generated in the cytoplasm can easily penetrate through membranes, thus also reaching the apoplast.

NO is an important signaling molecule in plants. It is involved in the control of various physiological processes, such as gravity responses, growth or flowering, regulation of stomatal aperture, xylem formation, stress response and adaptation, and also in defense responses. NO-activated signaling pathways, by analogy with animal systems, involve a number of secondary messengers, such as cyclic guanosine monophosphate, cyclic ADP-ribose, or Ca^{2+} (reviewed by Neill et al. 2003, 2008; Arasimowicz and Floryszak-Wieczorek 2007). In this pathway in animals, NO reacts with ferrous ion of the heme group in a soluble guanylate cyclase. This binding triggers conformational changes in the soluble guanylate cyclase resulting in 300-fold increase in the enzymatic activity (Denninger and Marletta 1999). There is a great probability that the same process is taking place in plants as application of NO donors causes a dramatic increase of the cyclic guanosine monophosphate level in tobacco (Durner et al. 1998).

NO can also evoke cellular responses by structural modification of proteins via direct addition of either an NO molecule or an NO-derived molecule to peptide amino acids. The most common are cysteine S-nitrosylation (Hess et al. 2005) and tyrosine nitration (Monteiro 2002). It is now well established in animal systems that NO is part of a universal redox-based signaling mechanism and S-nitrosylation could be regarded as an important posttranslational modification, fulfilling roles similar to phosphorylation. Knowledge of plant proteins regulated by S-nitrosylation or tyrosine nitration is actually just emerging (Lindermayr et al. 2005), and only a few examples have been definitely demonstrated. These include S-nitrosylation of glyceraldehyde 3-phosphate dehydrogenase (Lindermayr et al. 2005), methionine adenosyltransferase (Lindermayr et al. 2006), and metacaspase (Belenghi et al. 2007), and peroxiredoxin II E-stimulated tyrosine nitration (Romero-Puertas et al. 2007).

Many physiological processes and defense reactions seem to be regulated by NO, although the exact mechanisms of the actions of NO in most of the cases are barely known. NO burst was observed during differentiation of xylem cells (Gabaldón et al. 2005). One of the potential NO targets seem to be wall-resident peroxidase participating in the lignification of secondary cell walls (Ferrer and Ros Barcelo 1999). NO also affects cellulose metabolism in tomato. The cellulose content increased in response to picomolar concentrations of NO, while nanomolar concentrations had the opposite effect. Those data may suggest that NO is also involved in the regulation of primary wall synthesis (Correa-Arangunde et al. 2008).

Treatment of potato leaflets with NO, resulting in the enhanced synthesis and accumulation of callose, demonstrated its involvement in the wound healing response (París et al. 2007). NO also plays a significant role in plant defense responses (Delledonne et al. 1998). Interestingly, increased NO generation during plant-pathogen interactions might not be a plant-specific phenomenon. Recently, it was demonstrated that in response to cellobiose, originating from plant walls, pathogenic *Streptomyces* species can initiate NO burst releasing NO to the plant apoplast (Johnson et al. 2008). Finally, NO seems to play a crucial role during hypersensitive response, including programmed cell death. During programmed cell death a tightly regulated interplay between NO and various ROS species is observed. Usually, like in soybean, both NO and other ROS are required to initiate programmed cell death (Delledonne et al. 2001). In some cases, also independent action of NO causes programmed cell death (Clarke et al. 2000). The timing of the NO burst seems to be a critical point in the induction of programmed cell death. Oxidative burst activates a burst of NO generation, while constant ROS production does not affect NO synthesis (de Pinto et al. 2006).

References

Aldington S, Fry S (1993) Oligosaccharins. Adv Bot Res 19:1–101

Alvarez ME, Pennell RI, Meijer PJ, Ishikawa A, Dixon RA, Lamb C (1998) Reactive oxygen intermediates mediate a systemic signal network in the establishment of plant immunity. Cell 92:773–784

Anderson CM, Wagner TA, Perret M, He ZH, He D, Kohorn BD (2001) WAKs: cell wall-associated kinases linking the cytoplasm to the extracellular matrix. Plant Mol Biol 47:197–206

Arasimowicz M, Floryszak-Wieczorek J (2007) Nitric oxide as a bioactive signaling molecule in plant stress responses. Plant Sci 172:876–887

Baier R, Schiene K, Kohring B, Flaschel E, Niehaus K (1999) Alfalfa and tobacco cells react differently to chitin oligosaccharides and *Sinorhizobium meliloti* nodulation factors. Planta 210:157–164

Baluška F, Salaj J, Mathur J, Braun M, Jasper F, Šamaj J, Chua N-H, Barlow PW, Volkmann D (2000) Root hair formation: F-actin-dependent tip growth is initiated by local assembly of profilin-supported F-actin meshworks accumulated within expansin-enriched bulges. Dev Biol 226:618–632

Baluška F, Šamaj J, Wojtaszek P, Volkmann D, Menzel D (2003) Cytoskeleton-plasma membrane-cell wall continuum in plants. Emerging links revisited. Plant Physiol 133:482–491

Barthou H, Petitprez M, Briere C, Souvre A, Alibert G (1999) RGD-mediated membrane-matrix adhesion triggers agarose-induced embryoid formation in sunflower protoplasts. Protoplasma 206:143–151

Belenghi B, Romero-Puertas MC, Vercammen D, Brackenier A, Inzé D, Delledonne M, Van Breusegem F (2007) Metacaspase activity of *Arabidopsis thaliana* is regulated by *S*-nitrosylation of a critical cysteine residue. J Biol Chem 282:1352–1358

Berger D, Altmann T (2000) A subtilisin-like serine protease involved in the regulation of stomatal density and distribution in *Arabidopsis thaliana*. Genes Dev 14:1119–1131

Bergmann DC, Sack FD (2007) Stomatal development. Annu Rev Plant Biol 58:163–181

Bethke PC, Badger MR, Jones RL (2004) Apoplastic synthesis of nitric oxide by plant tissues. Plant Cell 16:332–341

Bhat RA, Miklis M, Schmelzer E, Schulze-Lefert P, Panstruga R (2005) Recruitment and interaction dynamics of plant penetration resistance components in a plasma membrane microdomain. Proc Natl Acad Sci U S A 102:3135–3140

Bienert GP, Møller ALB, Kristiansen KA, Schulz A, Møller IM, Schjoerring JK, Jahn TP (2007) Specific aquaporins facilitate the diffusion of hydrogen peroxide across membranes. J Biol Chem 282:1183–1192

Bindschedler LV, Dewdney J, Blee KA, Stone JM, Asai T, Plotnikov J, Denoux C, Hayes T, Gerrish C, Davies DR, Ausubel FM, Paul Bolwell G (2006) Peroxidase-dependent apoplastic oxidative burst in Arabidopsis required for pathogen resistance. Plant J 47:851–863

Bisseling T (1999) The role of plant peptides in intercellular signaling. Curr Opin Plant Biol 2:365–368

Bloomfield G, Pears C (2003) Superoxide signaling required for multicellular development of *Dictyostelium*. J Cell Sci 116:3387–3397

Bolwell GP, Wojtaszek P (1997) Mechanisms for the generation of reactive oxygen species in plant defense – a broad perspective. Physiol Mol Plant Pathol 51:347–366

Borner GH, Lilley KS, Stevens TJ, Dupree P (2003) Identification of glycosylphosphatidylinositol-anchored proteins in Arabidopsis. A proteomic and genomic analysis. Plant Physiol 132:568–577

Braam J (2005) In touch: plant responses to mechanical stimuli. New Phytol 165:373–388

Braam J, Davies RW (1990) Rain-, wind-, and touch-induced expression of calmodulin and calmodulin-related genes in Arabidopsis. Cell 60:357–364

Brett C, Waldron K (1996) Physiology and biochemistry of plant cell walls, 2nd ed. Chapman & Hall, London

Brownlee C (2002) Role of the extracellular matrix in cell-cell signaling: paracrine paradigms. Curr Opin Plant Biol 5:396–401

Canny MJ (1995) Apoplastic water and solute movement: new rules for an old space. Annu Rev Plant Physiol Plant Mol Biol 46:215–236

Canut H, Carrasco A, Galaud J, Cassan C, Bouyssou H, Vita N, Ferrara P, Pont-Lezica R (1998) High affinity RGD-binding sites at the plasma membrane of *Arabidopsis thaliana* link the cell wall. Plant J 16:63–71

Chinchilla D, Bauer Z, Regenass M, Boller T, Felix G (2006) The Arabidopsis receptor kinase FLS2 binds flg22 and determines specificity of flagellin perception. Plant Cell 18:465–476

Clark SE (2001) Cell signaling at the shoot meristem. Nat Rev Mol Cell Biol 2:276–284

Clark SE, Williams RW, Meyerowitz EM (1997) The CLAVATA1 gene encodes a putative receptor kinase that controls shoot and floral meristem size in Arabidopsis. Cell 89:575–585

Clarke A, Desikan R, Hurst RD, Hancock JT, Neill SJ (2000) NO way back: nitric oxide and programmed cell death in *Arabidopsis thaliana* suspension cultures. Plant J 24:667–677

Correa-Arragunde N, Lombardo C, Lamattina L (2008) Nitric oxide: an active nitrogen molecule that modulates cellulose synthesis in tomato roots. New Phytol 179:386–396

Cosgrove DJ (2005) Growth of the plant cell wall. Nat Rev Mol Cell Biol 6:850–861

Côé F, Hahn MG (1994) Oligosaccharins: structures and signal transduction. Plant Mol Biol 26:1379–1411

Creelman RA, Mullet JE (1997) Oligosaccharins, brassinolides, and jasmonates: nontraditional regulators of plant growth, development, and gene expression. Plant Cell 9:1211–1223

Darvill AG, Augur C, Bergmann C, Carlson RW, Cheong J-J, Eberhard S, Hahn MG, Ló V-M, Marfa V, Meyer U, Mohnen D, O'Neill MA, Spiro MD, van Halbeek H, York WS, Albersheim P (1992) Oligosaccharins – oligosaccharides that regulate growth, development and defense responses in plants. Glycobiology 2:181–198

Decreux A, Messiaen J (2005) Wall-associated kinase WAK1 interacts with cell wall pectins in a calcium-induced conformation. Plant Cell Physiol 46:268–278

Decreaux A, Thomas A, Spies B, Brasseur R, Van Cutsem P, Messiaen J (2006) In vitro characterization of the homogalacturonan-binding domain of the wall-associated kinase WAK1 using site-directed mutagenesis. Phytochemistry 67:1068–1079

Deeks MJ, Hussey PJ, Davies B (2002) Formins: intermediates in signal-transduction cascades that affect cytoskeletal reorganization. Trends Plant Sci 7:492–498

De Jong AJ, Cordewener J, Lo Schiavo F, Terzi M, Vandekerckhove J, Van Kammen A, De Vries SC (1992) A carrot somatic embryo mutant is rescued by chitinase. Plant Cell 4:425–433

De Jong AJ, Heidstra R, Spaink HP, Hartog MV, Meijer EA, Hendriks T, Lo Schiavo F, Terzi M, Bisseling T, Van Kammen A, De Vries SC (1993) *Rhizobium* lipooligosaccharides rescue a carrot somatic embryo mutant. Plant Cell 5:615–620

Delledonne M, Xia Y, Dixon RA, Lamb C (1998) Nitric oxide functions as a signal in plant disease resistance. Nature 394:585–588

Delledonne M, Zeier J, Marocco A, Lamb C (2001) Signal interactions between nitric oxide and reactive oxygen intermediates in the plant hypersensitive disease resistance response. Proc Natl Acad Sci U S A 98:13454–13459

Denninger JW, Marletta MA (1999) Guanylate cyclase and the NO·/cGMP signaling pathway. Biochim Biophys Acta 1411:334–350

de Pinto MC, Paradiso A, Leonetti P, De Gara L (2006) Hydrogen peroxide, nitric oxide and cytosolic ascorbate peroxidase at the crossroad between defence and cell death. Plant J 48:784–795

DeWitt G, Richards J, Mohnen D, Jones AM (1999) Comparative compositional analysis of walls with two different morphologies: archetypical versus transfer-cell-like. Protoplasma 209:238–245

DeYulia GJ, Cárcamo JM, Bórquez-Ojeda O, Shelton CC, Golde DW (2005) Hydrogen peroxide generated extracellularly by receptor-ligand interaction facilitates cell signaling. Proc Natl Acad Sci U S A 102:5044–5049

D'Haeze W, Holsters M (2002) Nod factor structures, responses, and perception during initiation of nodule development. Glycobiology 12:79R–105R

Ding B, Kwon MO, Warnberg L (1996) Evidence that actin filaments are involved in controlling the permeability of plasmodesmata in tobacco mesophyll. Plant J 10:157–164

Durner J, Wendehenne D, Klessig DF (1998) Defense gene induction in tobacco by nitric oxide, cyclic GMP, and cyclic ADPribose. Proc Natl Acad Sci U S A 95:10328–10333

Farmer EE, Moloshok TD, Saxton MJ, Ryan CA (1991) Oligosaccharide signaling in plants. Specificity of oligouronide-enhanced plasma membrane protein phosphorylation. J Biol Chem 266:3140–3145

Farrokhi N, Whitelegge JP, Brusslan JA (2008) Plant peptides and peptidomics. Plant Biotechnol J 6:105–134

Felix G, Regenass M, Boller T (1993) Specific perception of subnanomolar concentrations of chitin fragments by tomato cells: Induction of extracellular alkalinization, changes in protein phosphorylation and establishment of a refractory state. Plant J 4:307–316

Felix G, Duran JD, Volko S, Boller T (1999) Plants have a sensitive perception system for the most conserved domain of bacterial flagellin. Plant J 18:265–276

Ferrer MA, Ros Barcelo A (1999) Differential effects of nitric oxide on peroxidase and H_2O_2 production by the xylem *of Zinnia elegans*. Plant Cell Environ 22:891–897

Fleischer A, O'Neill MA, Ehwald R (1999) The pore size of non-graminaceous plant cell walls is rapidly decreased by borate ester cross-linking of the pectic polysaccharide rhamnogalacturonan II. Plant Physiol 121:829–838

Fleming A (2005) The plant extracellular matrix and signaling. Fleming A Intercellular communication in plants. Blackwell, Oxford, pp 85–108

Fleming AJ, McQueen-Mason S, Mandel T, Kuhlemeier C (1997) Local expression of expansin induces the entire process of leaf development and modifies leaf shape. Science 276:1415–1418

Fletcher JC, Brand U, Running MP, Simon R, Meyerowitz EM (1999) Signaling of cell fate decisions by CLAVATA3 in Arabidopsis shoot meristems. Science 283:1911–1914

Foreman J, Demidchik V, Bothwell JH, Mylona P, Miedema H, Torres MA, Linstead P, Costa S, Brownlee C, Jones JD, Davies JM, Dolan L (2003) Reactive oxygen species produced by NADPH oxidase regulate plant cell growth. Nature 422:442–446

Gabaldón C, Gómez Ros LV, Pedreño MA, Ros Barceló A (2005) Nitric oxide production by the differentiating xylem of *Zinnia elegans*. New Phytol 165:121–130

Gens JS, Fujiki M, Pickard BG (2000) Arabinoglactan protein and wall-associated kinases in a plasmalemmal reticulum with specialized vertices. Protoplasma 212:115–134

Gisel A, Barella S, Hempel FD, Zambryski PC (1999) Temporal and spatial regulation of symplastic trafficking during development in *Arabidopsis thaliana* apices. Development 126:1879–1889

Golomb L, Abu-Abied M, Belausov E, Sadot E (2008) Different subcellular localizations and functions of Arabidopsis myosin VIII. BMC Plant Biol 8:3

Gouget A, Senchou V, Govers F, Sanson A, Barre A, Rouge P, Pont-Lezica R, Canut H (2006) Lectin receptor kinases participate in protein–protein interactions to mediate plasma membrane-cell wall adhesions in Arabidopsis. Plant Physiol 140:81–90

Green PB (1999) Expression of pattern in plants: combining molecular and calculus-based biophysical paradigms. Am J Bot 86:1059–1076

Gustin MC, Albertyn J, Alexander M, Davenport K (1998) MAP kinase pathways in the yeast *Saccharomyces cerevisiae*. Microbiol Mol Biol Rev 62:1264–1300

Haffani Y, Silva-Gagliardi N, Sewter S, Aldea M, Zhao Z, Nakhamchik A, Cameron R, Goring D (2006) Altered expression of PERK receptor kinases in Arabidopsis leads to changes in growth and floral organ formation. Plant Signal Behav 1:251–260

Hara K, Kajita R, Torii KU, Bergmann DC, Kakimoto T (2007) The secretory peptide gene *EPF1* enforces the stomatal one-cell-spacing rule. Genes Dev 21:1720–1725

Hardham AR, Jones DA, Takemoto D (2007) Cytoskeleton and cell wall function in penetration resistance. Curr Opin Plant Biol 10:342–348

Hardham AR, Takemoto D, White RG (2008) Rapid and dynamic subcellular reorganization following mechanical stimulation of Arabidopsis epidermal cells mimics responses to fungal and oomycete attack. BMC Plant Biol 8:63

Hayashi T, Harada A, Sakai T, Takagi A (2006) Ca^{2+} transient induced by extracellular changes in osmotic pressure in Arabidopsis leaves: differential involvement of cell wall-plasma membrane adhesion. Plant Cell Environ 29:661–672

Heese A, Hann DR, Gimenez-Ibanez S, Jones AME, He K, Li J, Schroeder JI, Peck SC, Rathjen JP (2007) The receptor-like kinase SERK3/BAK1 is a central regulator of innate immunity in plants. Proc Natl Acad Sci U S A 104:12217–12222

Heinisch JJ, Lorberg A, Schmitz HP, Jacoby JJ (1999) The protein kinase C-mediated MAP kinase pathway involved in the maintenance of cellular integrity in *Saccharomyces cerevisiae*. Mol Microbiol 32:671–680

Heinlein M, Epel BL (2004) Macromolecular transport and signaling through plasmodesmata. Int Rev Cytol 235:93–164

Hématy K, Höfte H (2008) Novel receptor kinases involved in growth regulation. Curr Opin Plant Biol 11:321–328

Hématy K, Sado P-E, Van Tuinen A, Rochange S, Desnos T, Balzergue S, Pelletier S, Renou J-P, Höfte H (2007) A receptor-like kinase mediates the response of *Arabidopsis* cells to the inhibition of cellulose synthesis. Curr Biol 17:922–931

Henry CA, Jordan JR, Kropf DL (1996) Localized membrane-wall adhesions in *Pelvetia* zygotes. Protoplasma 190:39–52

Hess DT, Matsumoto A, Kim SO, Marshall HE, Stamler JS (2005) Protein S-nitrosylation: purview and parameters. Nat Rev Mol Cell Biol 6:150–166

Hu W-J, Lung J, Harding SA, Popko JL, Ralph J, Stokke DD, Tsai C-J, Chiang VL (1999) Repression of lignin biosynthesis promotes cellulose accumulation and growth in transgenic trees. Nat Biotechol 17:808–812

Im K-H, Cosgrove DJ, Jones AM (2000) Subcellular localization of expansin mRNA in xylem cells. Plant Physiol 123:463–470

Ingber DE (2003) Tensegrity. I. Cell structure and hierarchical systems biology. J Cell Sci 116:1157–1173

Jeong S, Trotochaud AE, Clark SE (1999) The Arabidopsis CLAVATA2 gene encodes a receptor-like protein required for the stability of the CLAVATA1 receptor-like kinase. Plant Cell 11:1925–1934

Johnson EG, Sparks JP, Dzikovski B, Crane BR, Gibson DM, Loria R (2008) Plant-pathogenic *Streptomyces* species produce nitric oxide synthase-derived nitric oxide in response to host signals. Chem Biol 15:43–50

Jones JDG, Dangl JL (2006) The plant immune system. Nature 444:323–329

Jun JH, Fiume E, Fletcher JC (2008) The CLE family of plant polypeptide signaling molecules. Cell Mol Life Sci 65:743–755

Kiba A, Sugimoto M, Toyoda K, Ichinose Y, Yamada T, Shiraishi T (1998) Interaction between cell wall and plasma membrane via RGD motif is implicated in plant defense responses. Plant Cell Physiol 39:1245–1249

Knox JP, Linstead PJ, King J, Cooper C, Roberts K (1990) Pectin esterification is spatially regulated both within cell walls and between developing tissues of root apices. Planta 181:512–521

Kobe B, Kajava AV (2001) The leucine-rich repeat as a protein recognition motif. Curr Opin Struct Biol 11:725–732

Kohorn BD (2000) Plasma membrane-cell wall contacts. Plant Physiol 124:31–38

Kohorn BD (2001) WAKs: cell wall associated kinases. Curr Opin Cell Biol 13:529–533

Kohorn BD, Kobayashi M, Johansen S, Riese J, Huang LF, Koch K, Fu S, Dotson A, Byers N (2006) An Arabidopsis cell wall-associated kinase required for invertase activity and cell growth. Plant J 46:307–316

Kondo T, Sawa S, Kinoshita A, Mizuno S, Kakimoto T, Fukuda H, Sakagami Y (2006) A plant peptide encoded by CLV3 identified by in situ MALDI-TOF MS analysis. Science 313:845–848

Kutschera U (1995) Tissue pressure and cell turgor in axial plant organs: Implications for the organism theory of multicellularity. J Plant Physiol 146:126–132

Kutschera U (2008) The growing outer epidermal wall: design and physiological role of a composite structure. Ann Bot 101:615–621

Lang-Pauluzzi I, Gunning BES (2000) A plasmolytic cycle: the fate of cytoskeletal elements. Protoplasma 212:174–185

Lease KA, Walker JC (2006) The Arabidopsis unannotated secreted peptide database, a resource for plant peptidomics. Plant Physiol 142:831–838

Lerouge P, Roche P, Faucher C, Maillet F, Truchet G, Prome JC, Denarie J (1990) Symbiotic host-specificity of *Rhizobium meliloti* is determined by a sulphated and acylated glucosamine oligosaccharide signal. Nature 344:781–784

Leshem Y, Seri L, Levine A (2007) Induction of phosphatidylinositol 3-kinase-mediated endocytosis by salt stress leads to intracellular production of reactive oxygen species and salt tolerance. Plant J 51:185–197

Levin DE (2005) Cell wall integrity signaling in *Saccharomyces cerevisiae*. Microbiol Mol Biol Rev 69:262–291

Levy A, Erlanger M, Rosenthal M, Epel BL (2007) A plasmodesmata-associated beta-1,3-glucanase in Arabidopsis. Plant J 49:669–682
Lindermayr C, Saalbach G, Durner J (2005) Proteomic identification of S-nitrosylated proteins in *Arabidopsis*. Plant Physiol 137:921–930
Lindermayr C, Saalbach G, Bahnweg G, Durner J (2006) Differential inhibition of *Arabidopsis* methionine adenosyltransferases by protein S-nitrosylation. J Biol Chem 281:4285–4291
Lintilhac PM, Vesecky TB (1984) Stress-induced alignment of division plane in plant tissues grown in vitro. Nature 307:363–364
Lynch TM, Lintilhac PM (1997) Mechanical signals in plant development: a new method for single cell studies. Dev Biol 181:246–256
Madsen EB, Madsen LH, Radutoiu S, Olbryt M, Rakwalska M, Szczyglowski K, Sato S, Kaneko T, Tabata S, Sandal N, Stougaard J (2003) A receptor kinase gene of the LysM type is involved in legume perception of rhizobial signals. Nature 425:637–640
Majewska-Sawka A, Nothnagel EA (2000) The multiple roles of arabinogalactan proteins in plant development. Plant Physiol 122:3–10
Marshall JG, Dumbroff EB (1999) Turgor regulation via cell wall adjustment in white spruce. Plant Physiol 119:313–319
Mathieu Y, Guern J, Spiro MD, O'Neill MA, Kates KA, Darvill A, Albersheim P (1998) The transient nature of the oligogalacturonide-induced ion fluxes of tobacco cells is not correlated with fragmentation of the oligogalacturonides. Plant J 16:305–311
Matsubayashi Y (2003) Ligand-receptor pairs in plant peptide signaling. J Cell Sci 116:3863–3870
Matsubayashi Y, Sakagami Y (2006) Peptide hormones in plants. Annu Rev Plant Biol 57:649–674
McQueen-Mason SJ, Rochange F (1999) Expansins in plant growth and development: an update on an emerging topic. Plant Biol 1:19–25
Mellersh DG, Heath MC (2001) Plasma membrane-cell wall adhesion is required for expression of plant defense responses during fungal penetration. Plant Cell 13:413–424
Monteiro HP (2002) Signal transduction by protein tyrosine nitration: competition or cooperation with tyrosine phosphorylation-dependent signaling events? Free Radic Biol Med 33:765–773
Moscatiello R, Mariani P, Sanders D, Maathuis FJM (2006) Transcriptional analysis of calcium-dependent and calcium-independent signaling pathways induced by oligogalacturonides. J Exp Bot 57:2847–2865
Müller R, Bleckmann A, Simon R (2008) The receptor kinase CORYNE of *Arabidopsis* transmits the stem cell-limiting signal CLAVATA3 independently of CLAVATA1. Plant Cell 20:934–946
Nadeau JA, Sack FD (2002) Control of stomatal distribution on the *Arabidopsis* leaf surface. Science 296:1697–1700
Nakagawa N, Sakurai N (2001) Cell wall integrity controls expression of endoxyloglucan transferase in tobacco BY2 cells. Plant Cell Physiol 42:240–244
Nam KH, Li J (2002) BRI1/BAK1, a receptor kinase pair mediating brassinosteroid signaling. Cell 110:203–212
Neill S, Desikan R, Hancock J (2002) Hydrogen peroxide signaling. Curr Opin Plant Biol 5:388–395
Neill SJ, Desikan R, Hancock JT (2003) Nitric oxide signaling in plants. New Phytol 159:11–35
Neill S, Barros R, Bright J, Desikan R, Hancock J, Harrison J, Morris P, Ribeiro D, Wilson I (2008) Nitric oxide, stomatal closure, and abiotic stress. J Exp Bot 59:165–176
Niklas KJ (1992) Plant biomechanics. An engineering approach to plant form and function. University of Chicago Press, Chicago
Ogawa M, Shinohara H, Sakagami Y, Matsubayashi Y (2008) *Arabidopsis* CLV3 peptide directly binds CLV1 ectodomain. Science 319:294
Panteris E, Galatis B (2005) The morphogenesis of lobed plant cells in the mesophyll and epidermis: organization and distinct roles of cortical microtubules and actin filaments. New Phytol 167:721–732
París R, Lamattina L, Casalongué CA (2007) Nitric oxide promotes the wound-healing response of potato leaflets. Plant Physiol Biochem 45:80–86

Pilling E, Höfte H (2003) Feedback from the wall. Curr Opin Plant Biol 6:611–616

Popper ZA (2008) Evolution and diversity of green plant cell walls. Curr Opin Plant Biol 11:286–292

Radford JE, Vesk M, Overall RL (1998) Callose deposition at plasmodesmata. Protoplasma 201:30–37

Reise J, Ney J, Kohorn BD (2003). WAKs: cell wall associated kinases. Rose J The plant cell wall. Blackwell, Oxford, pp. 223–236

Ridley BL, O'Neill MA, Mohnen D (2001) Pectins: structure, biosynthesis, and oligogalacturonide-related signaling. Phytochemistry 57:929–967

Rinne PLH, van der Schoot C (1998) Symplasmic fields in the tunica of the shoot apical meristem coordinate morphogenetic events. Development 125:1477–1485

Romero-Puertas MC, Laxa M, Mattè A, Zaninotto F, Finkemeier I, Jones AME, Perazzolli M, Vandelle E, Dietz KJ, Delledonne M (2007) S-Nitrosylation of peroxiredoxin II E promotes peroxynitrite-mediated tyrosine nitration. Plant Cell 19:4120–4130

Roy S, Watada AE, Wergin WP (1997) Characterization of the cell wall microdomain surrounding plasmodesmata in apple fruit. Plant Physiol 114:539–547

Sampedro J, Cosgrove DJ (2005) The expansin superfamily. Genome Biol 6:242.1–242.11

Schindler M, Meiners S, Cheresh DA (1989) RGD-dependent linkage between plant cell wall and plasma membrane: Consequences for growth. J Cell Biol 108:1955–1965

Schmelzer E (2002) Cell polarization, a crucial process in fungal defence. Trends Plant Sci 7:411–415

Seifert GJ, Roberts K (2007) The biology of arabinogalactan proteins. Annu Rev Plant Biol 58:137–161

Senchou V, Weide R, Carrasco A, Bouyssou H, Pont-Lezica R, Govers F, Canut H (2004) High affinity recognition of a *Phytophthora* protein by *Arabidopsis* via an RGD motif. Cell Mol Life Sci 61:502–509

Shiu SH, Bleecker AB (2001) Receptor-like kinases from *Arabidopsis* form a monophyletic gene family related to animal receptor kinases. Proc Natl Acad Sci U S A 98:10763–10768

Shpak ED, McAbee JM, Pillitteri LJ, Torii KU (2005) Stomatal patterning and differentiation by synergistic interactions of receptor kinases. Science 309:290–293

Somerville C, Bauer S, Brininstool G, Facette M, Hamann T, Milne J, Osborne E, Paredez A, Persson S, Raab T, Vorwerk S, Youngs H (2004) Toward a system approach to understanding plant cell walls. Science 306:2206–2211

Staehelin C, Schultze M, Kondorosi E, Mellor RB, Boiler T, Kondorosi A (1994a) Structural modifications in *Rhizobium meliloti* Nod factors influence their stability against hydrolysis by root chitinases. Plant J 5:319–330

Staehelin C, Granado J, Müller J, Wiemken A, Mellor RB, Felix G, Regenass M, Broughton WJ, Boller T (1994b) Perception of Rhizobium nodulation factors by tomato cells and inactivation by root chitinases. Proc Natl Acad Sci U S A 91:2196–2200

Stöhr C, Ullrich WR (2002) Generation and possible roles of NO in plant roots and their apoplastic space. J Exp Bot 53:2293–2303

Suzuki K, Amino S-I, Takeuchi Y, Komamine A (1990) Differences in the composition of the cell walls of two morphologically different lines of suspension-cultured *Catharanthus roseus* cells. Plant Cell Physiol 31:7–14

Timmers ACJ, Auriac M-C, Truchet G (1999) Refined analysis of early symbiotic steps of the Rhizobium-Medicago interaction in relationship with microtubular cytoskeleton rearrangements. Development 126:3617–3628

Truchet G, Roche P, Lerouge P, Vasse J, Camut S, De Billy F, Prome JC, Denarie J (1991) Sulphated lipo-oligosaccharide signals of *Rhizobium meliloti* elicit root nodule organogenesis in alfalfa. Nature 351:670–673

Turner A, Wells B, Roberts K (1994) Plasmodesmata of maize root tips: structure and composition. J Cell Sci 107:3351–3361

Van Breusegem F, Bailey-Serres J, Mittler R (2008) Unraveling the tapestry of networks involving reactive oxygen species in plants. Plant Physiol 147:978–984 van Hengel AJ, Tadesse Z, Immerzeel P, Schols H, van Kammen A, de Vries SC (2001) N-acetylglucosamine and glucosamine-containing arabinogalactan proteins control somatic embryogenesis. Plant Physiol 125:1880–1890

Wendehenne D, Pugin A, Klessig DF, Durner J (2001) Nitric oxide: comparative synthesis and signaling in animal and plant cells. Trends Plant Sci 6:177–183

Williams M, Freshour G, Darvill A, Albersheim P, Hahn M (1996) An antibody Fab selected from a recombinant phage display library detects deesterified pectic polysaccharide rhamnogalacturonan II in plant cells. Plant Cell 8:673–685

Wojtaszek P (1997) Oxidative burst: an early plant response to pathogen infection. Biochem J 322:681–692

Wojtaszek P (2000) Genes and plant cell walls: a difficult relationship. Biol Rev 75:437–475

Wojtaszek P, Baluška F, Kasprowicz A, Łuczak M, Volkmann D (2007) Domain-specific mechanosensory transmission of osmotic and enzymatic cell wall disturbances to the actin cytoskeleton. Protoplasma 230:217–230

Wyatt SE, Carpita NC (1993) The plant cytoskeleton-cell wall continuum. Trends Cell Biol 3:413–417

Xia Y (2005) Peptides as signals. Fleming A Intercellular communication in plants. Blackwell. Blackwell, Oxford, pp 27–48

Yamasaki H (2005) The NO world for plants: achieving balance in an open system. Plant Cell Environ 28:78–84

Zambryski P, Crawford K (2000) Plasmodesmata: gatekeepers for cell-to-cell transport of developmental signals in plants. Annu Rev Cell Dev Biol 16:393–421

Zhou J, Wang B, Li Y, Wang Y, Zhu L (2007) Responses of *Chrysanthemum* cells to mechanical stimulation require intact microtubules and plasma membrane-cell wall adhesion. J Plant Growth Regul 26:55–68

Zonia L, Munnik T (2004) Osmotically induced cell swelling versus cell shrinking elicits specific changes in phospholipid signals in tobacco pollen tubes. Plant Physiol 134:813–823

Ionic Loops and Rebounds: Oxygen-Deprivation Signaling in Plants

Chalivendra C. Subbaiah

Abstract In flooded soils, oxygen is a serious limiting factor for plant growth and survival. However, as evident from their evolutionary diversification and ecological success, plants have acquired the ability to sense and adapt to oxygen deficits. Here, I attempt a synthesis of recent developments in oxygen-signaling research, in particular, the role of transient ionic (calcium) and redox (reactive oxygen species) changes in the perception of oxygen deprivation by plant cells. An emerging theme from this analysis is that the oxygen-signaling network comprises both positive and negative feedback loops that augment or moderate these cellular perturbations encoding the stress. Further, plant tolerance to oxygen deprivation seems to depend on the ability of cells to regulate these signaling circuits and rapidly attain homeostasis.

1 Introduction

Life evolved on the earth under strongly reducing conditions until the appearance of photoautotrophs and the release of oxygen via photolysis of water about two billion years ago (Papagiannis 1984). An immediate challenge posed by the nascent oxygenation of the earth to the then-existing organisms was that they needed to evolve antioxidant pathways and survive the strongly oxidizing environment. As the oxygen level rose, the ancient microbes were also under a massive but positive selection pressure to radically modify their metabolism. In particular, the acquisition of oxidative phosphorylation as a far more efficient (and switchable) alternative for fermentative pathways by the pioneering aerobes allowed a rapid explosion and elaboration of eukaryotic life in the Pre-Cambian period. The aerobic life style also entailed acquiring oxygen-sensitive switches that can modulate the expression of key metabolic genes appropriately, as oxygen levels fluctuated.

C.C. Subbaiah (✉)
Department of Biology, 1878 Campus Delivery, Colorado State University Fort Collins, CO 80521-1878, USA
e-mail: schalive@lamar.colostate.edu

F. Baluška and S. Mancuso (eds.), *Signaling in Plants*, Signaling and Communication in Plants, 195
DOI: 10.1007/978-3-540-89228-1_10, © Springer-Verlag Berlin Heidelberg 2009

Green plants continue to be the major contributors of oxygen, which is as much as 21% of the present Earth's atmosphere. At the same time, being sessile, plants are exposed to hypoxia (partial lack of oxygen) or even anoxia (total lack of oxygen) when they are flooded. Plants also experience oxygen-deprivation stress during their normal growth and development owing to poor diffusion of oxygen in tissues with closely packed cells. However, even terrestrially acclimated plant species can survive more than a few hours of submergence, indicating that plants possess exquisite oxygen-sensing mechanisms as well as robust response pathways. In addition to cell-autonomous signaling cascades (e.g., steps leading to the induction of glycolytic-fermentative pathway genes), multicellular plants possess mechanisms to perceive and mount coordinated responses to remote hypoxic stress signals (e.g., aerenchyma induction in aerial tissues following submergence of only roots; Mühlenbock et al. 2007). Adaptation to falling oxygen levels in the short term involves rapid but transient gene expression changes driving immediate metabolic adjustments (e.g., minimizing energy consumption while optimizing conservation, a switch from tricarboxylic acid cycle-oxidative phosphorylation to glycolytic-fermentative pathways or activating ionic and redox homeostasis pathways). Tolerance to chronic submergence requires, in addition, long-lasting alterations in gene expression culminating in anatomical and developmental modifications (e.g., growth retardation or "quiescence," self-pruning, formation of aerenchyma and adventitious roots) that would allow reestablishment of near-normal metabolism. Developmental programs of seeds, fruits, and tubers (other metabolically intensive and bulky tissues as well) are cued to the hypoxia prevailing in their inner cell layers to ensure successful perennation (Porterfield et al. 1999; Rolletscheck et al. 2002). Judicious activation of cell death pathways is also a part of the plant survival arsenal under energy crisis (Subbaiah et al. 2000; Gunawardena et al. 2001; Mühlenbock et al. 2007).

In addition to species-specific differences, responses to oxygen deficit vary even within a species, according to the magnitude (hypoxia versus anoxia), duration, and location (cell, tissue, or organ) of the stress (Okimoto et al. 1980; Subbaiah et al. 2000; Pang et al. 2006; reviewed in Geigenberger 2003; Bailey-Serres and Chang 2005; Bailey-Serres and Voesenek 2008). We are only beginning to understand how these stress variations are configured as well as how the multitude of response patterns are orchestrated by plants. Perturbations in a variety of signaling molecules, including ions, metabolites, macromolecular complexes, and hormones, have been observed following oxygen deprivation in plant cells, indicating combinations and permutations of these signals may encode diverse manifestations of oxygen deficit. Further, the diversity of signals reflects the intricate physiological ramifications of what is generally perceived as "predominantly an energy stress." The potential involvement of a wide gamut of signaling pathways may also mean a certain level of network robustness that ensures an adaptive response. On the other hand, the paucity of naturally occurring (or laboratory-derived) mutants that are impaired in oxygen signaling implies that the core pathway may be less redundant and (obviously) essential for survival.

2 Intracellular Signaling of Oxygen Deprivation

In addition to tissue-dependent or stress-intensity-dependent responses, plant cells seem to respond distinctly to internal or local oxygen deficits as against systemic signals of hypoxia or anoxia. It is not completely clear how plant cells encode and decode these two sources of stress. Recently there has been considerable progress in elucidating the role of second messengers, in particular, Ca^{2+} and reactive oxygen species (ROS), in perceiving oxygen deprivation within plant cells (reviewed in Bailey-Serres and Chang 2005; Subbaiah and Sachs 2007). The key questions still to be answered are how these ionic or redox perturbations are initiated, intercepted, and translated into gene expression changes. The following is a still-emerging scenario of how plant cells perceive oxygen deficits.

2.1 Is the Mitochondrion the Site of Oxygen Sensing in Plant Cells?

Since mitochondria are the primary consumers of oxygen, variations in oxygen availability are presumably sensed first in these organelles, leading to a retrograde regulation of nuclear genes (Subbaiah et al. 1994a). Inhibitors of the mitochondrial electron transport chain (mtETC) can trigger hypoxia-responsive genes under normoxia (Nie and Hill 1997; Yu et al. 2001; Subbaiah et al. 2006; Umbach AL, Zarkovic J, Yu J, Ruckle ME, McIntosh L, Hock JJ, Bingham, Subbaiah CC, White SJ, George RM, Rhoads DM, unpublished data), indicating that hypoxia-induced changes in the electron carriers or even the electron flux can act as the primary signal of oxygen deprivation in plant cells. In fact, a functional respiratory electron transport chain is required for the induction of hypoxic genes, as evidenced by genetic and transgenic analyses in well-characterized nonplant model systems (Kwast et al. 1999; Taylor et al. 1999; Guzy et al. 2005). In turn, hypoxia-induced alterations in gene expression are directed to fine-tune mitochondrial and cellular functions in response to changing cellular energy capacity and demands. For example, transcripts and proteins for tricarboxylic acid cycle enzymes or components of the mtETC are downregulated following oxygen deprivation and rapidly recovered upon reoxygenation. Prolonged anoxia leads to the induction of genes involved in cell lysis and death (Chang et al. 2000; Tsuji et al. 2000; Millar et al. 2004; Branco-Price et al. 2005; Gonzali et al. 2005; Huang et al. 2005; Ahsan et al. 2007), presumably triggered by events in mitochondria (Arpagaus et al. 2002; Virolainen et al. 2002). An attractive hypothesis supported by phylogenetic and experimental evidence is that the endosymbiotically originated mitochondria, similar to their pathogenic bacterial progenitors, determine the fate of their host cells. Frade and Michaelidis (1997) proposed that the eukaryotic programmed cell death is a consequence of acquiring aerobic metabolism via the endosymbiotic process and is regulated by cellular ATP levels. Several key enzymes of the cell death machinery, such as paracaspases and metacaspases belonging to the caspase-like protease superfamily, have homologs in *Alphaproteobacteria* (the prokaryotic ancestors of mitochondria), but not in Archaea, progenitors of present-day eukaryotes (Koonin and Aravind 2002). Thus, the centrality of the mitochondrion in sensing oxygen and the retrograde regulation of the cell fate by this organelle appear to be rooted in its prokaryotic lineage.

How is oxygen sensed in mitochondria? Except in some bacteria (see below), there is no evidence, so far, for the direct sensing of oxygen by a receptor-like protein in plants or other eukaryotes. In *Escherichia coli* and mammalian cells, proteins with Per–Arnt–Sim (PAS) domains (also known as the LOV domain for light, oxygen, voltage sensing) [e.g., *E. coli* Aer protein or hypoxia-induced factor (HIF) in mammals] are bona fide oxygen sensors (as indicated by genetic analysis) and link cellular energy status with oxygen-dependent gene expression (reviewed in Taylor and Zhulin 1999; Taylor 2007). These proteins do not bind oxygen but detect changes in redox status, proton motive force, or simply electron flux in the electron transport chain and interpret these perturbations as alterations in the energy/oxygen status (Taylor 2007). Nevertheless, PAS-domain proteins containing heme (e.g., FixL of *Bradyrhizobium japanicum*, NifL of *Azotobacter vinelandii*, DOS of *Mycobacterium tuberculosis*) bind oxygen and thus directly sense its availability (reviewed in Gilles-Gonzalez and Gonzalez 2005). This information is encoded as the differential autophosphorylation of an attached histidine kinase. Via a phospho-relay mechanism characteristic to all two-component signal transduction systems, these sensor proteins then activate a downstream transcription factor and thereby regulate hypoxic gene expression (Gong et al. 1998). In addition to a variety of PAS domains, other heme-based oxygen-sensing domains have been discovered recently in bacteria (e.g., the SCHIC domain of *Rhodobacter sphaeroides*; Moskvin et al. 2007). Many of these domains are phylogenetically conserved and, in some cases, recruited to carry out a different function (Taylor and Zhulin 1999; Moskvin et al. 2007). For example, the PAS-domain-containing proteins are well represented also in plants, some of which are key photoreceptors. In the *Arabidopsis* genome, there are 20 annotated genes that encode proteins with one or more PAS domains. Many of them belong to phytochrome or phototropin families, while others are functionally uncharacterized. An interesting possibility is that one or more of them could be located in mitochondria and function as an oxygen sensor, although it is not essential for a protein to be present inside the organelle to sense hypoxia signals generated in the mitochondrion. For example, HIF, the global regulator of hypoxic gene expression in mammalian cells, is extramitochondrial in its distribution and shuttles between the cytosol and the nucleus (Taylor and Zhulin 1999; Berchner-Pfannschmidt et al. 2008). Nonetheless, hypoxic stabilization and nuclear localization of HIF is mediated by ROS signals generated in mitochondria (Guzy et al. 2005). As discussed later, mobile signals are also initiated in mitochondria of plant cells upon oxygen deprivation and they are critical for oxygen-responsive gene expression.

2.2 *Cytosolic Calcium Perturbations Are Essential for Anoxic Gene Induction As Well As Postsubmergence Survival*

In maize cultured cells, oxygen deprivation triggers an immediate and reversible rise in cytosolic Ca^{2+} (Subbaiah et al. 1994a, 1998). These changes localize around mitochondria (Subbaiah et al. 1998) and are abolished by a putative mitochondrial Ca^{2+} channel blocker, ruthenium red (RR). RR also represses the induction of anoxia-responsive genes, alcohol dehydrogenase 1 (*adh1*; Subbaiah et al. 1994b, 1998), and sucrose synthase (*sh1*) in maize seedlings and cultured cells. Maize seedlings treated with RR are more sensitive to anoxic stress than *adh1* mutants, indicating that RR inhibited ionic fluxes act on the upstream events leading to gene expression changes (Subbaiah et al. 1994b). Ca^{2+} addition antagonizes RR effects, further supporting the inference that the ionic changes probably initiated in mitochondria signal anoxic responses in maize (Subbaiah et al. 1994a, b, 1998). Cytosolic Ca^{2+} changes are also an essential component of hypoxia signaling in *Arabidopsis* (Sedbrook et al. 1996; Chung and Ferl 1999) and have been implicated in the perception of oxygen deprivation in rice, barley, and arrowhead, as well (Nakazono et al. 2000; Tsuji et al. 2000; Tamura et al. 2001; Baxter-Burrell et al. 2002; Nie et al. 2006). While indicating that the pathway is conserved among plants, these studies also hint at the species- and organ-specific differences in Ca^{2+} signatures as well as stores contributing to these changes. The full complexity of Ca^{2+} signaling in oxygen perception still remains to be explored.

2.3 *Oxygen-Deprivation-Induced Cytosolic Ca^{2+} Signals Originate in Mitochondria*

Corroborative evidence for the role of mitochondria as an origin of the cytosolic Ca^{2+} signal came from the measurement of mitochondrial Ca^{2+} dynamics in response to anoxia (Subbaiah et al. 1998). In maize cells, mitochondria release Ca^{2+} immediately after the cessation of oxygen supply. A treatment with methomyl, an uncoupler that triggers mitochondrial Ca^{2+} release, induces the *adh1* gene in normoxic CMS-T maize seedlings (Kuzmin et al. 2004). There is growing evidence for the critical role of ionic changes originating in mitochondria in eliciting anoxic gene expression changes in other species as well. An anoxia-responsive hemoglobin gene (*Hb1*) is induced in barley aleurone cells by pharmacological disruption of mitochondrial ATP synthesis or the mtETC under aerated conditions (Nie and Hill 1997) and the same inhibitors are known to induce Ca^{2+} release from mitochondria in normoxic cells (Subbaiah et al. 1998). Anoxic induction of barley *Hb1* and *Adh1* is inhibited by RR in a calcium-dependent manner (Nie et al. 2006) and *Adh1* and *Pdc1*: *pyruvate decarboxylase1* are regulated likewise in rice (Nakazono et al. 2000; Tsuji et al. 2000). The anoxic suppression of rice AOX1a messenger RNA in a Ca^{2+} -dependent manner (i.e., a reversal of suppression by the Ca^{2+} channel blocker RR and the prevention of RR action in the presence of Ca^{2+}; Tsuji et al. 2000), also underscores the link between mitochondrial Ca^{2+} dynamics, electron transport, and altered nuclear gene expression under oxygen deprivation.

2.4 *ROS Induction May Be the Trigger for Mitochondrial Ca^{2+} Efflux in Oxygen-Deprived Plant Cells*

How is mitochondrial Ca^{2+} efflux initiated in oxygen-starved plant cells? In maize, mitochondria release their matrix calcium within 10 min of oxygen deprivation –well before membrane depolarization sets in (which occurs only after 30 min of uninterrupted anoxia). It follows that the Ca^{2+} efflux during early anoxia is regulated and not mediated by a passive leakage of the ion (Subbaiah et al. 1998). H^+/Ca^{2+} antiport is implicated in mediating anoxic mitochondrial Ca^{2+} release, on the basis of the cytosolic pH changes that are known to accompany oxygen deprivation (Gibbs and Greenway 2003). However, mitochondrial matrix pH may be unaffected by anoxia as reported in *Artemia* embryos (Kwast and Hand 1996) and, further, the vacuole appears to be the major proton sink in oxygen-deprived plant cells (Carystinos et al. 1995; Gibbs and Greenway 2003). Recent studies indicate that oxygen deprivation may trigger a rapid burst of ROS in mitochondria (Branco-Price et al. 2006; Subbaiah et al. 2006). A surge in mitochondrial ROS is essential for the hypoxia-mediated mitochondrial Ca^{2+} release and cytosolic Ca^{2+} increase in rat myocytes (Waypa et al. 2002). Although these details are yet to be worked out in plant mitochondria, the plot regarding the role of ROS in oxygen sensing in plants is thickening.

An oxidative burst under limiting oxygen levels may sound paradoxical but the induction of ROS is now recognized as an early event and an essential component of oxygen-deprivation signaling in many species. In yeast and mammalian cells, ROS are generated in mitochondria and released into the cytosol when electrons are backed up in the mtETC under oxygen deprivation. A buildup of ROS in the cytosol is required for hypoxia-responsive gene expression (Dirmeier et al. 2002; Guzy et al. 2005). In mammalian cells, complex III of the mtETC is the primary site of ROS generation under hypoxia and RNA interference suppression of the complex III assembly abrogates the ROS burst as well as hypoxic gene induction (Waypa et al. 2002; Guzy et al. 2005).

2.4.1 Role of Alternative Oxidase and Mitochondrial ROS in Oxygen Sensing in Plant Cells

The mtETC is a source of ROS and can potentially act as a site of oxygen sensing in plant cells as well. In fact, inhibitors of the mtETC are known to induce both ROS and hypoxic genes (Maxwell et al. 1999; Yao et al. 2002; Nie and Hill 1997;

Subbaiah et al. 2006, unpublished data). However, in nature, i.e., in the absence of respiratory poisons, the oxygen-signaling network is responsive to even mild alterations in oxygen availability. Hypoxic genes are induced at much higher concentrations than the $K_m(O_2)$ of cytochrome oxidase (Paul and Ferl 1991; Bucher et al. 1994), indicating a low-affinity system could be a more likely sensor. Plant mitochondria have the alternative oxidase (AOX), an additional but non-proton-pumping terminal oxidase that directly oxidizes ubiquinol and reduces oxygen. This energy dissipation pathway is used by plants to prevent the generation of semiubiquinone radical (i.e., via partial oxidation of ubiquinol) and consequent generation of ROS when complex III or complex IV is not fully functional (reviewed in Rhoads et al. 2006). However, even under a mild decrease in oxygen concentration, AOX is likely to be inhibited, owing to its much lower affinity for oxygen (K_m = 10–25 μM; Ribas-Carbo et al. 1994) than that of cyclooxygenase (K_m = 140 nM; Millar et al. 1994). In fact, AOX activity is completely abolished under prolonged hypoxia (Szal et al. 2003). An inhibition of AOX activity, a built-in antioxidant pathway of the mtETC, can instantly raise mitochondrial ROS levels (Popov 2003). The induction of ROS, in turn, may initiate mitochondrial Ca^{2+} efflux into the cytosol (Waypa et al. 2002) and ultimately trigger gene expression changes in the nucleus.

The above proposal was tested in *Arabidopsis* suspension-cultured cells using a pharmacological approach (Subbaiah et al. 2006, unpublished data). Two commonly used AOX activity inhibitors, salicyl hydroxamic acid and *n*-propyl gallate, induced messenger RNA accumulation as well as the activity of ADH, a hypoxia-inducible gene, under ambient oxygen levels. These inhibitors or hypoxia also induced AOX1a message, a commonly observed response to increased ROS levels. In fact, AOX inhibitors caused a rapid induction of mitochondrial ROS levels in treated cells, indicating that an inhibition of AOX activity leads to an increase in ROS and the consequent induction of the ADH gene under hypoxia. RR inhibited the induction of ADH by AOX inhibitors, suggesting that ADH induction by ROS is promoted by mitochondrial Ca^{2+} release. However, RR did not affect AOX1a message levels, indicating that AOX induction is primarily a direct response to increased ROS levels and does not involve Ca^{2+} fluxes (Subbaiah CC, Shah N, Rhoads DM, unpublished data). The pattern of AOX expression under oxygen deprivation merits a little elaboration. The gene is induced under mild and short-term hypoxia (Klok et al. 2002; Subbaiah et al. 2006) but is inhibited both at transcript and at protein levels by prolonged hypoxia or anoxia (Szal N et al. 2003; Liu et al. 2005; (Subbaiah CC, Shah N, Rhoads DM, unpublished data). This pattern is consistent with the ROS-mitigating action of AOX as well as its high K_m for oxygen, since the protein becomes ineffectual as the mitochondrial oxygen concentration decreases to low micromolar levels with no immediate prospect of restoration to normoxia. In addition, the repression of AOX expression under prolonged or severe oxygen deprivation seems to be directed and not part of a general decrease in transcriptional and translational activities that occurs under severe energy deficit. In rice seedlings, RR restores the levels of AOX message under prolonged submergence, while Ca^{2+} addition prevents RR action, indicating that the anoxia-induced Ca^{2+} perturbations negatively regulate AOX expression under prolonged oxygen deprivation (Tsuji et al. 2000). In turn, AOX inhibition and ROS induction seem to influence mitochondrial Ca^{2+} dynamics in a feedback loop, thus enhancing Ca^{2+} and ROS release from mitochondria into the cytosol.

2.4.2 Hypoxic ROS Burst at the Plasma Membrane and Potential Cross-Talk with Mitochondrial Signals

The participation of ROS in hypoxia signaling in plant cells was first documented in the Bailey-Serres laboratory, while characterizing an *Arabidopsis* insertional mutant that shows constitutive ADH activity (i.e., high ADH activity even under normoxia; Baxter-Burrell et al. 2002). The inactivated gene encodes ROPGAP4, a negative regulator of the Rho-like GTPases of plants (ROP) signaling pathway. ROP is known to interact with a NADPH oxidase (NOX; encoded by *rboh* genes) localized on the plasma membrane and trigger ROS production (Wong et al. 2007). Pharmacological treatments and ROS measurements indicated that ADH expression is proportional to ROS levels in the wild type and *ropgap4* mutant seedlings. Further, ROPGAP4 induction under prolonged hypoxia and its action as an attenuator of ROS production (via its inhibition of ROP) prevent unregulated oxidative stress and tissue death (Baxter-Burrell et al. 2002). The participation of NOX proteins in hypoxia signaling is also suggested by the hypoxia-inducibility of genes encoding them (Bailey-Serres and Chang 2005). Hypoxia or anoxia is the predominant abiotic stress that induces the gene family. Out of ten genes that encode NOX proteins in *Arabidopsis*, eight are expressed in roots (plant parts most likely to experience oxygen deprivation in nature, either by submergence or by soil compaction) and five are induced by oxygen deprivation (reviewed in Sagi and Fluhr 2006). Further, the induction of *rhob* genes appears to be regulated by an autocatalytic feedforward loop, i.e., the induction of a transcription factor, Zat12, by an initial rise in ROS and consequent transcriptional activation of *rhob* genes (Davletova et al. 2005) which may lead to a continued induction of ROS. Plant NOX proteins bind Ca^{2+} and are dependent on stimulus-mediated elevation of cytosolic Ca^{2+} for activation (Sagi and Fluhr 2001; Ogasawara et al. 2008). In turn, ROS catalyzed by NOX proteins can activate Ca^{2+} influx via nonselective cation channels (Demidchik and Maathuis 2007) and thus in a positive feedback loop (Kadota et al. 2004; Ogasawara et al. 2008; Takeda et al. 2008; reviewed in Sagi and Fluhr 2006) can enhance both ionic and redox signals at the plasma membrane.

A key question is how the ROP or NOX is initially activated under oxygen deprivation. One possible scenario calls for the interaction of mitochondrially originated signals with the molecular players at the plasma membrane. NOX is activated upon its binding to ROP in a Ca^{2+}-dependent manner (Wong et al. 2007). Even the hypoxic induction of ROP seems to involve cytosolic Ca^{2+} elevation (Baxter-Burrell et al. 2002), indicating that a release of the ion from mitochondria may trigger the events at the plasma membrane.

2.4.3 Blast from the Past?: Why Oxidative Stress Is an Inducer of Anaerobic Genes

It is now clear that plants, along with other eukaryotes, use ROS as second messengers in hypoxia perception. Paradoxically, most aerobes (including prokaryotes) also show the induction of hypoxic genes in response to other sources of oxidative stress even under fully oxygenated conditions (Desikan et al. 2001; Abercrombie et al. 2008; Chang et al. 2006; Pouysségur and Mechta-Grigoriou 2006). External oxidants, such as ozone, can directly act on components of the mitochondrial electron transport chain (mtETC) (e.g., cytochrome *c* oxidase) and induce ROS generation in mitochondria (Tiwari et al. 2002; Ederli et al. 2006). It is also possible that this response is a relic of the adaptive process by early aerobic organisms to highly fluctuating oxygen concentrations. Under hypoxia, ROS are generated owing to the partial reduction or incomplete oxidation of the ETC. Cells seem to perceive even nonhypoxic sources of oxidative stress as impending hypoxia and induce anaerobic genes (particularly, those encoding glycolytic-fermentative pathway enzymes) as a preemptive measure.

2.4.4 Ionic and Redox Homeostasis Under Oxygen Deprivation: A Role for Calcium

An imminent danger to cellular integrity under energy-deprived conditions is unregulated ion fluxes, particularly of Ca^{2+} and protons. In addition to mediating adaptive gene expression responses, oxygen-deprivation-induced cytosolic Ca^{2+} changes are critical to bring about ionic homeostasis (Subbaiah and Sachs 2003). Recent work suggests that Ca^{2+} changes under oxygen deprivation can potentially impact the redox homeostasis as well. As discussed in Sect. 2.4.2, cytosolic Ca^{2+} changes seem to be important in initiating ROS production particularly at the plasma membrane. At the same time, a continued cytosolic Ca^{2+} elevation may serve as a negative-feedback mechanism to modulate the duration and the intensity of oxidative burst at the plasma membrane. Such a feedback mechanism is proposed on the basis of the inhibitory effect of high Ca^{2+} levels on the interaction between NOX and ROP proteins (Wong et al. 2007). In addition, the H_2O_2-scavenging enzyme catalase is shown to bind and be regulated by the ubiquitous calcium-binding protein, calmodulin (Yang and Poovaiah 2002). Further, the Ca^{2+}-dependent downregulation of AOX under prolonged hypoxia (Tsuji et al. 2000) may also act as a switch for the induction of other antioxidant systems in mitochondria or elsewhere that can function under falling oxygen levels.

3 Conclusions

Oxygen is an ancient signal molecule that prompted the evolution of many biochemical pathways and gene-regulatory switches prevalent in present-day organisms, including plants. In the past decade there have been significant advances in our understanding of ionic, metabolic, and molecular components of plant oxygen sensing. However, the information is sketchy and a unified picture is still emerging. The mitochondrion appears to be the site where primary (ionic and redox) signals of oxygen deprivation are generated and, thus, central to oxygen sensing.

Acknowledgement I thank Barry Taylor (Loma Linda University, CA, USA) for a critical reading of the manuscript.

References

Abercrombie J, Halfhill M, Ranjan P, Rao M, Saxton A, Yuan J, Stewart C (2008) Transcriptional responses of *Arabidopsis thaliana* plants to As (V) stress. BMC Plant Biol 8:87

Ahsan N, Lee D-G, Lee S-H, Kang KY, Bahk JD, Choi MS, Lee I-J, Renaut J, Lee B-H (2007) A comparative proteomic analysis of tomato leaves in response to waterlogging stress. Physiol Planta 131:555–570

Arpagaus S, Rawyler AJ, Braendle RA (2002) Occurrence and characteristics of the mitochondrial permeability transition in plants. J Biol Chem 277:1780–1787

Bailey-Serres J, Chang R (2005) Sensing and signalling in response to oxygen deprivation in plants and other organisms. Ann Bot (Lond) 96:507–518

Bailey-Serres J, Voesenek LACJ (2008) Flooding stress: acclimations and genetic diversity. Annu Rev Plant Biol 59:313–39

Baxter-Burrell A, Yang Z, Springer PS, Bailey-Serres J (2002) RopGAP4-dependent rop GTPase rheostat control of Arabidopsis oxygen deprivation tolerance. Science 296:2026–2028

Berchner-Pfannschmidt U, Frede S, Wotzlaw C, Fandrey J (2008) Imaging of the hypoxia-inducible factor pathway: insights into oxygen sensing. Eur Respir J 32:210–217

Branco-Price C, Kawaguchi R, Ferreira RB, Bailey-Serres J (2005) Genome-wide analysis of transcript abundance and translation in Arabidopsis seedlings subjected to oxygen deprivation. Ann Bot 96:647–660

Branco-Price C, Chang R, Ferreira R, Bailey-Serres J (2006) Mitochondria-initiated signaling in Arabidopsis under low oxygen. Poster presented at Plant Biology 2006, American Society of Plant Biologists, Boston, 5–9 Aug 2006, abstract P36049

Bucher M, Brandle R, Kuhlemeier C (1994) Ethanolic fermentation in transgenic tobacco expressing Zymomonas mobilis pyruvate decarboxylase. EMBO J 13:2755–2763

Carystinos GD, Heather MR, Monroy AF, Rajinder DS, Poole RJ (1995) Vacuolar H^+-translocating pyrophosphatase is induced by anoxia or chilling in seedlings of rice. Plant Physiol 108:641–649

Chang W, Small DA, Toghrol F, Bentley WE (2006) Global transcriptome analysis of *Staphylococcus aureus* response to hydrogen peroxide. J Bacteriol 188:1648–1659

Chang WWP, Huang L, Shen M, Webster C, Burlingame AL, Roberts JKM (2000) Patterns of protein synthesis and tolerance of anoxia in root tips of maize seedlings acclimated to a low-oxygen environment, and identification of proteins by mass spectrometry. Plant Physiol 122:295–317

Chung H-J, Ferl RJ (1999) *Arabidopsis* alcohol dehydrogenase expression in both shoots and roots is conditioned by root growth environment. Plant Physiol 121:429–436

Davletova S, Schlauch K, Coutu J, Mittler R (2005) The zinc-finger protein Zat12 plays a central role in reactive oxygen and abiotic stress signaling in *Arabidopsis*. Plant Physiol 139:847–856

Demidchik V, Maathuis FJM (2007) Physiological roles of nonselective cation channels in plants: from salt stress to signalling and development. New Phytol 175:387–404

Desikan R, Mackerness SAH, Hancock JT, Neill SJ (2001) Regulation of the *Arabidopsis* transcriptome by oxidative stress. Plant Physiol 127:159–172

Dirmeier R, O'Brien KM, Engle M, Dodd A, Spears E, Poyton RO (2002) Exposure of yeast cells to anoxia induces transient oxidative stress. Implications for the induction of hypoxic genes. J Biol Chem 277:34773–34784

Ederli L, Morettini R, Borgogni A, Wasternack C, Miersch O, Reale L, Ferranti F, Tosti N, Pasqualini S (2006) Interaction between nitric oxide and ethylene in the induction of alternative oxidase in ozone-treated tobacco plants. Plant Physiol 142:595–608

Frade JM, Michaelidis TM (1997) Origin of eukaryotic programmed cell-death - a consequence of aerobic metabolism. BioEssays 19:827–832

Geigenberger P (2003) Response of plant metabolism to too little oxygen. Curr Opin Plant Biol 6:247–256

Gibbs J, Greenway H (2003) Review: mechanisms of anoxia tolerance in plants. I. Growth, survival and anaerobic catabolism. Funct Plant Biol 30:1–47

Gilles-Gonzalez MA, Gonzalez G (2005) Heme-based sensors: defining characteristics, recent developments, and regulatory hypotheses. J Inorg Biochem 99:1–22

Gong W, Hao B, Mansy SS, Gonzalez G, Gilles-Gonzalez MA, Chan MK (1998) Structure of a biological oxygen sensor: a new mechanism for heme-driven signal transduction. Proc Natl Acad Sci U S A 95:15177–15182

Gonzali S, Loreti E, Novi G, Poggi A, Alpi A, Perata P (2005) The use of microarrays to study the anaerobic response in *Arabidopsis*. Ann Bot 96:661–668

Gunawardena A, Pearce DM, Jackson MB, Hawes CR, Evans DE (2001) Characterisation of programmed cell death in aerenchyma formation. Planta 212:205–214

Guzy RD, Hoyos B, Robin E, Chen H, Liu L, Mansfield KD, Simon MC, Hammerling U, Schumacker PT (2005) Mitochondrial complex III is required for hypoxia-induced ROS production and cellular oxygen sensing. Cell Metab 1:401–408

Huang S, Greenway H, Colmer TD, Millar AH (2005) Protein Synthesis by rice coleoptiles during prolonged anoxia: implications for glycolysis, growth and energy utilization. Ann Bot 96:703–715

Kadota Y, Goh T, Tomatsu H, Tamauchi R, Higashi K, Muto S, Kuchitsu K (2004) Cryptogein-induced initial events in tobacco BY-2 cells: pharmacological characterization of molecular relationship among cytosolic Ca^{2+} transients, anion efflux and production of reactive oxygen species. Plant Cell Physiol 45:160–170

Klok EJ, Wilson IW, Wilson D, Chapman SC, Ewing RM, Somerville SC, Peacock WJ, Dolferus R, Dennis ES (2002) Expression profile analysis of the low-oxygen response in Arabidopsis root cultures. Plant Cell 14:2481–2494

Koonin EV, Aravind L (2002) Origin and evolution of eukaryotic apoptosis: the bacterial connection. Cell Death Differ 9:394–404

Kuzmin EV, Karpova OV, Elthon TE, Newton KJ (2004) Mitochondrial respiratory deficiencies signal up-regulation of genes for heat shock proteins. J Biol Chem 279:20672–20677

Kwast KE, Hand SC (1996) Oxygen and pH regulation of protein synthesis in mitochondria from *Artemia franciscana* embryos. Biochem J 313:207–213

Kwast KE, Burke PV, Staahl BT, Poyton RO (1999) Oxygen sensing in yeast: evidence for the involvement of the respiratory chain in regulating the transcription of a subset of hypoxic genes. Proc Natl Acad Sci U S A 96:5446–5451

Liu F, VanToai T, Moy LP, Bock G, Linford LD, Quackenbush J (2005) Global transcription profiling reveals comprehensive insights into hypoxic response in *Arabidopsis*. Plant Physiol 137:1115–1129

Maxwell DP, Wang Y, McIntosh L (1999) The alternative oxidase lowers mitochondrial reactive oxygen production in plant cells. Proc Natl Acad Sci U S A 96:8271–8276

Millar AH, Bergersen FJ, Day DA (1994) Oxygen affinity of terminal oxidases in soybean mitochondria. Plant Physiol Biochem 32:847–852

Millar AH, Trend AE, Heazlewood JL (2004) Changes in the mitochondrial proteome during the anoxia to air transition in rice focus around cytochrome-containing respiratory complexes. J Biol Chem 279:39471–39478

Moskvin OV, Kaplan S, Gilles-Gonzalez M-A, Gomelsky M (2007) Novel heme-based oxygen sensor with a revealing evolutionary history. J Biol Chem 282:28740–28748

Mühlenbock P, Plaszczyca M, Mellerowicz E, Karpinski S (2007) Lysigenous aerenchyma formation in *Arabidopsis* is controlled by *LESION SIMULATING DISEASE1*. Plant Cell 19:3819–3830
Nakazono M, Tsuji H, Li YH, Saisho D, Arimura S, Tsutsumi N, Hirai A (2000) Expression of a gene encoding mitochondrial aldehyde dehydrogenase in rice increases under submerged conditions. Plant Physiol 124:587–598
Nie X, Hill RD (1997) Mitochondrial respiration and hemoglobin gene expression in barley aleurone tissue. Plant Physiol 114:835–840
Nie X, Durnin DC, Igamberdiev AU, Hill RD (2006) Cytosolic calcium is involved in the regulation of barley hemoglobin gene expression. Planta 223:542–549
Ogasawara Y, Kaya H, Hiraoka G, Yumoto F, Kimura S, Kadota Y, Hishinuma H, Senzaki E, Yamagoe S, Nagata K, Nara M, Suzuki K, Tanokura M, Kuchitsu K (2008) Synergistic activation of the Arabidopsis NADPH oxidase AtrbohD by Ca^{2+} and phosphorylation. J Biol Chem 283:8885–8892
Okimoto R, Sachs MM, Porter EK, Freeling M (1980) Patterns of polypeptide synthesis in various maize organs under anaerobiosis. Planta 150:89–94
Pang JY, Newman I, Mendham N, Zhou M, Shabala S (2006) Microelectrode ion and O_2 fluxes measurements reveal differential sensitivity of barley root tissues to hypoxia. Plant Cell Environ 29:1107–1121
Papagiannis MD (1984) Life-related aspects of stellar evolution. Orig Life 14:43–50
Paul A-L, Ferl RJ (1991) Adh1 and Adh2 regulation. Maydica 36:129–134
Popov VN (2003) Possible role of free oxidation processes in the regulation of reactive oxygen species production in plant mitochondria. Biochem Soc Trans 31:1316–1317
Porterfield DM, Kuang A, Smith PJS, Crispi ML, Musgrave ME (1999) Oxygen-depleted zones inside reproductive structures of Brassicaceae: implications for oxygen control of seed development. Can J of Bot 77:1439–1446
Pouysségur J, Mechta-Grigoriou F (2006) Redox regulation of the hypoxia-inducible factor. Biol Chem 387:1337–1346
Rhoads DM, Umbach AL, Subbaiah CC, Siedow JN (2006) Mitochondrial ROS: contribution to oxidative stress and inter-organellar signaling. Plant Physiol 141:357–366
Ribas-Carbo M, Berry JA, Azcon-Bieto J, Siedow JN (1994) The reaction of the plant mitochondrial cyanide-resistant alternative oxidase with oxygen. Biochim Biophys Acta 118:205–212
Rolletschek H, Borisjuk L, Koschorreck M, Wobus U, Weber H (2002) Legume embryos develop in a hypoxic environment. J Exp Bot 53:1099–1107
Sagi M, Fluhr R (2001) Superoxide production by plant homologues of the gp91(phox) NADPH oxidase: modulation of activity by calcium and by tobacco mosaic virus infection. Plant Physiol 126:1281–1290
Sagi M, Fluhr R (2006) Production of reactive oxygen species by plant NADPH oxidases. Plant Physiol 141:336–340
Sedbrook JC, Kronebusch PJ, Borisy GG, Trewavas AJ, Masson PH (1996) Transgenic AEQUORIN reveals organ-specific cytosolic Ca^{2+} responses to anoxia and *Arabidopsis thaliana* seedlings. Plant Physiol 111:243–257
Subbaiah CC, Sachs MM (2003) Molecular and cellular adaptations of maize to flooding stress. Ann Bot 91:119–127
Subbaiah CC, Sachs MM (2007) Responses to oxygen deprivation and potential for enhanced flooding tolerance in Maize. In: Bennitzen J, Hake S (eds) The maize handbook, 2nd edn, Springer-Verlag, Berlin Heidelberg
Subbaiah CC, Bush DS, Sachs MM (1994a) Elevation of cytosolic calcium precedes anoxic gene expression in maize suspension-cultured cells. Plant Cell 6:1747–1762
Subbaiah CC, Zhang J-K, Sachs MM (1994b) Involvement of intracellular calcium in anaerobic gene expression and survival of maize seedlings. Plant Physiol 105:369–376
Subbaiah CC, Bush DS, Sachs MM (1998) Mitochondrial contribution to the anoxic Ca^{2+} signal in maize suspension-cultured cells. Plant Physiol 118:759–771

Subbaiah CC, Kollipara KP, Sachs MM (2000) A Ca^{2+}-dependent cysteine protease is associated with anoxia-induced root tip death in maize. J Exp Bot 51:721–730
Subbaiah CC, Shah N, Rhoads DM (2006) Role of reactive oxygen species and alternative oxidase in hypoxia signaling in Arabidopsis. Poster presented at Plant Biology 2006, American Society of Plant Biologists, Boston, 5–9 Aug 2006, abstract P36053
Szal B, Jolivet Y, Hasenfratz-Sauder MP, Dizengremel P, Rychter AM (2003) Oxygen concentration regulates alternative oxidase expression in barley roots during hypoxia and post-hypoxia. Physiol Plant 119:494–502
Takeda S, Gapper C, Kaya H, Bell E, Kuchitsu K, Dolan L (2008) Local positive feedback regulation determines cell shape in root hair cells. Science 319:1241–1244
Tamura S, Kuramochi H, Ishizawa K (2001) Involvement of calcium ion in the stimulated shoot elongation of arrowhead tubers under anaerobic conditions. Plant Cell Physiol 42:717–722
Taylor BL (2007) Aer on the inside looking out: paradigm for a PAS–HAMP role in sensing oxygen, redox and energy. Mol Microbiol 65:1415–1424
Taylor BL, Zhulin IB (1999) PAS domains: internal sensors of oxygen, redox potential, and light. Microbiol Mol Biol Rev 63:479–506
Taylor BL, Zhulin IB, Johnson MS (1999) Aerotaxis and other energy-sensing behavior in bacteria. Annu Rev Microbiol 53:103–128
Tiwari BS, Belenghi B, Levine A (2002) Oxidative stress increased respiration and generation of reactive oxygen species, resulting in ATP depletion, opening of mitochondrial permeability transition, and programmed cell death. Plant Physiol 128:1271–1281
Tsuji H, Nakazono M, Saisho D, Tsutsumi N, Hirai A (2000) Transcript levels of the nuclear-encoded respiratory genes in rice decrease by oxygen deprivation: evidence for involvement of calcium in expression of the alternative oxidase 1a gene. FEBS Lett 471:201–204
Umbach AL, Zarkovic J, Yu J, Ruckle ME, McIntosh L, Hock JJ, Bingham, Subbaiah CC, White SJ, George RM, Rhoads DM is the full list of authors and their initials.
Virolainen E, Blokhina O, Fagerstedt K (2002) Ca^{2+}-induced high amplitude swelling and cytochrome c release from wheat (*Triticum aestivum* L.) mitochondria under anoxic stress. Ann Bot 90:509–516
Waypa GB, Marks JD, Mack MM, Boriboun C, Mungai PT, Schumacker PT (2002) Mitochondrial reactive oxygen species trigger calcium increases during hypoxia in pulmonary arterial myocytes. Circ Res 91:719–726
Wong HL, Pinontoan R, Hayashi K, Tabata R, Yaeno T, Hasegawa K, Kojima C, Yoshioka H, Iba K, Kawasaki T, Shimamoto K (2007) Regulation of rice NADPH oxidase by binding of Rac GTPase to its N-terminal extension. Plant Cell 19:4022–4034
Yao N, Tada Y, Sakamoto M, Nakayashiki H, Park P, Tosa Y, Mayama S (2002) Mitochondrial oxidative burst involved in apoptotic response in oats. Plant J 30:567–579
Yang T, Poovaiah BW (2002) Hydrogen peroxide homeostasis: activation of plant catalase by calcium/calmodulin. Proc Natl Acad Sci U S A 99:4097–4102
Yu J, Nickels R, McIntosh L (2001) A genome approach to mitochondrial-nuclear communication in *Arabidopsis*. Plant Physiol Biochem 39:345–353

Signaling in Plant Gravitropism

Patrick H. Masson, John Stanga, Carolyn Neal, Laura Vaughn Katherine Baldwin, and Gengxiang Jia

Abstract Most plant organs use gravity as a guide for growth. Gravity sensing occurs mainly in specialized cells, named "statocytes," through the sedimentation of starch-filled plastids (amyloplasts). This movement triggers a pathway that leads to lateral auxin transport across the stimulated organ, responsible for curvature at the elongation zone. In this review, we discuss our current understanding of the molecular mechanisms that mediate gravity sensing and signal transduction in the statocytes of higher-plant organs. We also briefly describe the machinery that governs the transport of auxin across and along the stimulated organ, and refer to "Auxin and the Communication Between Plant Cells" by Peter Nick (this volume) for a discussion of the mechanisms that mediate cellular responses to auxin. We end this chapter by placing gravitropism within the larger context of complex behavioral plant growth responses to combinations of directional cues that are pervasive within a plant's environment.

1 Introduction

Gravitropism, which defines the ability of many plant and fungal organs to direct growth at a defined angle from the gravity vector (named "gravity set point angle"), is an important growth response that modulates the morphogenesis of these organisms. A long history of research in this field has shown that most plant organs use gravity as a guide for growth (reviewed in Boonsirichai et al. 2002). However, each organ interprets this directional cue in its own way, which also varies depending on

P.H. Masson (✉)
University of Wisconsin-Madison, 3262 Genetics-Biotechnology Center Building, 425 Henry Mall, Madison, WI 53706, USA
e-mail: phmasson@wisc.edu

J. Stanga, C. Neal, L. Vaughn, K. Baldwin, and G. Jia
Laboratory of Genetics, University of Wisconsin-Madison, 425G Henry Mall, Madison, WI 53706, USA

F. Baluška and S. Mancuso (eds.), *Signaling in Plants*, Signaling and Communication in Plants, 209
DOI: 10.1007/978-3-540-89228-1_11, © Springer-Verlag Berlin Heidelberg 2009

the organ's developmental stage and physiological status. In germinating seedlings, gravitropism dictates upward shoot growth and downward root orientation, thereby allowing these organs to quickly reach environments that allow them to fulfill their primary functions: photosynthesis, gas exchange, and reproduction for shoots; water and nutrient uptake and anchorage for roots. As the plant grows, secondary organs are initiated and grow along a vector that reflects their own interpretation of the gravity vector. Lateral branches of shoots and roots tend to grow at a defined angle from the gravity vector, distinct from the vertical or horizontal orientations. Leaves will display an angle from gravity that may vary depending on the time of day, the environmental conditions, and the age of the plant. At the extreme range of the spectrum, specialized lateral roots of mangrove plants (*Rhizophora mangle*), called "pneumatophores," display a negative gravitropic response (upward vertical growth) that brings them above the muddy surface of intertidal swamps, allowing them to aerate the root system. Conversely, the upward growth habit of peanut flowers is reverted upon fertilization, which triggers the downward outgrowth of a specialized organ at the basis of the fertilized ovule, the gynophore. The growth of this organ drives the developing fruits into the ground, a process that is needed for proper fruit development and maturation. Also, many plants develop specialized stems named "stolons" that tend to grow horizontally outward, either above or below the soil surface, producing new clones at their tip. This process permits a rapid colonization of the surroundings.

These disparate examples of plant organs with distinct gravity-guided growth habits testify to the existence of organ-specific developmental programs that dictate the direction of growth relative to gravity in accordance with the distinct functional constraints associated with these organs (Firn and Digby 1997). Overall, these responses combine with other responses to environmental cues such as light, touch, gradients in humidity, oxygen, ions, chemicals, and temperature to give a plant its final form, which is adapted to the characteristics of its surroundings, and contributes to its fitness within the environment (Galen et al. 2004).

To develop a gravitropic response, plant organs have to sense changes in their orientation within the gravity field and transduce the corresponding physical information into a physiological or biochemical signal. As discussed later, this physiological signal seems to be found in the form of a lateral gradient of auxin across the stimulated organ, which dictates differential cellular elongation on opposite flanks of the corresponding elongation zone, responsible for tip curvature (reviewed in Trewavas 1992; Fig. 1).

In this chapter, we review the molecular mechanisms that modulate the gravitropic response of higher-plant organs with special emphasis on the early phases of gravity sensing and signal transduction. We briefly describe the molecular mechanisms that mediate the polar transport of auxin in a gravity-dependent manner and discuss possible interactions between gravitropism and the tropic responses to other environmental stimuli. We would like to refer the reader to "Auxin and the Communication Between Plant Cells" by Peter Nick (this volume) for a more thorough discussion of mechanisms that mediate the polar transport of, and specific cellular responses to, auxin.

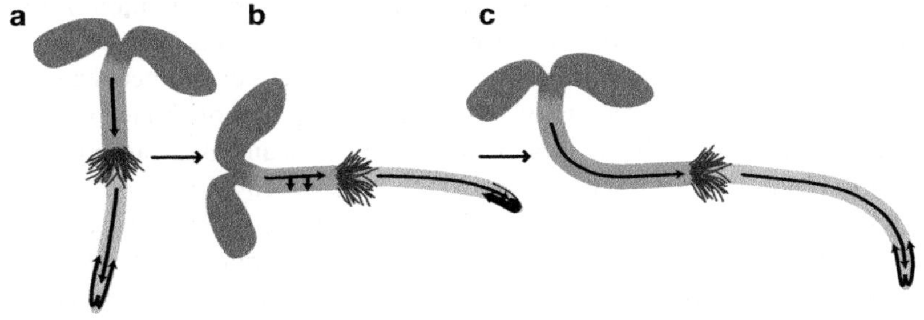

Fig. 1 Polar auxin transport and gravitropism in a young dicot seedling. **a** In a vertically oriented seedling, auxin, which is mainly synthesized in the shoot apex, is transported basipetally toward the root. There, auxin is transported acropetally toward the tip, where it accumulates in a region that overlaps with the quiescent center and upper cap. In that region, auxin is redistributed symmetrically to peripheral tissues (lateral cap), then transported basipetally through lateral cap and epidermal cells toward the elongation zone, where it functions as an inhibitor of cell elongation. **b** Upon gravistimulation, auxin is transported laterally toward the bottom flank of the hypocotyl, where it inhibits cell elongation, promoting an upward curvature. In roots, gravistimulation promotes a lateral transport of auxin downward across the cap. The corresponding lateral auxin gradient is then transported basipetally toward the elongation zone, where it promotes downward tip curvature. **c** At the end of the gravitropic response, the tips of both hypocotyl and root have reoriented into a vertical position, and the transport of auxin has resumed according to its pregravistimulation pattern. In these panels, the direction of auxin transport is indicated by *arrows* whose thicknesses illustrate the amounts of auxin transported in the corresponding streams

2 Gravity Sensing Occurs Mainly Within Specialized Cells of Roots and Shoots

2.1 Are Sedimenting Amyloplasts Responsible for Gravity Susception?

Higher-plant organs capable of gravitropism contain specialized cells with starch-filled plastids that sediment upon plant reorientation within the gravity field. The starch-statolith hypothesis, proposed at the beginning of the twentieth century, postulates that amyloplast sedimentation might be the primary mechanism for gravity sensing in plants (Haberlandt 1900; Nemec 1900). In roots, the primary site for gravity sensing is located at the center of the cap, within the upper tiers of columella cells (Blancaflor et al. 1998). These cells are highly polarized, with a peripheral endoplasmic reticulum (ER) that lines the bottom and sides of the cell and a nucleus at the top. Except for a highly dynamic network of actin microfilaments, the rest of the cytoplasm is largely deprived of organelles, allowing amyloplasts to sediment toward the new bottom upon reorientation within the gravity field. In above-ground organs, the statocytes, which also contain sedimentable amyloplasts, form the endodermis or bundle-sheath parenchyma that surrounds the vasculature (reviewed in Sack 1991; Boonsirichai et al. 2002; Blancaflor and Masson 2003).

Various ablation studies have confirmed the important contribution of the columella cells in roots and of the endodermis in hypocotyls and inflorescence stems for gravity sensing and signal transduction. In roots, these experiments consisted of analyzing the gravitropic sensitivity and curvature response of ablated organs missing specific cell types owing to treatment with a laser beam (Blancaflor et al. 1998), heavy-ion microbeams (Tanaka et al. 2002), or targeted expression of a protein synthesis inhibitor (diphtheria toxin A; Tsugeki and Fedoroff 1999). Treated roots missing the central columella cells of the cap displayed dramatic alterations in their ability to sense a gravity stimulus, whereas roots that lost the lateral cap cells were unable to transmit the signal from the site of perception to the site of response.

Physical ablation of the statocytes in above-ground organs is less feasible considering their embedded location and extent within these organs. However, genetic screens allowed the identification of radial patterning mutants that lack a normal endodermis, including the *Arabidopsis sgr1/scr* and *sgr7/shr* mutants (Fukaki et al. 1998). In these mutants, hypocotyls and inflorescence stems display altered gravitropism, whereas root gravitropism remains normal (Fukaki et al. 1998). Similarly, the agravitropic phenotype displayed by the stems of *Shidare-asagao (weeping)* mutant of morning glory (*Pharbitis nil*) is due to a mutation in an ortholog of *Arabidopsis SCR* (Kitazawa et al. 2005). Together, these observations are consistent with a role for the endodermal cells in the sensation or transduction of gravity signals by above-ground organs.

That amyloplast sedimentation within the statocytes might contribute to gravity sensing, as originally proposed by Haberlandt (1900) and Nemec (1900), is supported by a number of experimental results. For instance, early centrifugation studies identified the root tip, including the cap, as the main site for gravity sensing in roots, although they also suggested the existence of a secondary site within the distal elongation zone (Haberlandt 1908; Poff and Martin 1989). Furthermore, starch-deficient mutants have been shown to display altered gravitropism, although the remnant response has been the source of substantial controversy (Caspar and Pickard 1989; Kiss et al. 1999). Importantly, acquisition of gravisensitivity by embryonic roots of germinating flax seeds was shown to coincide with the development of mature amyloplasts within the cap statocytes (Ma and Hasenstein 2006).

A number of *Arabidopsis thaliana* mutants with defects in shoot gravitropism display characteristics compatible with the starch-statolith hypothesis. Before describing these mutants, we should first mention that the statocytes of hypocotyls and stems (endodermal cells) differ from those of roots in that they contain large vacuoles that appose all organelles, including the amyloplasts, onto the cellular cortex. In contrast, only small vacuoles can be found in the root columella cells, which do not restrict the movement of amyloplasts within the cytoplasm. In the shoot statocytes, amyloplast sedimentation has to occur within transvacuolar strands (Saito et al. 2005).

Interestingly, several mutants with altered shoot gravitropism are defective in vacuolar biogenesis or function. The *sgr2*, *sgr3*, *sgr4/zig*, and *grv2* mutants of *Arabidopsis* appear to affect vacuolar membrane trafficking, resulting in abnormal vacuolar structures in the endodermis and several other cell types (Kato et al. 2002; Yano et al. 2003). The *SGR2* gene encodes a phospholipase-like protein that localizes to the vacuole and some smaller organelles that remain uncharacterized, whereas *ZIG* encodes a soluble *N*-ethylmaleimide-sensitive factor attachment protein receptor (SNARE) protein involved in vesicular trafficking to the vacuole (Kato et al. 2002; Morita et al. 2002). *SGR3* encodes a syntaxin that localizes to the prevacuolar and vacuolar compartments (Yano et al. 2003). *GRV2/KAM2* encodes a J-domain protein that is peripherally associated with membranes of the endosome, shares similarities with the RME-8 protein of *Caenorhabditis elegans* previously implicated in endocytosis, and is required for proper formation of the endosomes that traffic proteins to the vacuoles (Silady et al. 2004, 2007; Tamura et al. 2007). Importantly, all four mutations affect the ability of amyloplasts to sediment to the bottom of the endodermal statocytes upon gravistimulation and result in altered shoot gravitropism. Altogether, these exciting results suggest that vacuole biogenesis and/or integrity are necessary for normal graviperception through amyloplast sedimentation.

Another important test of the starch-statolith hypothesis involves the guided displacement of amyloplasts within the statocytes, using magnetophoresis. Because starch has diamagnetic properties, high-gradient magnetic fields (HGMFs) exert a ponderomotive force on them. Therefore, Kuznetsov and Hasenstein (1996) built a device that allowed them to apply HGMFs in close proximity of plant organs. As expected, these HGMFs promoted both the magnetophoretic displacement of amyloplasts within the statocytes of flax and *Arabidopsis* root tips, and a lateral curvature in a direction predicted by this displacement. Neither displacement nor tip curvature could be observed in a starchless mutant of *Arabidopsis*, suggesting that the HGMFs regulate tip curvature through their ponderomotive effect on starch-filled amyloplasts within the statocytes (Kuznetsov and Hasenstein 1996).

Similar applications of HGMFs in proximity of hypocotyls and inflorescence stems resulted in both a localized displacement of the amyloplasts within the statocytes and a curvature at the site of stimulation as long as the field was applied within the elongation zone (Weise et al. 2000). When applied to wild-type or dark-grown *lazy-2* mutant hypocotyls of tomato, these HGMFs promoted a curvature in the direction opposite to that of amyloplast displacement. However, when applied to red-light-treated *lazy-2* hypocotyls, they promoted a curvature in the direction of amyloplast displacement. Because *lazy-2* hypocotyls display negative gravitropism in darkness and positive gravitropism upon exposure to red light, the results of these magnetrophoretic experiments support the contention that amyloplast displacement within the statocytes of higher-plant organs is sufficient to trigger a physiological signal that leads to a curvature response (Kuznetsov and Hasenstein 1997; Hasenstein and Kuznetsov 1999).

2.2 *Could Alternative Mechanisms Also Contribute to Gravity Sensing?*

Despite the volley of evidence supporting the starch-statolith hypothesis, several observations suggest another mechanism of gravity perception may also exist. Quite relevant to this contention is the observation that starch-deficient mutants still display a remnant gravitropic response despite lacking starch in their statocytes and showing no evidence of amyloplast sedimentation upon gravistimulation (Caspar and Pickard 1989). In the absence of amyloplast sedimentation, how could these mutants perceive gravity? Could an alternative gravity-sensing mechanism function in these mutants, and possibly the wild type as well?

To answer these fascinating questions, it seems appropriate to remind the reader of early centrifugation experiments suggesting a contribution of the distal elongation zone to gravity perception in roots (Haberlandt 1908; see above). Similarly, root-cap ablation experiments suggested the existence of a secondary site of gravity perception, away from the cap, because gravistimulated decapped roots resumed a gravitropic curvature (1) after return to the vertical if the cap of an unstimulated root was applied back to their tip (Keeble et al. 1929) and (2) if auxin was applied to the severed tip of horizontally positioned roots (Geiger-Huber and Huber 1945; Pilet 1953). Unfortunately, taken individually, these experiments suffer from potential side effects associated with diverse mechanostimuli that accompany these harsh treatments.

Recent studies have been designed to overcome the difficulties and potential artifacts associated with the earlier cap-ablation investigations discussed above. These new investigations seem to support the existence of an alternative mechanism of gravity sensing in the distal elongation zone of roots. Amongst these, work initiated in the Evans laboratory at Ohio State University relies on a customized system, the ROTATO, to provide a constant gravistimulus to responding seedlings. This device carries a motorized rotating stage that holds the responding seedling and is controlled by real-time, high-resolution image analysis software that is designed to maintain any predetermined segment of a graviresponding root tip at a defined angle from the vertical. When a subapical region of the root tip, such as the distal elongation zone, is chosen as a stationary target for ROTATO, the feedback device keeps on rotating the stage to maintain the distal elongation zone at a predefined angle from the vertical. Surprisingly, when this experiment was carried out, the device had to continue rotating the stage even after the cap had reached a vertical orientation (Wolverton et al. 2002b). Such continued rotation would not be expected if the root cap were the only gravity-sensing organ in roots because a cap-derived gravitropic signal should have been interrupted as soon as the cap reached a vertical orientation. This surprising result suggests that cells within the distal elongation zone might also contribute to gravity sensing in roots, although to a lesser extent than the root cap (Wolverton et al. 2002b).

The existence of a secondary site of gravity perception in the root distal elongation zone is consistent with the distinct physiological properties of its cells, and the fact that its contribution to early phases of gravitropism appears independent of an auxin gradient generated across the cap (reviewed in Evans and Ishikawa 1997). It may also help explain some unexpected results that have recently been reported in the literature. For instance, it has been demonstrated that decapped roots of maize will respond to gravistimulation if exposed to inhibitors of actin polymerization or myosin activity. This observation is consistent with the existence of a mechanism for gravity sensing outside the cap (Mancuso et al. 2006). Similarly, utilization of a novel computer-vision platform allowed detection of a bidirectional movement of the root gravitropic curvature during the initial 2–3h of a gravitropic response. The acropetal component of the movement is not expected if one considers the root cap as the only site of gravity sensing in roots (Miller et al. 2007).

If a secondary site of gravity sensing exists in the root distal elongation zone, the perception mechanism must differ from that described by the starch-statolith model. Indeed, cells in this region do not contain sedimenting amyloplasts; therefore, other hitherto uncharacterized organelles may serve as statoliths within those cells, or the whole protoplast may serve as a gravity susceptor. Interestingly, the gravitational-pressure model, which originated from studies on gravistimulus-induced changes in the polarity of cytoplasmic streaming in large *Chara* internodal cells, postulates that plant cells perceive gravity by sensing the weight of the protoplast on its cell wall. Differential tension between the upper and lower sides of the cell would activate receptors at the plasma membrane, possibly mechanosensitive ion channels, which in turn would trigger a transduction pathway that ultimately leads to a response (Staves 1997). While this system seems to apply to the large internodal cells of *Chara*, it remains debatable whether it can also function in the smaller cells of higher plants. Yet, rice roots decreased their rate of gravitropic curvature when exposed to media of increasing density, a behavior that is consistent with the gravitational-pressure model, and cannot be easily reconciled with the starch-statolith hypothesis (Staves et al. 1997).

3 Gravity Signal Transduction in the Statocytes

3.1 What Is the Gravity Receptor?

Even though experimental evidence supports a role for amyloplast sedimentation in gravity sensing within the statocytes, the molecular mechanisms that convert the corresponding information into a physiological process remain poorly understood. Several models have been proposed to explain this process. First, the highly polarized structure of the root statocytes, with amyloplasts that sediment on top of the peripheral ER, led to the suggestion that changes in the distribution of forces exerted by sedimenting amyloplasts on resisting structures such as the ER or the actin cytoskeleton might promote the opening of mechanosensitive ion channels in ER or plasma membranes, triggering gravity signal transduction within the statocytes (Sievers et al. 1984, 1989; Volkmann and Baluska 1999; Yoder et al. 2001). Unfortunately, these models seem oversimplified because connections between amyloplasts and the ER may not be needed for full gravitropism (Wendt et al. 1987) and because disruption of the actin cytoskeleton network with pharmacological agents enhances gravitropism in roots, hypocotyls, and inflorescence stems, rather than inhibiting it (Yamamoto and Kiss 2002; Hou et al. 2003, 2004). The latter observation is important because it suggests that the actin cytoskeleton is an important modulator of gravity sensing in the statocytes, where it may serve as a negative regulator of the process, possibly ensuring a fast termination of the curvature response as soon as the columella has reached its gravity set point angle (Hou et al. 2004).

A possible involvement of mechanosensitive, Ca^{2+}-selective, ion channels in gravity signal transduction has been suggested by a number of investigators (Sievers et al. 1984, 1989; Pickard and Ding 1993; Volkmann and Baluska 1999; Yoder et al. 2001), and several lines of experimental evidence have supported this contention. For instance, pharmacological agents that interfere with the activity of such channels, of calmodulin (a Ca^{2+}-dependent regulatory protein) or of Ca^{2+}-ATPases, all inhibit gravitropism, as do Ca^{2+} chelators (reviewed in Sinclair and Trewavas 1997; Fasano et al. 2002; Monshausen et al. 2008). Also, the levels of calmodulin are high in the root tip (Allan and Trewavas 1985; Stinemetz et al. 1987), where they correlate with the organ's responsiveness to gravity (Stinemetz et al. 1987), and increase upon gravistimulation (Sinclair et al. 1996). Gravistimulation may also change the abundance of calmodulin protein across monocot pulvini because calmodulin messenger RNA is recruited into polysomes on the lower side upon gravistimulation (Heilmann et al. 2001).

A possible involvement of Ca^{2+} in gravity signal transduction is also consistent with recent observations that suggest a role for inositol 1,4,5-trisphosphate ($InsP_3$) in this process. Indeed, this compound has been implicated in triggering Ca^{2+} release from intracellular stores in plant cells as it does in animals (Wang 2004). The pulvini of monocots constitute an interesting model for the study of gravitropism because they provide large amounts of easily dissectible material for biochemical and cell-biological analyses of the response. Taking advantage of these interesting properties, Imara Perera and her collaborators at North Carolina State University were able to observe fluctuations in $InsP_3$ levels between upper and lower flanks of pulvini within the first few minutes of gravistimulation. These fluctuations were followed by a sustained increase on the lower flank accompanied by an increase in phosphatidylinositol phosphate kinase activity (Perera et al. 1998, 1999, 2001). Similar increases in $InsP_3$ levels were also observed in *Arabidopsis thaliana* inflorescence stems within 15min of gravistimulation, supporting the universality of this response. When $InsP_3$ levels were altered by transgenic expression of a human inositol polyphosphate 5-phosphatase enzyme, *Arabidopsis* inflorescence stems displayed dramatic alterations in their ability to respond to gravistimulation (Perera et al. 2006).

Altogether, these results are consistent with an involvement of Ca^{2+} as a second messenger in the transduction of gravity signals in plant organs; therefore, one should be able to observe changes in cytosolic Ca^{2+} levels within the statocytes upon gravistimulation. Despite previous unsuccessful attempts at identifying such Ca^{2+} fluctuations (Sedbrook et al. 1996; Legue et al. 1997; Fasano et al. 2002), recent experiments have demonstrated a small biphasic Ca^{2+} transient following gravistimulation of *Arabidopsis thaliana* seedlings, using the transgenic aequorin Ca^{2+} reporter system (Plieth and Trewavas 2002; Toyota et al. 2008). This Ca^{2+} response seems to originate from the hypocotyls and petioles, although its cellular source has not been established (Toyota et al. 2008). The first peak lasts a few seconds after the onset of gravistimulation, and is reminiscent of responses to mechanostimulation, whereas the second peak is more sustained, and lasts for a period proportional to the strength of the stimulus (angle of reorientation within the gravity field) (Plieth and Trewavas 2002). Importantly, pharmacological studies have pointed to an involvement of both mechanosensitive ion channels and intracellular-store channels in these responses (Toyota et al. 2008). Unfortunately, the aequorin-based luminescence responses observed in these studies were very small, implying that the corresponding Ca^{2+} responses may have been restricted to only a few cells, or may have been highly localized within the graviresponding cells. Furthermore, the absence of a signal within the roots cannot be interpreted as reflecting a true absence of Ca^{2+} response in that organ considering the sensitivity of the Ca^{2+} detection system used in these studies. In fact, it is worthwhile noting that the high levels of calmodulin found in the root-cap statocytes would allow the perception of even small changes in cytosolic Ca^{2+} levels within those cells (Sinclair et al. 1996).

While the studies discussed above suggest the involvement of Ca^{2+} as a second messenger in gravitropic signaling, the mechanosensitive channels predicted to be responsible for such responses have not been identified. Unfortunately, many of the mechanosensitive ion channels identified in animals and yeast do not have clear orthologs in plants. However, plant orthologs to bacterial small-conductance mechanosensitive ion channels (MscS) have been identified (Haswell and Meyerowitz 2006). Initial studies on two of the ten MscL-like (*MSL*) channel genes of *Arabidopsis* (*MSL2* and *MSL3*) suggest that the corresponding proteins may also function as mechanosensitive ion channels in plants. These channels localize within the inner envelope of plastids, and they regulate plastid size, shape, and possibly division (Haswell and Meyerowitz 2006). Importantly, some of the other members of the *Arabidopsis MSL* gene family are predicted to associate with the plasma membrane, and are being investigated for potential roles in mechano- and gravisensing (Haswell and Meyerowitz 2006).

In addition to the MSL proteins, another Ca^{2+} channel has recently been identified in *Arabidopsis* and proposed to function in mechanotransduction. The *Arabidopsis* plasma-membrane-associated MCA1 protein was shown to rescue a *mid1* mutant of yeast, which lacks a Ca^{2+}-permeable stretch-activated channel. *MCA1* overexpression in *Arabidopsis* also resulted in increased cytosolic Ca^{2+} responses to mechanostimulation, and knockout mutant roots displayed an altered ability to respond to mechanostimulation (Nakagawa et al. 2007). Hence, several families of membrane-associated proteins have now been identified as potential stretch-activated channels in plants, and future work should investigate a possible role for these proteins in gravity signal transduction.

The reader should be alerted here to the fact that gravity sensing does not necessarily require the involvement of mechanosensitive ion channels as receptors of amyloplast sedimentation within the statocytes. Indeed, a ligand-receptor model of gravity signal transduction has recently been proposed to explain gravity sensing in *Chara* rhizoids (Limbach et al. 2005). In this single-cell gravitropic system, the statoliths ($BaSO_4$-filled vacuoles) have to sediment on specific subapical areas of the plasma membrane in order for that cell to develop a curvature (Braun 2002). Experiments involving short-term exposure to the microgravity environment created by parabolic flights have suggested that a simple contact between the statoliths and the sensitive membrane, rather than pressure, might be sufficient for gravisensing in this system (Limbach et al. 2005). The authors concluded that a membrane-associated receptor present at the subapical region of the rhizoid may be able to interact with a ligand carried at the surface of the sedimenting statoliths. Interaction would trigger a pathway that leads to the curvature observed in this system. It is possible that a similar ligand-receptor interaction model is also functional in the statocytes of higher plants. In this case, the ligand (or receptor) would have to be carried at the surface of sedimenting amyloplasts, whereas the receptor (or ligand) would be carried by sensitive membranes the plastids sediment upon, such as those of the ER that lines the root statocytes. Ligand-receptor interaction would trigger a pathway that could also involve Ca^{2+}, $InsP_3$, and/or pH as second messengers, eventually leading to the cellular polarization needed to redirect auxin transport downward. The recent identification of outer-envelope, plastid-associated proteins as gravity signal transducers in *Arabidopsis thaliana* may allow us to test this interesting new model in the future (see Sect. 4.1).

3.2 pH Changes May Also Contribute to Gravity Signal Transduction

Another inorganic ion that may participate in gravity signal transduction within the statocytes is H^+. Experiments in which researchers traced the pH of the statocytes of maize pulvini and *Arabidopsis* roots during a graviresponse detected substantial alkalinization of the cytoplasm within seconds of a stimulus (Scott and Allen 1999; Fasano et al. 2001; Johannes et al. 2001; Boonsirichai et al. 2003; Hou et al. 2004; Young et al. 2006). In maize pulvini, gravity-induced alkalinization of the statocytes occurred only in a region of the cells where amyloplasts sediment, and it did not occur in the surrounding parenchyma cells (Johannes et al. 2001). In *Arabidopsis* root statocytes, the characteristics of the gravity-induced cytoplasmic pH fluctuations differed depending on the source of the data. Although Scott and Allen (1999) reported distinct kinetics of cytoplasmic alkalinization between upper and lower statocytes in layer 2 of the *Arabidopsis* root-cap columella, and an acidification of tier 3 cells. Fasano et al. (2001) observed a similar cytoplasmic alkalinization of all statocytes in these two tiers, which peaked after 1–2min of gravistimulation, and returned back to the baseline within 8–12min. These fast pH responses to gravistimulation were substantially attenuated in a starch-deficient mutant, and they were accompanied by an acidification of the apoplast, suggesting the involvement of plasma-membrane or vacuolar H^+ transporters (Fasano et al. 2001; Li et al. 2005). Importantly, interfering with these pH responses by inducing the release of preloaded caged protons in the root statocytes (Fasano et al. 2001) or by treatment with agents that alkalinize the apoplast (Scott and Allen 1999) delayed the gravitropic response. Also, treating *Arabidopsis* seedlings with agents that destabilize actin microfilaments resulted in a sustained pH increase of root statocytes upon short periods of gravistimulation, along with expanded gravicurvature (Hou et al. 2004). Conversely, treating seedlings with agents that acidify the apoplast under conditions that do not interfere with root growth rate enhanced the gravitropic response (Scott and Allen 1999). Together, these data support a role for cytosolic pH in gravity signal transduction in both root and pulvinus statocytes, possibly via the facilitation of auxin transport through the statocytes (Fasano et al. 2002; Monshausen and Sievers 2002; Monshausen et al. 2008). Consistent with this model, *Arabidopsis* mutants with altered expression of a proton pyrophosphatase (AVP1) display altered auxin transport associated with modified expression and localization of the PIN1 auxin efflux facilitator (Li et al. 2005).

3.3 Gravistimulation Promotes the Formation of a Lateral Auxin Gradient Across Stimulated Organs

The Cholodny-Went theory proposes that gravistimulation promotes the formation of a lateral gradient of auxin across the stimulated organ, leading to differential cell elongation on opposite flanks, responsible for tip curvature (Cholodny 1928; Went 1928; Fig. 1). In roots, the lateral gradient of auxin has to be transmitted from the cap to the elongation zone, where a curvature response develops. In shoot organs, the gradient forms right at the site of tip curvature. Several lines of experimental evidence support this long-held model of gravitropism. First, mutations that affect auxin synthesis, transport, or response have been shown to affect gravitropism (reviewed in Muday and Rahman 2008). Second, physiological experiments following the fate of exogenously supplied radioactive auxin demonstrated its lateral transport toward the bottom flank of gravistimulated shoots (Parker and Briggs 1990) or roots (Young et al. 1990), and showed that this redistribution precedes the development of a gravitropic curvature (Parker and Briggs 1990). Third, lateral gradients in free auxin concentration were detected across gravistimulated pulvini, coleoptiles, and gynophores, whose polarities were compatible with the direction of curvature response (Kaufman et al. 1995; Moctezuma and Feldman 1999; Philippar et al. 1999; Long et al. 2002). Similarly, the recent use of reporter constructs carrying an auxin-responsive promoter that drives the expression of a reporter gene (encoding either β-glucuronidase or the green fluorescent protein) in transgenic plants allowed indirect detection of lateral auxin gradients across gravistimulated roots, hypocotyls, or inflorescence stems, with accumulation at the bottom flank (Li et al. 1991; Rashotte et al. 2001; Friml et al. 2002; Boonsirichai et al. 2003; Ottenschlager et al. 2003; Hou et al. 2004; Nadella et al. 2006; Young et al. 2006; Lewis et al. 2007; Harrison and Masson 2008a).

To better understand how this lateral auxin gradient is established upon gravistimulation, let us first describe the molecular mechanisms that modulate polar auxin transport in plants.

3.4 *Auxin Influx and Efflux Carrier Complexes Mediate Polar Auxin Transport*

As discussed in "Auxin and the Communication Between Plant Cells" by Peter Nick (this volume), auxin, which is mainly synthesized in young shoot tissues, uses a cell-to-cell transport system that functions in the tip to base direction in shoots. When auxin reaches the root, it is transported through the central cylinder into the root tip, where it adds to an uncharacterized pool of locally synthesized auxin, forming an auxin-maximum center that overlaps with the quiescent center and top layers of the root-cap columella (reviewed in Wolverton et al. 2002a; Muday and Rahman 2008). There, auxin is redistributed laterally to peripheral tissues, then transported basipetally through lateral-cap and epidermal cell files toward the elongation zone, where it inhibits cell elongation.

Individual auxin-transporting cells express a combination of plasma-membrane-associated auxin influx and efflux carriers. Cytoplasmic auxin, found mainly in the ionic form, cannot easily cross the membrane. On the other hand, a substantial fraction of the auxin present in the acidic environment of the apoplast is found in the protonated form and can diffuse through the plasma membrane. Hence, apoplastic auxin is taken up by transporting cells both by diffusion through the plasma membrane and via transmembranous auxin influx carriers that belong mainly to the AUX1 family of auxin/H^+symporters (Swarup et al. 2004).

To exit the cell, auxin requires the intervention of auxin efflux carrier complexes that contain PIN proteins and members of the multidrug-resistant (MDR)/P-glycoprotein (PGP) family of ATP-binding cassette transporters (Blakeslee et al. 2007). The PIN and PGP proteins seem to be parts of independent auxin transport complexes that may also interact in a tissue-specific manner (Blakeslee et al. 2007). Importantly, the polar localization of PIN proteins at the plasma membrane of transporting cells predicts the direction of auxin flow (reviewed in Muday and Rahman 2008).

Considering the molecular basis of polar auxin transport, how could gravistimulation promote a lateral transport of auxin across a plant organ, leading to its accumulation at the bottom flank, as suggested by Cholodny and Went 80 years ago?

3.5 *The Arabidopsis PIN3 Auxin Efflux Facilitator May Be an Important Modulator of Lateral Auxin Transport Across Gravistimulated Plant Organs*

A member of the *Arabidopsis PIN* gene family, *PIN3*, appears to be a key player in gravity signal transduction within both root and shoot statocytes. In roots, *PIN3* is expressed in cells of the pericycle and within the top two cell layers of the columella - precisely those that contribute most to gravity sensing in that organ (see Sect. 2.1; (Blancaflor et al. 1998)). In above-ground organs, *PIN3* is expressed in the endodermal cells of stems and hypocotyls where its protein localizes mainly at the inner lateral plasma membrane, with some association with the basal membrane in some cells (Friml et al. 2002).

In the columella cells of the cap, PIN3 localization is highly dynamic, and varies depending on the root's orientation within the gravity field. In vertically oriented roots, PIN3 is localized symmetrically on all sides of the root-cap statocytes; however, upon root reorientation within the gravity field, PIN3 quickly relocalizes, accumulating in the new lower membrane within 2min (Friml et al. 2002). This rapid relocalization of an efflux facilitator should lead to a preferential transport of auxin toward the bottom flank of the cap, creating a lateral auxin gradient across the gravistimulated root (Friml et al. 2002). Consistent with a key role for PIN3 in gravitropism, *pin3* mutant seedlings display altered root and hypocotyl gravitropism, even though gravity-induced PIN3 relocalization has not yet been documented in gravistimulated hypocotyl or stem statocytes (Friml et al. 2002; Harrison and Masson 2008a).

It is interesting to note here that the second messengers associated with early phases of gravity signal transduction (intermediates and products of the phosphatidylinositol pathway, Ca^{2+}, pH; see Sects. 3.1, 3.2) have been implicated in the control of vesicular trafficking in both animal and plant systems (Stevenson et al. 2000; Reddy 2001; Wenk and De Camilli 2004; Li et al. 2005); therefore, at least some of them may be involved, directly or indirectly, in controlling the localization of PIN3 within the statocytes.

3.6 In Roots, the Gravity-Induced Lateral Auxin Gradient Has To Be Transmitted to the Elongation Zone for a Curvature Response To Develop

In shoots, the lateral gradient of auxin generated upon gravistimulation is directly responsible for the curvature response at that site of the organ, as discussed in Sect. 2.1. In roots, however, a lateral gradient generated across the cap has to be transported to the elongation zone in order to promote tip curvature. This transport occurs through the basipetal auxin transport stream (reviewed in Muday and Rahman 2008). As discussed in Sect. 3.4, auxin transport requires both auxin influx and efflux carrier complexes. The basipetal transport of auxin in roots is mediated by the AUX1 influx carrier, which functions as an auxin/H^+ symporter, and the PIN2 auxin efflux facilitator. *AUX1* is expressed in the lateral-cap and epidermal cells of the root tip where its protein localizes uniformly at the plasma membrane (Swarup et al. 2001). AUX1 is also expressed in the protophloem cells of the root central cylinder, where its basal localization contributes to the acropetal auxin flow toward the tip (Swarup et al. 2001). *PIN2*, on the other hand, is expressed in the lateral-cap and epidermal cells of the root elongation zone where it is essentially located in the basal-side plasma membrane, directing basipetal auxin transport (Müller et al. 1998). It is also expressed in the cortical cells of the root tip, where its apical localization suggests a contribution to auxin recycling toward the tip (Friml 2003). As expected, mutations in both *AUX1* and *PIN2* eliminate the gravitropic response (Bennett et al. 1996; Chen et al. 1998; Luschnig et al. 1998; Müller et al. 1998; Utsuno et al. 1998). Furthermore, the agravitropic phenotype of *aux1* mutants can be rescued by constructs that express wild-type *AUX1* specifically in the lateral cap and apical four to five epidermal cells of the root elongation zone, indicating that its contribution to the basipetal auxin transport stream is sufficient for its role in root gravitropism (Swarup et al. 2005).

The MDR4 ATP-binding cassette transporter also contributes to the basipetal auxin transport stream in *Arabidopsis* root tips. Its expression expands basally farther from the tip than that of PIN2, suggesting that this MDR transporter may take over the basipetal flow of auxin in basal regions of the root tip where PIN2 has ceased to function. In agreement with this model, *mdr4* mutant roots display an accelerated gravitropic curvature relative to the wild type, accompanied by an increased ability to develop a lateral auxin gradient upon gravistimulation. It was hypothesized that decreased basipetal auxin transport on the basal side of the elongation zone allows for auxin accumulation at the distal elongation zone, where the curvature response occurs (Lewis et al. 2007). This interpretation is in line with similar explanations of the accelerated gravitropic and phototropic responses displayed by *mdr1* single and *mdr1 pgp19* double mutant hypocotyls. Indeed, the *MDR1* and *PGP19* genes contribute to basipetal auxin transport in hypocotyls and inflorescence stems, along with PIN1, and the corresponding mutations result in decreased basipetal auxin transport, thereby increasing the potential for gravity- or light-induced lateral auxin transport in these organs (Noh et al. 2003).

3.7 Auxin-Induced Changes in the Cycling and Stability of Auxin Transporters May Also Contribute to the Formation of a Lateral Auxin Gradient Across the Gravistimulated Root

In this chapter, the discussion has, so far, focused mainly on how gravity susception through amyloplast sedimentation within the statocytes is transduced into the formation of a lateral auxin gradient across the stimulated organ, leading to tip curvature. Yet, recent studies of auxin transport in roots suggest that the process may be more complex. Indeed, auxin influx and efflux carriers are continuously cycled between the plasma membrane and intracellular compartments, and PIN2 is no exception (see "Auxin and the Communication Between Plant Cells" by Peter Nick, this volume). Importantly, some of the endocytosed PIN2 protein is ubiquitinated and targeted for degradation by the proteasome in an auxin-sensitive manner: increased auxin levels stabilize PIN2 at the plasma membrane (Abas et al. 2006). Hence, in gravistimulated root tips, the initial lateral auxin gradient that is generated across the cap, then transmitted to the distal elongation zone, has the potential of affecting both the process of cellular elongation and the association of PIN2 with the basal membrane of transporting lateral-cap and epidermal cells. The increased levels of auxin found at the lower flank of gravistimulated roots allow for higher PIN2 levels within the plasma membrane at the basal side of the transporting cells, whereas lower auxin levels at the upper flank result in increased PIN2 internalization and degradation by the proteasome (Abas et al. 2006). This differential abundance of plasma-membrane-associated PIN2 protein between upper and lower flanks of gravistimulated roots is likely to enhance the steepness of the auxin gradient across the stimulated distal elongation zone. Furthermore, the lower levels of auxin reaching the upper flank of the elongation zone should result in accelerated cell elongation, a process that was actually observed in the distal elongation zone through careful kinematic studies of graviresponding *Arabidopsis* roots (Ishikawa and Evans 1997; Chavarria-Krauser 2006). Because this region of the *Arabidopsis* root also serves as a secondary site for gravity sensing, it will be interesting to investigate the possibility that PIN2 internalization and degradation might also be directly controlled by gravity sensing, explaining the remnant gravitropic response displayed by *pgm* mutant roots in the absence of amyloplast sedimentation within the statocytes (Wolverton et al. 2002b).

4 How Can We Uncover New Gravity Signal Transducers?

4.1 Genetics to the Rescue: Genes That Contribute to Gravity Signal Transduction in the Statocytes

Genetics has helped in the identification and initial characterization of proteins that contribute to gravity signal transduction within the statocytes. Mutations in the *ARG1/RHG* and *ARL2* loci of *Arabidopsis thaliana* were shown to result in specific defects in root and hypocotyl gravitropism without significant alterations in root-growth responses to phytohormones and polar auxin transport inhibitors, starch accumulation in the statocytes, or phototropic response (Fukaki et al. 1997; Sedbrook et al. 1999; Boonsirichai et al. 2003; Guan et al. 2003; Harrison and Masson 2008a). Both genes were shown to encode paralogous proteins that contain a J domain at their N terminus, a hydrophobic region and a C-terminal coiled-coil domain that shares similarity with coiled coils found in a number of microtubule- and microfilament-associated proteins (Sedbrook et al. 1999; Guan et al. 2003). The J domain present in both proteins suggests they function as molecular chaperones, in association with HSP70-like proteins, to modulate the folding, targeting, or macromolecular assembly of specific gravity signal transducer(s) within the statocytes (Boonsirichai et al. 2003; Young et al. 2003). While *ARG1* is expressed ubiquitously in plants, expression of wild-type ARG1 in the statocytes of *arg1–2* mutant roots is sufficient to restore wild-type root gravitropism, whereas its expression in the endodermis under the control of the *SCR* promoter allows specific rescue of hypocotyl gravitropism; hence, the functional domain of *ARG1* expression for gravitropism is the statocyte, suggesting a role for this protein in early phases of gravity signal transduction (Boonsirichai et al. 2003). *ARL2*, on the other hand, is expressed specifically in the statocytes, confirming its role in gravity signal transduction (Harrison and Masson 2008a).

A combination of biochemical and cell-biological experiments revealed that both ARG1 and ARL2 are peripheral membrane proteins that associate with both the plasma membrane and components of the vesicular trafficking pathway (Boonsirichai et al. 2003; Harrison and Masson 2008a). Interestingly, both proteins are needed for gravity-induced PIN3 relocalization to the lower membrane of the statocytes and for the establishment of a lateral auxin gradient across the root cap upon gravistimulation (Harrison and Masson 2008a). Taken together, these data strongly suggest that ARG1 and ARL2 modulate the cycling of PIN3 and other membrane-associated proteins between the plasma membrane and intracellular organelles within the statocytes. This hypothesis is in agreement with genetic studies that indicate that *ARG1* and *ARL2* function in the same genetic pathway as *PIN3* in roots (Harrison and Masson 2008a, b).

As mentioned above, mutations in *ARG1* and *ARL2* affect all physiological outputs of gravity signal transduction in the root statocytes, including cytoplasmic alkalinization, (atleast for ARG1) PIN3 relocalization, and the lateral transport of auxin to the bottom flank of a stimulated organ. However, a mutation in one of two genes that encode adenosine kinase (*ADK1*), which was shown to specifically affect root gravitropism by interfering with PIN3 expression in the statocytes, its relocalization in response to gravistimulation, and the lateral transport of auxin across gravistimulated root caps, did not affect the gravity-induced cytoplasmic alkalinization of the statocytes (Young et al. 2006). These interesting results could indicate that the gravity signal transduction pathway within the statocytes is linear, with the cytoplasmic alkalinization step preceding that of PIN3 relocalization. Alternatively, the pathway could be branched, with cytoplasmic alkalinization forming an ADK1-independent branch, whereas PIN3 relocalization would form the ADK1-dependent branch. Both branches would converge to promote the lateral transport of auxin across the gravistimulated cap (Young et al. 2006; Harrison et al. 2008). It will be necessary to elucidate the mode of ADK1 action in gravitropism to resolve the linear or branched nature of the gravity signal transduction pathway in roots.

While *ARG1*, *ARL2*, and *PIN3* appear to function in the same genetic pathway in roots, *ARG1* (or *ARL2*) and *PGM* (gene encoding phosphoglucomutase, an enzyme involved in starch biosynthesis) do not: the corresponding double mutants (*arg1–2 pgm* and *arl2–3 pgm*) display strongly enhanced gravitropic defects relative to single mutants (Guan et al. 2003). This observation suggested that new early gravity signal transducers could be isolated through genetic screens for enhancers of the gravitropic defect associated with *arg1–2*. Such a screen was carried out, and allowed the identification of *modifier of arg1 1–1* (*mar1–1)* and *mar2–1*. While *mar1–1* displays a slight alteration in root gravitropism in a wild-type ARG1 background, *mar2–1* shows a wild-type gravitropism in such a background. However, both mutations completely obliterate the gravitropic response of either *arg1–2* or *arl2–3* mutant roots or hypocotyls. Hence, both *mar1–1* and *mar2–1* function as true genetic enhancers of *arg1–2* and *arl2–3*. Molecular cloning revealed mutations in two genes that encode components of the plastidic outer membrane protein import complex. Amazingly, these two mutations do not affect starch accumulation in the plastids, nor do they alter amyloplast sedimentation in the root cap statocytes (Stanga J, Boonsirichai K, Sedbrook J, Oregui M, and Masson PH, in preparation). These data suggest a role for this complex in gravity signal transduction, either directly through interactions between a cytoplasmic domain of the complex and another transducer in the statocytes whose location or activity might be modulated by *ARG1/ARL2*, or indirectly by targeting the insertion within the plastidic outer membrane of (a) protein(s) that might serve as a signal transducer (Stanga J et al., in preparation). Work is in progress to test these alternative hypotheses.

While roots constitute a good model for a genetic investigation of gravity signal transduction in plants, the inflorescence stem of *Arabidopsis* is also ideally suited for that purpose because it is possible to separate in time the initial events of gravity signal transduction from the ensuing events of lateral auxin transport and curvature response in this system. Indeed, early studies of inflorescence stem gravitropism in this plant demonstrated that a gravistimulus provided in the cold cannot promote the development of a lateral auxin gradient or a gravitropic curvature if the plant is maintained in the cold. However, a stem that is gravistimulated in the cold, then repositioned vertically before being transferred back to room temperature, can "remember" the stimulus in the cold, and develop an appropriate curvature response as long as the waiting period following stimulation is sufficiently short (less than 1h) (Fukaki et al. 1996; Wyatt et al. 2002; Nadella et al. 2006).

A clever genetic screen was designed to identify genes involved in early phases of gravity sensing or signal transduction, taking advantage of the temporal separation that exists between the phases of (1) gravity sensing and signal transduction in the cold and (2) auxin transport and curvature response at room temperature. Conditional mutations were isolated, which affect the ability of inflorescence stems stimulated in the cold to develop a tip curvature upon return to root temperature, while not altering the ability of mutant stems to respond like the wild type to a normal gravistimulus provided at room temperature (Wyatt et al. 2002; Nadella et al. 2006). Three classes of *gravity persistence signal* (*gps*) mutants were isolated in this screen: (1) class 1 mutants, such as *gps1*, show no stem bending in response to cold stimulation ("no-response" mutants); (2) class 2 mutants, such as *gps2*, have stems that bend in the wrong direction in response to a stimulus provided in the cold ("wrong-way" mutants); and (3) class 3 mutants, such as *gps3*, develop an exaggerated stem curvature response to a cold stimulus after return to root temperature ("overachiever" mutants) (Wyatt et al. 2002). Because auxin transport is inhibited in the cold (Nadella et al. 2006), and the Gps phenotypes are conditional upon the stimulus being provided in the cold, it was concluded that these mutants likely affect early stages of gravity sensing or signal transduction, rather than the later phases of lateral auxin transport or curvature response (Nadella et al. 2006).

Initial results in the molecular cloning and characterization of the *GPS* genes support the contention that these genes might function in early phases of gravity sensing or signal transduction. Indeed, *GPS1* encodes a P450-type enzyme that may contribute to flavonoid metabolism (Withers JW and Wyatt GE, unpublished data).

Because flavonoids are important regulators of auxin transport (Buer and Muday 2004), it seems likely that GPS1 modulates the transport of auxin in gravirespond-ing *Arabidopsis* seedlings. *GPS2* encodes a putative synaptobrevin/vesicle-associ-ated membrane protein vesicle SNARE that may regulate the transport of PIN auxin efflux facilitators or other regulatory molecules in cells of the inflorescence stem (McCallister J and Wyatt, SE unpublished results). Finally, *GPS3* encodes a puta-tive transcription factor with a B3 DNA-binding domain similar to that of auxin response factors. However, *GPS3* lacks the C-terminal dimerization domain found in auxin response factor proteins (Nadella V, and Wyatt, SE unpublished data).

In addition to the *GPS* genes discussed above, the *shoot gravitropic response 5*, *shoot gravitropic response 6*, and *shoot gravitropic response 9* (*sgr5*, *sgr6*, and *sgr9*) mutants have also been proposed to affect genes that function in the early phases of gravity signal transduction. Indeed, these mutations affect stem gravitro-pism without altering amyloplast sedimentation in the stem statocytes (Morita et al. 2006). While the function of *SGR6* remains unknown because the protein it encodes shares sequence similarity only with proteins of unknown function (Yano D et al., unpublished results), *SGR5* encodes a C2H2 zinc-finger-type transcription factor that is mainly expressed in the endodermis. Specific expression of wild-type *SGR5* in the endodermis of *sgr5–1* mutant plants rescues stem gravitropism to wild-type levels, supporting a role for *SGR5* in gravity signal transduction within the stem statocytes (Morita et al. 2006).

Mutations in *SGR9*, a locus that encodes a ring finger-type protein associated with the plastids, strongly enhance the gravitropic defect of *sgr5*, and alter the ability of mutant plastids to sediment in the statocytes. However, a preliminary report suggests that the gravitropic defect associated with *sgr5 sgr9* mutant stems can be partially rescued by drugs that inhibit the polymerization of actin. This rescue is accompanied by a restoration of amyloplast sedimentation in the stem statocytes. Hence, it appears likely that the plastid-associated SGR9 protein contributes to gravity susception by disrupting the polymerization of actin in the vicinity of amyloplasts, which facilitates their sedimentation within the statocytes (Tasaka M, Nakamura M, and Morita M (2008). Action dynamics involved in gravity per-ception in Arabidopsis inflorescence stem. Abstract, paper Number F11-0007-08 37th COSPAR Scientific Assembly 2008, Montreal, Canada).

4.2 Approaches Borrowed from Systems Biology Also Provide Clues to the Molecular Mechanisms That Govern Gravitropism

While genetics has been invaluable in our study of plant gravitropism, it has also been plagued by difficulties associated with functional redundancy and pleiotropy. To overcome such difficulties, researchers have tried to rely on alternative strategies derived from systems biology to identify new molecular mechanisms associated with gravitropism. Completion of sequencing of the *Arabidopsis* genome has allowed the development of global approaches to investigate genome-wide gene-expression changes that occur quickly after the onset of a gravistimulus. The first attempt at identifying global gene-expression changes in response to gravistimulation relied on an 8,300-gene microarray to identify some of the *Arabidopsis* genes whose expression in entire seedlings varies in response to gravistimulation. In this pioneering work, 1.7% of the genes analyzed changed in expression within 30min of gravistimulation, 39% of which were also responsive to gentle mechanostimulation. The promoters of many of these gravity-responsive genes contained a conserved motif that may function as a *cis*-acting regulatory element for expression responses to gravistimulation (Moseyko et al. 2002).

In this study, the transcripts were extracted from entire seedlings, with no attempt at separating organs that display opposite gravitropic responses such as roots and hypocotyls. Furthermore, only one third of the *Arabidopsis* genes were sampled for expression changes. To work around these difficulties, the Winter-Sederoff group at North Carolina State University chose to study global transcriptional responses to gravistimulation of *Arabidopsis* root tips, using the whole-genome Affymetrix ATH1 microarray (22,744 genes; Kimbrough et al. 2004). They developed a time-course study of transcriptional responses during the first hour of gravistimulation, and included a mechanostimulated control in their experiment (root tips from seedlings that had been gently rocked back and forth for 5s, before being returned to their original position). Using this careful approach, Kimbrough and his collaborators showed that a vast majority of the differentially expressed genes (96%, or 1,665 genes) respond transcriptionally to both gravitstimulation and gentle mechanostimulation. Only five of the differentially expressed genes found in this study were also found in the study of Moseyko et al. (2002), probably reflecting the differences in experimental conditions and tissues analyzed.

Of the differentially expressed root-tip genes identified by Kimbrough et al. (2004), only 65 showed evidence of upregulation specifically in response to gravistimulation, including five that were upregulated by a factor of 3 or more within only 2min of gravistimulation, and whose expression remained high after 30min of response (Kimbrough et al. 2004). Importantly, when three of these fast graviresponding genes were tested for gravity-induced expression changes in mutant root tips, none showed a response in *arg1–2* and *arl2–3* mutant root tips, whereas one retained some inducibility in *adk1–1* root tips. This interesting result is consistent with a key role for *ARG1* and *ARL2* in early phases of gravity signal transduction (Yester et al. 2006). It also supports the contention that *ADK1* is an important gravity signal transducer for some, but not all, root-tip responses to gravistimulation, as discussed in Sect. 4.1.

Because most signal transduction pathways include steps that involve the posttranslational modification of proteins, transcription-profiling studies are not likely to reveal all aspects of an organ's molecular responses to gravistimulation. Such studies should be coupled to large-scale investigations of protein levels, localization, stability, and/or modification in response to gravistimulation. A pioneering study by Sakamoto et al. (1993) used two-dimensional gel electrophoresis to demonstrate that gravistimulation promotes an increase in abundance of some proteins within 24h of gravistimulation, and a change in the phosphorylation status of other proteins. In a separate study, Kaufman and coworkers identified two soluble and two membrane-associated phosphoproteins that are differentially expressed between the lower versus the upper flanks of gravistimulated pulvini (Chang and Kaufman 2000; Chang et al. 2003).

Unfortunately, these initial studies did not attempt to identify the gravity-responsive proteins. Taking advantage of new technological developments in mass spectrometry and its utilization in protein identification (Li and Assmann 2000), Motoshi Kamada and his collaborators at the Aerospace Exploration Agency, in Tsukuba, Japan, identified 16 proteins, including some cytoskeleton-associated proteins and others related to the Ca^{2+} signaling pathway, as being differentially represented in *Arabidopsis thaliana* root tips upon 0.5–3h of gravistimulation, a time that was sufficient to allow both gravity sensing and signal transduction as well as initial curvature response. Other proteins were also shown to change their molecular weight upon gravistimulation, suggesting posttranslational modification (Kamada et al. 2005).

In an attempt to focus more on early phases of gravity signal transduction, Young et al. (2006) and Murthy et al. (2008) analyzed changes in protein profiles of *Arabidopsis* root tips upon 12min of gravistimulation. Fifty-seven protein spots were shown by two-dimensional gel electrophoresis to change in staining intensity after 12min of gravistimulation relative to an unstimulated control, and the corresponding proteins were identified by mass spectrometry. Fifty-three of these proteins were shown to respond specifically to gravistimulation, whereas the remaining four responded to both gravistimulation and a gentle mechanostimulus control. Interestingly, three of the graviresponding proteins were enzymes involved in the AdoMet methyl-donor cycle, including adenosine kinase, whose role in root gravitropism was subsequently investigated using reverse genetic strategies (see Sect. 4.1) (Young et al. 2006).

Six of the differentially represented protein spots identified by Murthy et al. (2008) were also included in the group of 16 differentially represented proteins identified by Kamada et al. (2005), demonstrating reproducibility of this molecular response. However, only one of the 57 differentially represented protein spots was found to derive from a gene shown to be transcriptionally regulated by gravistimulation (Kimbrough et al. 2004). This surprising result suggests that posttranscriptional regulatory processes may play important roles in gravity signal transduction. It should, however, be cautioned that these global transcript- and protein-profiling experiments were carried out under different conditions, possibly explaining the different results between them.

Taken together, these data, along with those of experiments that focused on analyzing differential protein representation/modification in *Arabidopsis* cell cultures subjected to either hypergravity or clinorotation (Wang et al. 2006; Barjaktarovic et al. 2007), provide fertile ground for further investigations on gravity sensing and signal transduction. They also provide a large population of molecular markers that should be very useful in our attempts to elucidate the branched pathways that lead to gravity signal transduction and the corresponding curvature response, as nicely illustrated in the analysis of ADK1's contribution to gravity signal transduction (see Sect. 4.1) (Young et al. 2006). Interestingly, the initial proteomic and expression profiling studies described above identified various metabolic pathways as being responsive to gravistimulation. Therefore, it will be important to carry out a comprehensive metabolomic screen in future investigations of plant gravitropism. Indeed, such strategies should lead to the identification of potentially important new signaling molecules in the gravitropic response pathways.

5 Do Other Phytohormones Contribute to Gravitropic Signaling?

Most plant hormones, including ethylene, cytokinins, brassinosteroids, abscisic acid, gibberellins, salicylic acid, and jasmonic acid, have been implicated in the modulation of gravitropism, mainly through indirect effects on auxin transport and/ or response, as discussed in "Auxin and the Communication Between Plant Cells" by Peter Nick in this volume (Philosoph-Hadas et al. 2005; Muday and Rahman 2008). However, one of them, cytokinin, has been proposed to contribute more directly to the gravitropic signal that is transmitted from the root cap to the elongation zones upon gravistimulation (Aloni et al. 2004). Roni Aloni and his collaborators at Tel Aviv University used a cytokinin-sensitive *ARR5-GUS* reporter to investigate gravity-dependent changes in cytokinin levels within *Arabidopsis* root tips. They found a fast asymmetrical induction of reporter expression within lateral-cap cells at the bottom side of gravistimulated roots. They also showed that application of cytokinin to one side of the elongation zone of vertical roots results in tip curvature toward the side of application. On the basis of these results, they concluded that gravistimulation might promote a lateral transport of cytokinin across the cap, generating a lateral gradient that, upon transmission to the elongation zone, could be responsible for the initial phases of gravitropic curvature.

One limitation of this model is that it currently lacks genetic support. Indeed, cytokinin-deficient mutants have not shown substantial alterations in their ability to develop a root gravitropic response (Aloni et al. 2004). It is also important to note that the expression of one of the genes that contribute to cytokinin biosynthesis in the root tip, *isopentenyl transferase 5* (*IPT5*), is sensitive to auxin (Miyawaki et al. 2004). Therefore, it is possible that the activation of *ARR5-GUS* expression on the bottom flank of gravistimulated caps might simply occur as a consequence of gravity-induced auxin accumulation at the bottom flank of the cap. Even though gravistimulation promoted earlier asymmetrical activation of *ARR5-GUS* expression at the bottom flank of the cap relative to *DR5-GUS* expression (Aloni et al. 2004), the timing of expression of these two reporters is not a conclusive way of determining which signal (auxin or cytokinin) appears first at the tip upon gravistimulation because both responses are at the end of their respective pathways, and possible lags in these pathways have not been investigated. Future work aimed at defining the effect on root gravitropism of mutations that affect cytokinin synthesis and response, and a careful investigation of cytokinin transport across the tip, would be needed before the contribution of cytokinin signaling to gravity signal transduction in roots can be ascertained.

6 Conclusion and Future Perspectives

In this chapter, we have reviewed some of the molecular mechanisms that govern gravity signal transduction in plants. Our understanding of the events that allow plant statocytes to transduce the information derived from amyloplast sedimentation into a physiological signal that dictates directional changes in auxin transport remains limited. However, recent studies have suggested a critical role for the machinery that modulates the activity of plasma-membrane-associated auxin transporters, and/or their cycling between plasma membrane and intracellular compartments. We also do not understand how similar signals (changes in orientation within the gravity field) are interpreted differently by plant organs, or by the same organ at different developmental times or under distinct environmental conditions.

In addition to the extreme behaviors displayed by the specialized organs described in Sect. 1, amazing interactions between distinct signal transduction pathways that lead to tropic responses have been uncovered. For instance, touch stimulation has been shown to inhibit root gravitropism, a process that is believed to condition obstacle avoidance (Massa and Gilroy 2003). This touch-induced inhibition of gravitropism occurs even though distinct cells sense the gravity and touch stimuli (columella and lateral-cap cells, respectively), and it involves a decrease in root gravitropic sensitivity, suggesting that a signal is transmitted from the lateral cap to the statocytes (Massa and Gilroy 2003). Preliminary evidence supports a role for intercellular Ca^{2+} signaling in this process (Monshausen et al. 2008). Because Ca^{2+} seems to modulate early phases of both gravity and touch transduction, it seems necessary for the columella cells to distinguish between the Ca^{2+} signatures associated with touch and gravity signaling, a discriminatory process that remains poorly characterized.

Similarly, both light and water gradients can promote tropic responses (phototropism, as discussed in "Signaling in Phototropism" by Brandon Celaya et al., this volume, and hydrotropism, respectively), while modulating gravitropism (Lu and Feldman 1997; Correll and Kiss 2002; Takahashi et al. 2002; Kiss et al. 2003; Takahashi et al. 2003; Cassab 2008; Mullen and Kiss 2008). In the case of hydrotropism, the modulation of gravitropism seems to once again involve a desensitization of the root statocytes to gravistimulation (Takahashi et al. 2003). Deciphering the molecular mechanisms that allow cross-talk between these pathways should constitute a research priority in the future.

Altogether, it will also be critical to establish the mechanisms that allow plant organs to develop complex growth behaviors in response to combinations of directional stimuli such as gravity, touch, light, gradients in humidity, ions, chemicals, oxygen, and temperature. The task seems daunting in view of the perceived complexity of the partially elucidated pathways that govern simple tropic responses to gravity, light, touch, and water gradients. Yet, powerful new tools derived from systems biology, including genetic, genomic, proteomic, metabolomic, cell-biological, and biochemical approaches, should lead the way toward exciting new discoveries in our understanding of these complex behavioral plant-growth responses to their environment. Only then will it become possible to engineer plants that are better adapted to defined ecological niches, allowing increased production of plant parts usable in agriculture, human or animal consumption, or for biomass production for conversion into biofuels.

Acknowledgements This work was made possible by grants from the National Science Foundation (MCB-0240084, IOS-0642865, and IOS-0821884), the National Aeronautics and Space Administration (NNA04CC71G), by University of Wisconsin-Madison College of Agriculture and Life Sciences Hatch funds, and by University of Wisconsin-Madison Graduate School Grant-in-Aid to P.H.M.

References

Abas L, Benjamins R, Malenica N, Paciorek T, Wirniewska J, Moulinier-Anzola J, Sieberer T, Friml J, Luschnig C (2006) Intracellular trafficking and proteolysis of the Arabidopsis auxin-efflux facilitator PIN2 are involved in root gravitropism. Nat Cell Biol 8:249–256

Allan E, Trewavas A (1985) Quantitative changes in calmodulin and NAD kinase during early cell development in the root apex of *Pisum sativum*. Planta 165:493–501

Aloni R, Langhans M, Aloni E, Ullrich C (2004) Role of cytokinin in the regulation of root gravitropism. Planta 220:177–182

Barjaktarovic Z, Nordheim A, Lamkemeyer T, Fladerer C, Madlung J, Hampp R (2007) Time-course of changes in amounts of specific proteins upon exposure to hyper-g, 2-D clinorotation, and 3-D random positioning of Arabidopsis cell cultures. J Exp Bot 58:4357–4363

Bennett M, Marchant A, Green H, May S, Ward S, Millner P, Walker A, Schulz B, Feldmann K (1996) *Arabidopsis AUX1* gene: a permease-like regulator of root gravitropism. Science 273:948–950

Blakeslee J, Bandyopadhyay A, Lee OR, Mravec J, Titapiwatanakun B, Sauer M, Makam S, Cheng Y, Bouchard R, Adamec J, Geisler M, Nagashima A, Sakai T, Martinoia E, Friml J, Peer W, Murphy A (2007) Interactions among PIN-FORMED and P-glycoprotein auxin transporters in Arabidopsis. Plant Cell 19:131–147

Blancaflor E, Masson P (2003) Plant gravitropism. Unraveling the ups and downs of a complex process. Plant Physiol 133:1677–1690

Blancaflor E, Fasano J, Gilroy S (1998) Mapping the functional roles of cap cells in the response of *Arabidopsis* primary roots to gravity. Plant Physiol 116:213–222

Boonsirichai K, Guan C, Chen R, Masson P (2002) Root gravitropism: an experimental tool to investigate basic cellular and molecular processes underlying mechanosensing and signal transmission in plants. Annu Rev Plant Biol 53:421–447

Boonsirichai K, Sedbrook J, Chen R, Gilroy S, Masson P (2003) ARG1 is a peripheral membrane protein that modulates gravity-induced cytoplasmic alkalinization and lateral auxin transport in plant statocytes. Plant Cell 15:2612–2625

Braun M (2002) Gravity perception requires statoliths settled on specific plasma membrane areas in characean rhizoids and protonemata. Protoplasma 219:150–159

Buer C, Muday G (2004) The *transparent testa4* mutation prevents flavonoid synthesis and alters auxin transport and the response of Arabidopsis roots to gravity and light. Plant Cell 16:1191–1205

Caspar T, Pickard B (1989) Gravitropism in a starchless mutant of *Arabidopsis*: implications for the starch-statolith theory of gravity sensing. Planta 177:185–197

Cassab G (2008) Other tropisms and their relationship to gravitropism. In: Gilroy S, Masson P (eds) Plant tropisms. Blackwell, Oxford, pp 123–140

Chang S, Kaufman P (2000) Effects of staurosporine, okadaic acid and sodium fluoride on protein phosphorylation in graviresponding oat shoot pulvini. Plant Physiol Biochem 38:315–323

Chang S, Cho M, Kim S-K, Lee J, Kirakosyan A, Kaufman P (2003) Changes in phosphorylation of 50 and 53 kDa soluble proteins in graviresponding oat (*Avena sativa*) shoots. J Exp Bot 54:1013–1022

Chavarria-Krauser A (2006) Quantification of curvature production in cylindrical organs, such as roots and hypocotyls. New Phytol 171:633–641

Chen R, Hilson P, Sedbrook J, Rosen E, Caspar T, Masson P (1998) The *Arabidopsis thaliana AGRAVITROPIC 1* gene encodes a component of the polar-auxin-transport efflux carrier. Proc Natl Acad Sci U S A 95:15112–15117

Cholodny H (1928) Beiträge zur hormonalen theorie von tropismen. Planta 6:118–134

Correll M, Kiss J (2002) Interactions between gravitropism and phototropism in plants. J Plant Growth Reg 21:89–101

Evans M, Ishikawa H (1997) Cellular specificity of the gravitropic motor response in roots. Planta 203:S115–S122

Fasano J, Massa G, Gilroy S (2002) Ionic signaling in plant responses to gravity and touch. J Plant Growth Reg 21:71–88

Fasano J, Swanson S, Blancaflor E, Dowd P, Kao T, Gilroy S (2001) Changes in root cap pH are required for the gravity response of the *Arabidopsis* root. Plant Cell 13:907–921

Firn R, Digby J (1997) Solving the puzzle of gravitropism – has a lost piece been found? Planta 203:S159–S163

Friml J (2003) Auxin transport – shaping the plant. Curr Opin Plant Biol 6:7–12

Friml J, Wisniewska J, Benkova E, Mendgen K, Palme K (2002) Lateral relocation of auxin efflux regulator PIN3 mediates tropism in *Arabidopsis*. Nature 415:806–809

Fukaki H, Fujisawa H, Tasaka M (1996) Gravitropic response of inflorescence stems in *Arabidopsis thaliana*. Plant Physiol 110:933–943

Fukaki H, Fujisawa H, Tasaka M (1997) The *RHG* gene is involved in root and hypocotyl gravitropism in *Arabidopsis thaliana*. Plant Cell Physiol 38:804–810

Fukaki H, Wysocka-Diller J, Kato T, Fujisawa H, Benfey P, Tasaka M (1998) Genetic evidence that the endodermis is essential for shoot gravitropism in *Arabidopsis thaliana*. Plant J 14:425–430

Galen C, Huddle J, Liscum E (2004) An experimental test of the adaptive evolution of phototropins: blue-light photoreceptors controlling phototropism in *Arabidopsis thaliana*. Evolution 58:515–523

Geiger-Huber M, Huber J (1945) Über die Ursache des gegensätzlichen geotrophischen Verhaltens von Spross und Wurzel. Experientia 1:26–28

Guan C, Rosen E, Boonsirichai K, Poff K, Masson P (2003) The *ARG1-LIKE2* (*ARL2*) gene of *Arabidopsis thaliana* functions in a gravity signal transduction pathway that is genetically distinct from the *PGM* pathway. Plant Physiol 133:100–112

Haberlandt G (1900) Über die Perzeption des geotropischen Reizes. Ber Dtsch Bot Ges 18:261–272

Haberlandt G (1908) Über die Verteilung der geotropischen Sensibilität in der Wurzel. Jahrb Wiss Bot 45:575–600

Harrison B, Masson P (2008a) ARL2, ARG1 and PIN3 define a gravity signal transduction pathway in root statocytes. Plant J 53:380–392

Harrison B, Masson P (2008b) Do ARG1 and ARL2 form an actin-based gravity-signaling chaperone complex in root statocytes? Plant Signal Behav 3:1–4

Harrison B, Morita M, Masson P, Tasaka M (2008) Signal transduction in gravitropism. In: Gilroy S, Masson P (eds) Plant tropisms. Blackwell, Oxford, pp 21–45

Hasenstein K, Kuznetsov O (1999) The response of *lazy-2* tomato seedlings to curvature-inducing magnetic gradients is modulated by light. Planta 208:59–65

Haswell E, Meyerowitz E (2006) MscS-like proteins control plastid size and shape in *Arabidopsis thaliana*. Curr Biol 16:1–11

Heilmann I, Shin J, Huang J, Perera I, Davies E (2001) Transient dissociation of polyribosomes and concurrent recruitment of calreticulin and calmodulin transcripts in gravistimulated maize pulvini. Plant Physiol 127:1193–1203

Hou G, Mohamalawari D, Blancaflor E (2003) Enhanced gravitropism of roots with a disrupted cap actin cytoskeleton. Plant Physiol 131:1360–1373

Hou G, Kramer V, Wang Y, Chen R, Perbal G, Gilroy S, Blancaflor E (2004) The promotion of gravitropism in Arabidopsis roots upon actin disruption is coupled with the extended alkalinization of the columella cytoplasm and a persistent lateral auxin gradient. Plant J 39:113–125

Ishikawa H, Evans M (1997) Novel software for analysis of root gravitropism: comparative response patterns of *Arabidopsis* wild-type and *axr1* seedlings. Plant Cell Environ 20:919–928

Johannes E, Collings D, Rink J, Allen N (2001) Cytoplasmic pH dynamics in maize pulvinal cels induced by gravity vector changes. Plant Physiol 127:119–130

Kamada M, Higashitani A, Ishioka N (2005) Proteomic analysis of root gravitropism. Biol Sci Space 19:148–154

Kato T, Morita M, Fukaki H, Yamauchi Y, Uehara M, Niihama M, Tasaka M (2002) SGR2, a phospholipase-like protein, and ZIG/SGR4, a SNARE, are involved in the shoot gravitropism of *Arabidopsis*. Plant Cell 14:33–46

Kaufman P, Wu L, Brock T, Kim D (1995) Hormones and the orientation of growth. Kluwer, Dordecht

Keeble F, Nelson M, Snow R (1929) Integration of plant behaviour, I. Separate geotropic stimulation of tip and stump in roots. Proc R Soc Lond B Biol Sci 105:493–498

Kimbrough J, Salinas-Mondragon R, Boss W, Brown C, Winter Sederoff H (2004) The fast and transient transcriptional network of gravity and mechanical stimulation in the Arabidopsis root apex. Plant Physiol 136:2790–2805

Kiss J, Edelmann R, Wood P (1999) Gravitropism of hypocotyls of wild-type and starch-deficient *Arabidopsis* seedlings in spaceflight studies. Planta 209:96–103

Kiss J, Correll M, Mullen J, Hangarter R, Edelmann R (2003) Root phototropism: how light and gravity interact in shaping plant form. Gravit Space Biol Bull 16:55–60

Kitazawa D, Hatakeda Y, Kamada M, Fujii N, Miyazawa Y, Hoshino A, Iida S, Fukaki H, Morita M, Tasaka M, Suge H, Takahashi H (2005) Shoot circumnutation and winding movements require gravisensing cells. Proc Natl Acad Sci U S A 102:18742–18747

Kuznetsov O, Hasenstein K (1996) Intracellular magnetophoresis of amyloplasts and induction of root curvature. Planta 198:87–94

Kuznetsov O, Hasenstein K (1997) Magnetophoretic induction of curvature in coleoptiles and hypocotyls. J Exp Bot 48:1951–1957

Legue V, Blancaflor E, Wymer C, Perbal G, Fantin D, Gilroy S (1997) Cytoplasmic free Ca2+ in *Arabidopsis* roots changes in response to touch but not gravity. Plant Physiol 114:789–800

Lewis D, Miller N, Splitt B, Wu G, Spalding E (2007) Separating the roles of acropetal and basipetal auxin transport on gravitropism with mutations in two Arabidopsis multidrug resistance-like ABC transporter genes. Plant Cell 19:1838–1850

Li J, Assmann S (2000) Mass spectrometry. An essential tool in proteome analysis. Plant Physiol 123:807–809

Li J, Yang H, Peer W, Richter G, Blakeslee J, Bandyopadhyay A, Titapiwantakun B, Undurraga S, Khodakovskaya M, Richards E, Krizek B, Murphy A, Gilroy S, Gaxiola R (2005) Arabidopsis H+-PPase AVP1 regulates auxin-mediated organ development. Science 310:121–125

Li Y, Hagen G, Guilfoyle T (1991) An auxin-responsive promoter is differentially induced by auxin gradients during tropisms. Plant Cell 3:1167–1175

Limbach C, Hauslage J, Schäfer C, Braun M (2005) How to activate a plant gravireceptor. Early mechanisms of gravity sensing studied in Characean rhizoids during parabolic flights. Plant Physiol 139:1030–1040

Long J, Zhao W, Rashotte A, Muday G, Huber S (2002) Gravity-stimulated changes in auxin and invertase gene expression in maize pulvinal cells. Plant Physiol 128:591–602

Lu Y-T, Feldman L (1997) Light-regulated root gravitropism: a role for, and characterization of, a calcium/calmodulin-dependent protein kinase homolog. Planta 203:S91–S97

Luschnig C, Gaxiola R, Grisafi P, Fink G (1998) EIR1, a root-specific protein involved in auxin transport, is required for gravitropism in *Arabidopsis thaliana*. Genes Dev 12:2175–2187

Ma Z, Hasenstein K (2006) The onset of gravisensitivity in the embryonic root of flax. Plant Physiol 140:159–166

Mancuso S, Barlow P, Volkmann D, Baluska F (2006) Actin turnover-mediated gravity response in maize root apices. Gravitropism of decapped roots implicates gravisensing outside of the root cap. Plant Signal Behav 1:52–58

Massa G, Gilroy S (2003) Touch modulates gravity sensing to regulate the growth of primary roots of *Arabidopsis thaliana*. Plant J 33:435–445

Miller N, Parks B, Spalding E (2007) Computer-vision analysis of seedling responses to light and gravity. Plant J 52:374–381

Miyawaki K, Matsumoto-Kitano M, Kakimoto T (2004) Expression of cytokinin biosynthetic isopentenyltransferase genes in Arabidopsis: tissue specificity and regulation by auxin, cytokinin, and nitrate. Plant J 37:128–138

Moctezuma E, Feldman L (1999) Auxin redistributes upwards in graviresponding gynophores of the peanut plant. Planta 209:180–186

Monshausen G, Sievers A (2002) Basipetal propagation of gravity-induced surface pH changes along primary roots of *Lepidium sativum* L. Planta 215:980–988

Monshausen G, Swanson S, Gilroy S (2008) Touch sensing and thigmotropism. In: Gilroy S, Masson P (eds) Plant tropisms. Blackwell, Oxford, pp 91–122

Morita M, Kato T, Nagafusa K, Saito C, Ueda T, Nakano A, Tasaka M (2002) Inovolvement of the vacuoles of the endodermis in the early process of shoot gravitropism in *Arabidopsis*. Plant Cell 14:47–56

Morita M, Sakaguchi K, Kiyose S, Taira K, Kato T, Nakamura M, Tasaka M (2006) A C2H2-type zinc finger protein, SGR5, is involved in early events of gravitropism in Arabidopsis inflorescence stems. Plant J 47:619–628

Moseyko N, Zhu T, Chang H-S, Wang X, Feldman L (2002) Transcription profiling of the early gravitropic response in *Arabidopsis* using high-density oligonucleotide probe microarrays. Plant Physiol 130:720–728

Muday G, Rahman A (2008) Auxin transport and the integration of gravitropic growth. In: Gilroy S, Masson P (eds) Plant tropisms. Blackwell, Oxford, pp 47–78

Mullen J, Kiss J (2008) Phototropism and its relationship to gravitropism. In: Gilroy S, Masson P (eds) Plant tropisms. Blackwell, Oxford, pp 79–90

Müller A, Guan C, Gälweiler L, Tanzler P, Huijser P, Marchant A, Parry G, Bennett M, Wisman E, Palme K (1998) *AtPIN2* defines a locus of *Arabidopsis* for root gravitropism control. EMBO J 17:6903–6911

Murthy N, Young L-S, Sabat G, Masson P (2008) Identification of gravity-regulated root tip proteins in Arabidopsis. (In preparation)

Nadella V, Shipp M, Muday G, Wyatt S (2006) Evidence for altered polar and lateral auxin transport in the *gravity persistent signal* (*gps*) mutants of Arabidopsis. Plant Cell Environ 29:683–690

Nakagawa Y, Katagiri T, Shinozaki K, Qi Z, Tatsumi H, Furuichi T, Kishigami A, Sokabe M, Kojima I, Sato S, Kato T, Tabata S, Iida K, Terashima A, Nakano M, Ikeda M, Yamanaka T, Iida H (2007) Arabidopsis plasma membrane protein crucial for Ca2+ influx and touch sensing in roots. Proc Natl Acad Sci U S A 104:3639–3644

Nemec B (1900) Über die Art der Wahrnehmung des Schwekraftreizes bei den Pflanzen. Ber Dtsch Bot Ges 18:241–245

Noh B, Bandyopadhyay A, Peer W, Spalding E, Murphy A (2003) Enhanced gravi- and phototropism in plant mdr mutants mislocalizing the auxin efflux protein PIN1. Nature 423:999–1002

Ottenschlager I, Wolff P, Wolverton C, Bhalereo R, Sandberg G, Ishikawa H, Evans M, Palme K (2003) Gravity-regulated differential auxin transport from columella to lateral root cap cells. Proc Natl Acad Sci U S A 100:2987–2991

Parker K, Briggs W (1990) Transport of indole-3-acetic acid during gravitropism in intact maize coleoptiles. Plant Physiol 94:1763–1769

Perera I, Heilmann I, Boss W (1998) Phosphoinositide signaling during gravity perception and response of maize pulvini. In: Plant Biol '98 – Abstract 107:46

Perera I, Heilmann I, Boss W (1999) Transient and sustained increases in inositol-1,4,5-trisphosphate precede the differential growth response in gravistimulated maize pulvini. Proc Natl Acad Sci U S A 96:5838–5843

Perera I, Heilmann I, Chang S, Boss W, Kaufman P (2001) A role for inositol 1,4,5-trisphosphate in gravitropic signaling and the retention of cold-perceived gravistimulation of oat shoot pulvini. Plant Physiol 125:1499–1507

Perera I, Hung C-Y, Brady S, Muday G, Boss W (2006) A universal role for inositol 1,4,5-trisphosphate-mediated signaling in plant gravitropism. Plant Physiol 140:746–760

Philippar K, Fuchs I, Luthen H, Hoth S, Bauer CS, Haga K, Thiel G, Ljung K, Sandberg G, Boettger M, Becker D, Hedrich R (1999) Auxin-induced K+ channel expression represents an essential step in coleoptile growth and gravitropism. Proc Natl Acad Sci U S A 96:12186–12191

Philosoph-Hadas S, Friedman H, Meir S (2005) Gravitropic bending and plant hormones. Vitam Horm 72:31–78

Pickard B, Ding J (1993) The mechanosensory calcium-selective ion channel: key component of a plasmalemmal control centre? Aust J Plant Physiol 20:439–459

Pilet P (1953) Auxines et amidon. IV. Essais d'interprtation du gotropisme des racines du *Lens culinaris* Medikus. Bull Soc Vaudoise Sci Nat 65:394–395

Plieth C, Trewavas A (2002) Reorientation of seedlings in the Earth's gravitational field induces cytosolic calcium transients. Plant Physiol 129:786–796

Poff K, Martin H (1989) Site of graviperception in roots: a re-examination. Physiol Plant 76:451–455

Rashotte A, DeLong A, Muday G (2001) Genetic and chemical reductions in protein phosphatase activity alter auxin transport, gravity response and lateral root growth. Plant Cell 13:1683–1696

Reddy A (2001) Molecular motors and their functions in plants. Int Rev Cytol 204:97–178

Sack F (1991) Plant gravity sensing. Intl Rev Cytol 127:193–252

Saito C, Morita M, Kato T, Tasaka M (2005) Amyloplasts and vacuolar membrane dynamics in the living graviperceptive cell of the Arabidopsis inflorescence stem. Plant Cell 17:548–558

Sakamoto K, Shiraishi H, Okada K, Shimura Y (1993). Proteins induced by physical stimuli in root tips of Arabidopsis thaliana seedlings. Plant Cell Physiol 34:297–304

Scott A, Allen N (1999) Changes in cytosolic pH within *Arabidposis* root columella cells play a key role in the early signaling pathway for root gravitropism. Plant Physiol 121:1291–1298

Sedbrook J, Kronebusch P, Borisy G, Trewavas A, Masson P (1996) Transgenic *AEQUORIN* reveals organ-specific cytosolic Ca2+ responses in *Arabidopsis thaliana* seedlings. Plant Physiol 111:243–257

Sedbrook J, Chen R, Masson P (1999) *ARG1* (Altered Response to Gravity) encodes a DnaJ-like protein that potentially interacts with the cytoskeleton. Proc Natl Acad Sci U S A 96:1140–1145

Sievers A, Behrens H, Buckhout T, Gradmann D (1984) Can a Ca2+ pump in the endoplasmic reticulum of the Lepidium root be the trigger for rapid changes in membrane potential after gravistimulation? J Plant Physiol 114:195–200

Sievers A, Kruse S, Kuo-Huang L, Wendt M (1989) Statoliths and microfilaments in plant cells. Planta 179:275–278

Silady R, Kato T, Lukowitz W, Sieber P, Tasaka M, Somerville C (2004) The *gravitropism defective 2* mutants of Arabidopsis are deficient in a protein implicated in endocytosis in *Caenorhabditis elegans*. Plant Physiol 136:3095–3103

Silady R, Ehrhardt D, Jackson K, Faulkner C, Oparka K, Somerville C (2007) The GRV2/RME-8 protein of Arabidopsis functions in the late endocytic pathway and is required for vacuolar membrane flow. Plant J 53:29–41

Sinclair W, Trewavas A (1997) Calcium in gravitropism: a re-examination. Planta 203:S85–S90

Sinclair W, Oliver I, Maher P, Trewavas A (1996) The role of calmodulin in the gravitropic response of *Arabidopsis thaliana agr-3* mutant. Planta 199:343–351

Staves M (1997) Cytoplasmic streaming and gravity sensing in *Chara* internodal cells. Planta 203:S79–S84

Staves M, Wayne R, Leopold A (1997) The effect of external medium on the gravitropic curvature of rice (*Oryza sativa*, Poaceae) roots. Am J Bot 84:1522–1529

Stevenson J, Perera I, Heilmann I, Persson S, Boss W (2000) Inositol signaling and plant growth. Trends Plant Sci 5:252–258

Stinemetz C, Kuzmanoff K, Evans M, Jarret H (1987) Correlations between calmodulin activity and gravitropic sensitivity in primary roots of maize. Plant Physiol 84:1337–1342

Swarup R, Friml J, Marchant A, Ljung K, Sandberg G, Palme K, Bennett M (2001) Localization of the auxin permease AUX1 suggests two functionally distinct hormone transport pathways operate in the *Arabidopsis* root apex. Genes Dev 15:2648–2653

Swarup R, Kargul J, Marchant A, Zadik D, Rahman A, Mills R, Yemm A, May S, Williams L, Millner P, Tsurumi S, Moore I, Napier R, Kerr I, Bennett M (2004) Structure-function analysis of the presumptive Arabidopsis auxin permease AUX1. Plant Cell 16:3069–3083

Swarup R, Kramer E, Perry P, Knox K, Leyser H, Haseloff J, Beemster G, Bhalerao R, Bennett M (2005) Root gravitropism requires lateral root cap and epidermal cells for transport and response to a mobile auxin signal. Nat Cell Biol 7:1057–1065

Takahashi N, Goto N, Okada K, Takahashi H (2002) Hydrotropism in abscisic acid, wavy, and agravitropic mutants of *Arabidopsis thaliana*. Planta 216:203–211

Takahashi N, Yamazaki Y, Kobayashi A, Higashitani A, Takahashi H (2003) Hydrotropism interacts with gravitropism by degrading amyloplasts in seedling roots of *Arabidopsis* and radish. Plant Physiol 132:805–810

Tamura K, Takahashi H, Kunieda T, Fuji K, Shimada T, Hara-Nishimura I (2007) Arabidopsis KAM2/GRV2 is required for proper endosome formation and functions in vacuolar sorting and determination of the embryo growth axis. Plant Cell 19:320–332

Tanaka A, Kobayashi Y, Hase Y, Watanabe H (2002) Positional effect of cell inactivation on root gravitropism using heavy-ion microbeams. J Exp Bot 53:683–687

Toyota M, Furuichi T, Tatsumi H, Sokabe M (2008) Cytoplasmic calcium increases in response to changes in the gravity vector in hypocotyls and petioles of Arabidopsis seedlings. Plant Physiol 146:505–514

Trewavas A (1992) What remains of the Cholodny–Went theory? Plant Cell Environ 15:759–794

Tsugeki R, Fedoroff N (1999) Genetic ablation of root cap cells in *Arabidopsis*. Proc Natl Acad Sci U S A 96:12941–12946

Utsuno K, Shikanai T, Yamada Y, Hashimoto T (1998) *AGR*, an *Agravitropic* locus of *Arabidopsis thaliana*, encodes a novel membrane-protein family member. Plant Cell Physiol 39:1111–1118

Volkmann D, Baluska F (1999) Actin cytoskeleton in plants: from transport networks to signaling networks. Microsc Res Tech 47:135–154

Wang X (2004) Lipid signaling. Curr Opin Plant Biol 7:329–336

Wang H, Zheng H, Sha W, Zeng R, Xia Q (2006) A proteomic approach to analysing responses of *Arabidopsis thaliana* callus cells to clinostat rotation. J Exp Bot 57:827–835

Weise S, Kuznetsov O, Hasenstein K, Kiss J (2000) Curvature in *Arabidopsis* inflorescence stems is limited to the region of amyloplast displacement. Plant Cell Physiol 41:702–709

Wendt M, Kuo-Huang L, Sievers A (1987) Gravitropic bending of cress roots without contact between amyloplasts and complexes of endoplasmic reticulum. Planta 172:321–329

Wenk M, De Camilli P (2004) Protein-lipid interactions and phosphoinositide metabolism in membrane traffic: insights from vesicle recycling in nerve terminals. Proc Natl Acad Sci U S A 101:8262–8269

Went F (1928) Wuchstoff und Wachstum. Rec Trav Bot Neerl 25:1–116

Wolverton C, Ishikawa H, Evans M (2002a) The kinetics of root gravitropism: dual motor and sensors. J Plant Growth Reg 21:102–112

Wolverton C, Mullen J, Ishikawa H, Evans M (2002b) Root gravitropism in response to a signal originating outside of the cap. Planta 215:153–157

Wyatt S, Rashotte A, Shipp M, Robertson D, Muday G (2002) Mutations in the gravity persistence signal loci in *Arabidopsis* disrupt the perception and/or signal transduction of gravitropic stimuli. Plant Physiol 130:1426–1435

Yamamoto K, Kiss J (2002) Disruption of the actin cytoskeleton results in the promotion of gravitropism in inflorescence stems and hypocotyls of *Arabidopsis*. Plant Physiol 128:669–681

Yano D, Sato M, Saito C, Sato M, Morita M, Tasaka M (2003) A SNARE complex containing SGR3/AtVAM3 and ZIG/VTI11 in gravity-sensing cells is important for Arabidopsis shoot gravitropism. Proc Natl Acad Sci U S A 100:8589–8594

Yester J, Kimbrough J, Salinas-Mondragon R, Masson P, Brown C, Winter Sederoff H (2006) Regulation of transcription in roots of Arabidopsis gravity mutants. Gravit Space Biol Bull 19:159–160

Yoder T, Zheng H, Todd P, Staehelin L (2001) Amyloplast sedimentation dynamics in maize columella cells support a new model for the gravity-sensing apparatus of roots. Plant Physiol 125:1045–1060

Young L, Evans M, Hertel R (1990) Correlations between gravitropic curvature and auxin movement across gravistimulated roots of *Zea Mays*. Plant Physiol 92:792–796

Young J, Barral J, Hartl U (2003) More than folding: localized functions of cytosolic chaperones. Trends Biochem Sci 28:541–547

Young L-S, Harrison B, Murthy N, Moffatt B, Gilroy S, Masson P (2006) Adenosine kinase modulates root gravitropism and cap morphogenesis in *Arabidopsis thaliana*. Plant Physiol 142:564–573

Signaling in Phototropism

R. Brandon Celaya, Ullas V. Pedmale, and Emmanuel Liscum

Abstract Land plants cope with the same environmental challenges as animals but have the added complication of being fixed to the ground. Thus, adaptability to variable environmental circumstances is essential to plant survival and fitness. A consequence of this condition is the necessity of plants to possess sophisticated sensors to adjust to changes. Plants take the input from their myriad of physiological sensors and respond physiologically. Among these responses are the tropisms, or directional growth responses that are oriented relative to a directional stimulus. Phototropism is among one of the best-studied tropic responses where plant tissues perceive and grow directionally upon perception of a directional light stimulus - positively, or towards the light source, in the case of shoots and negatively, or away from the light source, in the case of roots. From a historical perspective, the phototropic phenomenon has been known for hundreds of years (Whippo and Hangarter in *Plant Cell* 18:1110–1119, 2006). Yet, only in the past few decades has the phenomenon been carefully studied to the extent that the basis of this response has become clearer. While recent analyses have yielded detailed biochemical mechanisms for some of the phototropic receptors, a great deal remains unknown. In this review of phototropism in plants, the focus is on the growth of our understanding of phototropism from the simple observations of plant growth, to the initial physiological experiments, to the most recent detailed molecular mechanisms. With the advances in genetic and molecular tools we are now in a position to understand the nature of phototropic signaling and its regulation in great detail. Over the past few years we have come to learn much about the complex interplay of molecules,

R.B. Celaya and U.V. Pedmale
Christopher S. Bond Life Sciences Center and Division of Biological Sciences,
University of Missouri Columbia MO 65211, USA

E. Liscum (✉)
Emmanuel Liscum 371E C.S. Bond Life Sciences Center,
University of Missouri Columbia, MO 65211, USA
e-mail: liscume@missouri.edu

F. Baluška and S. Mancuso (eds.), *Signaling in Plants,* Signaling and Communication in Plants, 239
DOI: 10.1007/978-3-540-89228-1_12, © Springer-Verlag Berlin Heidelberg 2009

including the photoreceptors, accessory proteins, transcription factors, and effector molecules necessary to perceive the light cues, modulate signaling, activate gene transcription, and elicit physiological change. While some of these players are known, undoubtedly a role for many others will emerge in future studies and such advances will provide new avenues of research.

1 Phototropim: A "Simple" Physiological Response

A detailed historical perspective of phototropism has recently been published and is an excellent resource for further reading (Whippo and Hangarter 2006), as is a more classical perspective from the field (Sachs 1887). Here we shall summarize several of the key experimental results that have shaped the field as we see it today.

Early physiological analysis of phototropism was largely phenomenological with explanations varying from loss of water on the illuminated side of the seedling to "natural magic" explanations suggesting the seedlings were "rejoicing" in the presence of sunlight to the effects of the so-called vital force (Whippo and Hangarter 2006). It was not until the late 1870s and early 1880s that experiments performed by Julius von Wiesner (von Wiesner 1878; Sachs 1887), followed by Charles and Francis Darwin (Darwin 1880), gave some inkling as to how phototropism might occur. Original observations in *Avena sativa* coleoptiles, later reconfirmed, suggested the phototropic response occurred in distinct, fluence (a unit of light intensity per unit area; e.g., micromoles meter^{-2}) responsive phases (Thimann 1960; Zimmerman and Briggs 1963a, b). The first positive curvature occurs at low fluence levels and is proportionate to the amount of light to its peak curvature. After this point the responsiveness declines within a region where very little response is seen called the "neutral zone." With extended exposure to light, the seedling again appears responsive as it enters into the second positive curvature. It is the second positive curvature to which much of the attention has been devoted experimentally. As developments in the field progressed, the use of *Arabidopsis thaliana* for these physiological experiments grew, taking advantage of the vast genetic resources available. *Arabidopsis* seedlings were shown to have much the same response pattern for first and second positive phototropism as well as a zone of indifference (Janoudi and Poff 1990). Eventually, loss-of-function mutant screens were undertaken and began to identify strains that showed aberrant phototropic responsiveness (Khurana and Poff 1989). These initial studies provided the foundation that led to the identification and cloning of the genetic locus associated with a photoreceptor responsible for phototropic responsiveness initially called *NON-PHOTOTROPIC HYPOCOTYL 1* (*NPH1*) (Liscum and Briggs 1995), eventually renamed *PHOTOTROPIN 1* (*PHOT1*) (Briggs et al. 2001). Subsequently, a complementary DNA coding for a similar gene was identified originally called *NPH1-like 1* (*NPL1*) (Jarillo et al. 1998). This gene was further characterized for phototropin-mediated phenotypes (Jarillo et al. 2001; Kagawa et al. 2001) and was eventually renamed *PHOTOTROPIN*

2 (*PHOT2*) (Briggs et al. 2001). It is these photoreceptors that represent the starting point for the phototropic signal progression.

2 Point of Reception: The Phototropins and Other Photoreceptors

The phototropins represent a class of blue-light photoreceptors that play a key role in the phototropic response, though their involvement in a number of other processes has been described and will be considered further here. A great deal of emphasis has been placed on understanding how phototropin protein structure changes in response to blue-light absorption. In this context, the advances in our understanding of phototropin structure will be considered, as well as recent developments that illuminate the nature of how phototropin relocalization may represent a key first step in signal transduction.

2.1 The Many Aspects of LOV: LOV Domain Characteristics

Phototropins represent a class of blue-light-sensing proteins that contain two Per-Arnt-Sim (PAS) domains, as well as a serine/threonine kinase domain. The PAS domains in phot1 and phot2 are members of a smaller family with sensitivity to light, oxygen, and voltage (LOV). These domains have been studied in phototropins across a number of plant species (Kasahara et al. 2002) and have been shown to have high identity and similar spectrometric properties (Christie et al. 1999). The blue-light-sensing core of the LOV domains establishes a linkage between a canonical cysteine domain (GXNCRFLQ) and flavin mononucleotide (FMN) (Christie et al. 1998). Structurally, the LOV domains consist of four β-sheets and three α-helices. One of these α-helices, denoted α′A, appears to be a small "connector helix" that contains the key cysteine residue and, together with three of the β-sheets (C, D, and E), makes contact with the FMN (Crosson and Moffat 2001). Substitution of this critical cysteine residue with alanine has been shown to have substantial effects on the photophysical properties (Salomon et al. 2000; Swartz et al. 2001; Christie et al. 2002; Corchnoy et al. 2003; Guo et al. 2005) of the protein as well as physiological consequences for phototropin-mediated phenotypes (Christie et al. 2002; Cho et al. 2007). Moreover, results from such studies indicated a predominant role for the LOV2 domain over the LOV1 domain for both phot1 and phot2. An initial photochemical study of the LOV domains had shown that LOV2 has a quantum efficiency that is 10 times that of the LOV1 domain (Salomon et al. 2000). This difference has led to the assertion that the heightened sensitivity of the LOV2 domain explains the effects of the cysteine mutation within this domain (Christie et al. 2002).

The role of the LOV1 domain remained largely unclear until very recently. Initial experiments studying the dimerization qualities of the LOV domains of phot1a from *Avena sativa* through gel chromatography indicated a capacity of LOV1 but not of LOV2 to form dimers (Salomon 2004). Biophysical evidence of the dimerization capability of the LOV domains has come to light (Nakasako et al. 2004). This result is not entirely unexpected given that phototropin molecules in *Arabidopsis* (Reymond et al. 1992a, b; Cho et al. 2007) as well as other species (Reymond et al. 1992a, b) can cross-phosphorylate one another. There is some evidence that LOV2, under certain restrictive conditions of concentration and in the context of an additional phototropin peptide backbone, is also capable of dimerization (Eitoku et al. 2005). The limited LOV2 dimerization activity will be reconsidered during discussion of the linker region between the LOV domains and the protein kinase domain. A recent crystal structure analysis of the LOV1 domains from *Arabidopsis* phot1 and phot2 has demonstrated a dimerization event of these domains characterized by hydrophobic interactions of the A α-helix with several of the β-sheets, bonding between several β-sheets through van der Waals forces, and hydrogen bonding and a disulfide bridge with the Cys261 residues (Nakasako et al. 2008). It is worth noting that while LOV domains in *Neurospora* white-collar proteins (wc-1 and wc-2) have been shown to homodimerize as well as heterodimerize (Ballario et al. 1998), this has not proven to be the case in phototropin from the alga *Chlamydomanas* (Fedorov et al. 2003) owing in part to a lack of the key disulfide bridge forming cysteine (Nakasako et al. 2008). This may suggest the utility and thus retainment of phototropin dimerization may vary from species to species, though *Chlamydomanas* phototropin can rescue phototropin deficiency in *Arabidopsis* (Onodera et al. 2005). Interestingly though, not all phototropins appear to have LOV domains with the capacity for dimerization, a case particularly evident between phot1 and phot2 from *Arabidopsis* (Nakasako et al. 2004). This may represent a key difference in the signaling patterns of these two receptors. Altogether, these results suggest a dynamic and unique role for each of the LOV domains of phototropin. In some cases, LOV1 appears to function not only as an attenuator of phototropin activity but also as a site for dimerization, while the more photsensitive LOV2 appears to regulate phototropin activation after sensing blue light. Both domains are potential sites for protein-protein interactions.

2.2 A Circle of Light: The Photocycle of the Phototropin LOV Domains

LOV domains of phototropin undergo a multistep photocycle (Christie et al. 1999; Salomon et al. 2000; Corchnoy et al. 2003; Fedorov et al. 2003; Iwata et al. 2003; Neiß and Saalfrank 2003; Guo et al. 2005; Sato et al. 2005; Christie et al. 2007; Matsuoka et al. 2007) that eventually triggers dramatic changes in the quaternary structure of the protein (Swartz et al. 2002; Harper et al. 2003, 2004; Iwata et al. 2003; Nakasako et al. 2004). This cycle occurs via a number of intermediate steps and has been greatly sorted out through a number of studies combining mutants and biophysics. The changes in the protein structure subsequent to blue-light exposure and concomitant with the activation of the serine/threonine kinase are referred to as the "phototropin switch" (Crosson and Moffat 2002; Harper et al. 2003). The initial study (Swartz et al. 2001) of the LOV domain photocyle described three distinct phases [denoted as LOV^{X}_{n} where *X* indicates the state with dark (D), light-activated (L), signaling (S), and transition (T) and *n* denotes the absorption maximum for each state]: (1) in the ground state (LOV^{D}_{447}), FMN remains noncovalently linked to the phototropin protein; (2) in the excited state (LOV^{L}_{660}), very rapidly (within 30 ns) after blue-light exposure, spectroscopic studies suggest the presence of a flavin triplet state (Swartz et al. 2001); (3) finally, the formation of the cysteinyl adduct with the flavin C(4a) results in the presumed signaling form designated LOV^{S}_{390}. The cysteinyl adduct then decays back (approximately 60 s) into the LOV^{D}_{447} state (Salomon et al. 2000; Swartz et al. 2001; Kasahara et al. 2002). The initial steps of this process have been studied in great detail utilizing different LOV domains from both plant and algal phototropins.

The first proposed mechanism, based on the LOV2 domain from the fern *Adiantum*, suggested light leads to an alteration of the ground-state electron configuration within the isoalloxazine ring of FMN (Crosson and Moffat 2001). It was later found that *Adiantum* LOV2 showed a small 8° rotation of the FMN and only movement among the side chains of the residues within the FMN binding pocket and not in the protein backbone (Crosson and Moffat 2002). This alteration leads to a partial negative charge on N5, leading to the formation of a thiol anion which proceeds as a nucleophile and attacks C(4a), forming the cystienyl adduct. This mechanism was eventually further refined (Neiß and Saalfrank 2003; Dittrich et al. 2005; Sato et al. 2005; Matsuoka et al. 2007) to describe the mechanism in greater detail. Of particular interest has been the identification of the thiol species present and ascertaining the subsequent molecular steps. Early quantum calculations favored a radical combination of the excited flavin semiquinone and the thiol radical prior to intersystem crossing and recombination to form the adduct (Neiß and Saalfrank 2003). Subsequently, a mechanism was proposed (Fedorov et al. 2003) which, in contrast to others (Crosson and Moffat 2001; Swartz et al. 2001), suggested that C(4a) remains neutral during adduct formation and that instead a partial positive charge from C(5a) facilitates the negative charge on N5. This charge attracts the acidic hydrogen of the cysteine group, leaving the sulfur at a net negative charge. Because of the neutrality of C(4a), N5 pull will leave a partial positive charge facilitating combination with the sulfur. Still unclear was the thiol species participating in the adduct formation. Low-temperature infrared spectroscopy, used to distinguish between protonated and nonprotonated thiol groups, revealed a protonated sulfur group (Iwata et al. 2002, 2003). This finding has led to the hypothesis that the mechanism involves a protonated thiol radical (Matsuoka et al. 2007). While the mechanism has become clearer, other studies have begun focusing on the surrounding residues of the apoprotein and their role(s) in the photoreaction (Nozaki et al. 2004; Christie et al. 2007; Jones et al. 2007; Yamamoto et al. 2008). These findings have contributed to understanding how the decay of the active

LOV^S_{390} form is affected by mutation of surrounding residues (Christie et al. 2007; Yamamoto et al. 2008), as well as through base-catalyzed breakdown (Alexandre et al. 2007), and how adduct formation is transmitted to rest of the protein (Nozaki et al. 2004; Jones et al. 2007). It is this transmission of the light signal to the rest of the protein and its consequences for phototropin activity that we consider next.

2.3 *On the Hinge of Signaling: The Contributions of the J*α *Helix*

After photoactivation of the LOV domains, the events leading to phototropin activation had been unclear. Yet, the events following the formation of LOV^S_{390} must transmit changes to the rest of the protein to achieve a signaling state - they must switch on phototropin. Evidence of how the formation of LOV^S_{390} led to changes in the protein conformation came to light upon the description of an α-helix connecting the serine/threonine kinase with the LOV domains (Harper et al. 2003). This work along with subsequent findings (Harper et al. 2004) have shown that numerous changes in the conformation of the phototropin protein follow blue-light absorption. The molecular consequences of the Jα helix are becoming apparent through recently obtained biophysical results. Through utilization of ultrafast transient gating and transient lens florescence measurement techniques, several insights regarding the LOV2 domain's interaction with the linker peptide (including the Jα helix) have been obtained. The first of these studies showed that a small but detectable difference in the diffusion constants between the LOV2 and the LOV2-linker peptides (Eitoku et al. 2005). This suggested that upon blue-light absorption $LOV2^D_{447}$ proceeds to the adduct $LOV2^S_{390}$ state, but before decay to the ground state, a transition state ($LOV2^T_{390}$) is formed where the linker region is destabilized prior to signaling. As mentioned previously, LOV2 is typically monomeric in solution (Salomon et al. 2004), though recent studies suggest a dimerization is possible and is greatly facilitated by the presence of the linker region (Nakasako et al. 2004; Nakasone et al. 2006). It seems then that LOV2 undergoes a substantial change in its interaction with the Jα helix that it has been postulated to be repressed in the dark (Matsuoka and Tokutomi 2005). Subsequent studies have proposed a multistep transition state for the $LOV2^S_{390}$ adduct involving a precursor state where the hinge helix is diassociated (lose non-ionic contact with) from LOV2 (referred to as $LOV2^{T}_{390}{}^{pre}$) on a millisecond time scale followed by an unfolding of the helix to a presignaling state ($LOV2^T_{390}$) on the millisecond time scale (Nakasone et al. 2007). These findings have been supported energetically by a thermodynamic analysis of these intermediate states (Eitoku et al. 2007). As the underlying mechanism becomes clearer and clearer, other studies have focused on the identification of residues that are key in processing the LOV domain adduct formation to the C-terminal protein kinase domain to initiate signaling (Jones et al. 2007). While these findings point to investigation of the interaction of the kinase domain with LOV2, experiments to assess the functional roles of these residues have yet to be undertaken largely owing to the insolubility of full-length phototropin (Christie et al. 1998).

2.4 It Takes LOV and Kinase: Activation of Phosphorylation

Following the events of the "phototropin switch," the serine/threonine kinase in the C-terminal protein is activated. The phototropin kinase shares similarity with cyclic-AMP-dependent protein kinase A, cyclic-GMP-dependent protein kinase G, and phospholipids-dependent protein kinase C (AGC) group VIII kinases and phosphorylation motifs similar to protein kinase A targets (Salomon et al. 2003). Many of the aspects of phototropin kinase highlighted here have been nicely reviewed recently (Tokutomi et al. 2008). While, the kinase activation has been studied extensively, native substrates have yet to be identified other than the phototropins themselves. The phototropin protein kinase activity, like for other members of the AGC group, requires a crucial aspartate residue for phosphotransfer (Hanks and Hunter 1995). Loss of this residue negates autophosphorylation in phototropin expressed in insect cells (Christie et al. 2002), as well as phototropic responsiveness in lines solely expressing this mutated phototropin (*phot1-7*) (Cho et al. 2007). The autophosphorylation sites in oat (Salomon et al. 2003) and *Arabidopsis* (Inoue et al. 2008a, b; Sullivan et al. 2008) phot1 have been identified and their functional roles addressed physiologically, at least in part (Salomon et al. 2003; Inoue et al. 2008a, b; Sullivan et al. 2008). Despite the work done to find substrates of the phototropin kinases, no other bona fide targets have been found. A recent study has shown that the phot2 kinase is able to phosphorylate casein, even in the absence of the crucial "phototropin switch" Jα helix described previously. Although the implications of this finding remain unclear, it is possible that phototropin structural changes and kinase activity may be distinct (Matsuoka and Tokutomi 2005). Studies of the oat phototropin autophosphorylaton sites indicated that individual sites may become phosphorylated under increasing fluence levels (Salomon et al. 2003). The authors suggest this finding may indicate a role of accessory proteins or the effects of phototropin localization.

2.5 Signaling on the Run?: Implications of Phototropin Subcellular Movements

To this point the discussion has largely reflected the intramolecular changes within the phototropins. However, a growing field of inquiry is reflected in recent studies of phototropin localization and its potential physiological implications. These studies have largely revolved around experiments examining localization of an *Arabidopsis* phot1 protein tagged with a green fluorescent protein (GFP) reporter (Sakamoto and Briggs 2002) under a number of conditions (Sakamoto and Briggs 2002; Wan et al. 2007). The phot1 protein was first observed biochemically as an approximately 120 kDa protein in many plant species that shifted in its mobility upon blue-light treatment, suggesting a posttranslational modification (Gallagher et al. 1988; Reymond et al. 1992a, b; Palmer et al. 1993a, b; Short et al. 1993) that was later shown to reflect autophosphorylation (Knieb et al. 2005). In addition, it had previously been shown that the phot1 protein was cofractionated with plasma membrane proteins (Short et al. 1993). Studies with the phot1-GFP fusion protein confirmed that phot1 localizes to the plasma membrane in etiolated seedlings (Sakamoto and Briggs 2002). Interestingly though, following blue-light exposure, phot1 appeared to relocalize from the plasma membrane into the cytosol. These findings were subsequently followed up on in an attempt to understand the nature of this relocalization biochemically (Knieb et al. 2004). Recent work utilizing a tagged phot2 transgene has shed some light on this phenomenon. The phot2-GFP protein also appears to undergo relocalization into the cytosol upon blue-light exposure (Kong et al. 2006). The authors of this work further ascertained that it is the C-terminal kinase domain that is crucial for the plasma membrane localization though the kinase activity, per se, appears not to be essential (Kong et al. 2006). With respect to phot1, the C-terminal-associated localization has also been confirmed biochemically in lines lacking the N-terminal LOV1 domain but that contain the rest of the protein (Thomson et al. 2008). This subcellular relocalization was addressed in greater detail biochemically (Knieb et al. 2004) and in subsequent studies utilizing the phot1-GFP transgenic line (Wan et al. 2007). There are many possible explanations for this relocalization and its functional consequences. One possibility is that phototropin protein activity is downregulated through relocalization following blue-light exposure. This is supported by the finding that a C-terminal portion of phot2 represses phototropic responsiveness, though this appears to require a functional kinase (Kong et al. 2007). It is also possible that the relocalization may be a key step in activation of phototropic signaling as has been seen in the case of the brassinolide receptor BRI1 (Geldner et al. 2007), perhaps through a mechanism somewhat similar to the "signaling endosome" described in animal systems (Sorkin and Von Zastrow 2002) and hypothesized in plants (Raikhel and Hicks 2007). In any case, the study of phototropin movement may well lead to further insights into the initial steps of phototropin signaling.

3 Other Points of Light: Additional Photoreceptors and Phototropism

As described previously, the physiological action spectrum of phototropism indicated that blue light had the strongest effect in eliciting the response. The absorption spectrum of phototropin matches nicely the action spectrum for phototropism; however, because the blue-light-sensing cryptochromes also are active at these wavelengths, the involvement of phototropin in phototropism has been scrutinized (Ahmad et al. 1998; Lascève et al. 1999). Early studies suggested cryptochrome mutants were defective in their first positive response, implying a crucial sensing role (Ahmad et al. 1998). A more detailed physiological analysis of both first and second positive phototropic responses showed that single mutants of *CRYPTOCHROME 1* and *CRYPTOCHROME 2* show only a small diminishment in the magnitude of their response for the first or the second positive response, with no alteration in fluence responsiveness (Lascève et al. 1999). In contrast, the *cry1-304 cry2-1* double mutant had impaired second positive responsiveness, which implied a downstream role for the cryptochrome in phototropin signaling (Lascève et al. 1999; Whippo and Hangarter 2003). Yet the aforementioned finding must eventually be reconciled with recent work that suggests cryptochromes reduce levels of phot1 protein (Kang et al. 2008). This latter finding is consistent with the knowledge that crys and phots have distinct roles in the inhibition of hypocotyl growth in response to blue light (Cosgrove 1981).

Physiological studies of phototropism utilizing red-light treatments in combination with blue light, as well as phytochrome mutants, also clearly indicated a role for the red-light-sensing phytochromes in the first positive (Blaauw-Jansen 1959; Zimmerman and Briggs 1963a, b; Blaauw and Blaauw-Jansen 1964; Chon and Briggs 1966; Iino et al. 1984; Parker et al. 1989) and the second positive (Iino 1990; Harper et al. 2000; Stowe-Evans et al. 2001; Whippo and Hangarter 2004) phototropic responses in the shoots as well as an influence on root phototropism (Kiss et al. 2001; Ruppel et al. 2001; Correll et al. 2003). The influence of phytochrome on the first positive phototropic response leads to a dramatic amplification of the response (Konjevic et al. 1989; Parks et al. 1996; Janoudi et al. 1997), similar to what has been seen in the second positive response (Stowe-Evans et al. 2001; Whippo and Hangarter 2004). The amplification of the phototropic response by the phytochromes clearly implicates cross-talk in these signaling pathways. It was thus intriguing to find that phototropin can interact with phytochrome kinase substrate protein 1 (PKS1) (Lariquet et al. 2006), and that *pks* mutants show reduction in phototropic responsiveness (Lariguet et al. 2006; Boccalandro et al. 2007).

4 Other Phototropin Functions: Phototropins Do More Than Phototropism

To this point the focus of this chapter has largely been on the role that phototropins play in the phototropic response. Although it is clearly important to phototropism, a number of other physiological responses have been linked to the phototropins and a number of key studies have demonstrated the utility of these proteins in a number of processes (well reviewed in Christie 2007). Here we shall summarize these findings in the context of putative new players in the phototropin signaling pathway and their potential role in signal response.

The opening and closing of the pores on the leaf, called "stomata," is crucial for gas-exchange processes leading to photosynthesis and is affected by both light and hormones (Zeiger 1983; Assmann 1993; Kinoshita et al. 2001; Schroeder et al. 2001). Control of the guard cell aperture is regulated largely through the flow of potassium ions regulated by the action of an H^+-ATPase. Until recently, it was unclear as to how the blue-light induction of stomatal opening and the phototropins were linked to this proton pump. A link was established once a 14-3-3 protein that interacts with the proton pump was also found to interact with phosphorylated phototropin (Kinoshita et al. 2003). The importance of this interaction was resolved utilizing the *phot1 phot2* double mutant, which was shown to lack proton pumping (Ueno et al. 2005). Since the function of the 14-3-3 protein is thought largely to be one of a facilitator of protein-protein interaction, it remains unclear as to whether there are other components needed for stomatal responsiveness that are triggered by blue light and the phototropins. Recently a protein from *Vicia faba*, which was identified through a yeast-two-hybrid screen, was shown to interact with the N-terminal domain of phototropin (Emi et al. 2005). This protein, referred to as Vf-PHOT1a INTERACTING PROTEIN (VfPIP), appears to be a dynein light chain like protein. Interestingly there are no heavy dynein chain proteins in the *Arabidopsis* genome (Lawrence et al. 2001) and given their requirement for recruitment of dynein based protein movement (King 2000), the role of VfPIP is unclear.

While ionic influences on cell growth in plants are known, it is worth noting that blue light absorbed by phototropins can elicit a rapid increase in Ca^{2+} concentration in the cytosol (Harada et al. 2003; Stoelzle et al. 2003). While this may point to a possible secondary step in phototropin signaling, and recent evidence suggests this may still be the case (Chen et al. 2008), negligible changes in phototropic responsiveness upon chelation of Ca^{2+} may point to another role for calcium (Folta et al. 2003). Rapid ion changes in response to blue light have been linked to the phototropins (Babourina et al. 2002). This response, at least in the case of proton extrusion, appears to require an active phototropin kinase (Inoue et al. 2008a, b). Continued investigation in this field may indicate multiple pathways for this ionic influence.

Chloroplast movement is an adaptive response to changes in light conditions hypothesized to maximize light harvesting for photosynthesis under low to moderate light intensities, while also protecting chloroplasts from bleaching and photodamage under high light conditions (Wada et al. 2003). This dual functionality results in two separate phases: chloroplast accumulation and chloroplast avoidance, which appear to employ phot1 and phot2, respectively, likely as a consequence of their light sensitivities (Sakai et al. 2001). A number of interesting loci have been described that appear to be associated with phototropin-dependent chloroplast movements. One locus, *CHLOROPLAST UNUSUAL POSITIONING 1* (*CHUP1*), appears to encode a protein involved in both the accumulation and avoidance responses since mutations in *CHUP1* impair both responses (Oikawa et al. 2003). The CHUP1 protein is chloroplast-localized and interacts with F-actin (Oikawa et al. 2003). Mutations in another locus, *J-DOMAIN PROTEIN REQUIRED FOR CHLOROPLAST ACCUMULATION RESPONSE 1* (*JAC1*), lack proper chloroplast accumulation responses but have a normal avoidance response (Suetsugu et al. 2005). Interestingly the JAC1 protein contains an auxilin-clathrin uncoating domain (Suetsugu et al. 2005). It remains to be shown how these proteins and others (such as those encoded by the *PLASTID MOVEMENT IMPAIRED* (*PMI*) genes (DeBlasio et al. 2005; Luesse et al. 2006) function with respect to phototropins. Further studies will hopefully clarify whether these proteins receive cues directly or indirectly from phototropins, and what is nature of these processes.

Recently, the roles of phototropins in a number of other contexts throughout the stages of plant development have been explored. Among these are the leaf movement (separate from those associated with circadian rhythms; Engelmann 1992) and leaf flattening phenotypes. In the *phot1* mutant background, it has been observed that 7-day-old seedlings show a reduced petiole angle to the horizontal (Inoue et al. 2008a, b). This reduction along with deficiencies in leaf flattening suggest phototropins (and their interactors) play a role in the light capturing efficiency and thus overall growth efficiency (Inoue et al. 2008a, b). A study of the drought tolerance between phototropin mutants and those expressing a functional copy indicated a reduction in biomass for the mutants (Galen et al. 2004). These findings show the overall consequences for the plant as a whole based on the activity of phototropin in both the root and the aerial tissues.

5 Phototropin-Interacting Proteins: NPH3 and RPT2

Several loci involved in phototropism, in addition to PHOT1 and PHOT2, have been identified through loss-of-function genetic screens (Khurana and Poff 1989; Khurana et al. 1989; Liscum and Briggs 1995; Liscum and Briggs 1996). Two such mutants were strain JK218 (Khurana and Poff 1989) and *non-phototropic hypocotyl 3* (*nph3*) (Liscum and Briggs 1995). Subsequent genetic analysis demonstrated that these mutants are alleles of the same locus, now simply designated *NPH3* (Motchoulski and Liscum 1999). An independent screen identified three additional mutants *root phototropism 1*, *root phototropism 2*, and *root phototropism 3* (*rpt1*, *rpt2*, *rpt3*) (Okada and Shimura 1992). (While the names of the aforementioned mutants may imply organ-specific function of the encoded proteins, in fact phototropism is altered in all mutants in both hypocotyl and root.) *rpt1* was shown to be allelic to *PHOT1*, and *rpt3* allelic to *NPH3*, while *rpt2* mapped to a previously undescribed gene (Sakai et al. 2000). Further analysis identified RPT2 as a member of a family of 31 proteins in *Arabidopsis* that includes NPH3 (Liscum and Reed 2002). On the basis of these findings, additional members of this protein family have been designated NRL1-NRL29, for NPH3/RPT2-LIKE. At present little has been reported relative to the potential functions of other NRL family members (Cheng et al. 2007). It should be noted that in rice, a homolog of *NPH3* called *COLEOPTILE PHOTOTROPISM 1 (CPT1)* has been identified and many of the same physiological defects are observed in the *nph3* and *cpt1* mutants, suggesting conservation of function throughout the plant kingdom (Haga et al. 2005).

While both NPH3 and RPT2 function in phototropism (Inada et al. 2004) important differences exist between their functions. Firstly, *nph3* mutants show complete loss of phototropism, while *rpt2* mutants have diminished response at high fluence rates but normal responsiveness at low fluence rates (Sakai et al. 2000; Inada et al. 2004). These diverged phenotypes are likely explained by the findings that *NPH3* appears to be constitutively expressed, whereas *RPT2* is expressed highly only after light exposure (Sakai et al. 2000). The N-terminal region of

RPT2 appears to interact with phot1 LOV domains (Inada et al. 2004), as opposed to NPH3, which utilizes its C-terminal region as the phot1-interacting domain (Motchoulski and Liscum 1999). Lastly, RPT2 and NPH3 seem to have different effects on other phototropin-mediated phenotypes (Inada et al. 2004). In addition, the posttranslational status of RPT2 appears to lack the dynamic changes that NPH3 exhibits (Motchoulski and Liscum 1999; Pedmale and Liscum 2007). Recently a number of insights into the posttranslational state of NPH3 have revealed key posttranslational changes related to phototropic response. NPH3 has not only been shown to be a bona fide phosphoprotein, but these phosphorylation events are key to the phototropic response. The transition of NPH3 from the phosphorylated, inactive dark state to the dephosphorylated, active light state is required for phototropic response (Pedmale and Liscum 2007). This study utilized pharmacological tools such as phosphatase inhibitors to show a repression of phototropic responsiveness in *Arabidopsis* seedlings. Another recent study found a number of putative phosphorylation sites for NPH3, though single mutations of these serines to alanine appear to have no effect on phototropic response (Tsuchidamayama et al. 2008). Given the narrow scope of mutants assessed in this study though, it is difficult to ascertain the role of phosphorylation in phototropism from these findings. What remains clear is that the link between light-induced biochemical changes in NPH3 and phototropic responsiveness will be a continuing topic for further investigation. It also seems clear that while NPH3 and RPT2 may be similar in structure, their regulation and functions are at least in part diverged.

While the study of NPH3, RPT2, and the other NRL members has offered some new partners in the phototropin-mediated signaling, many questions still persist. Previously mentioned studies clearly indicate a posttranslational modification of NPH3, though there is clear dispute as to its function (Pedmale and Liscum 2007; Tsuchidamayama et al. 2008). Further, it remains to be determined what residues are key for the interaction of the C-terminal of NPH3 and the LOV domains of phot1 as well as how light affects these forms. While the interaction of phototropin and NPH3 and RPT2 is evident, the lack of biochemical activity for these genes leaves the issue of functionality for NPH3 and RPT2. One possibility is that these proteins act as a scaffold (Faux and Scott 1996; Pawson and Scott 1997; Tsunoda et al. 1998) for a larger complex and, here the functionality of the BTB/POZ domain (Albagli et al. 1995) must be further elucidated particularly given the tantalizing prospect of interaction with cullin3 (CUL3) and participation of ubiquitin-mediated, posttranslational effects. This point warrants further consideration as a wealth of examples in a number of systems (Furukawa et al. 2003; Geyer et al. 2003; Krek 2003; Pintard et al. 2003; Xu et al. 2003; Pintard et al. 2004; van den Heuvel 2004; Wilkins et al. 2004; Perez-Torrado et al. 2006), including *Arabidopsis* (Wang et al. 2004), have shown that BTB-CUL3 interactions can play a crucial role in regulation through recruitment of the ubiquitylation machinery. This leaves the pertinent question of the substrate for ubiquitylation for BTB^{NPH3}-CUL3.

6 The Role of Auxin

To this point we have focused on those factors that are physically or genetically linked to phototropin at the beginning of the signaling pathway. We now turn the emphasis towards the downstream events, particularly those related to response to auxin.

As early as 1880, Charles Darwin (1880) had suspected that a growth-influencing substance might be responsible for the physiological changes leading to phototropism. More than 100 years later, we now know the identity of this substance and a great deal about the influences it has on gene expression and growth. An exhaustive discussion or review of the history of auxin is beyond the scope of this chapter, but several key events related to phototropism are worth noting. Darwin (1880) noticed that covering the shoot tip of the seedling seemed to influence the curvature response, suggesting the location of the growth-influencing substance. Following Darwin's studies, a number of investigators began the search for this elusive substance. Eventually, three key experiments utilizing *Avena* coleoptiles proved critical in identifying the *Wuchsstoff* (growth substance).

The first experiments performed by Boysen-Jensen around 1913 showed that a barrier, in this case a mica plate, blocking movement of the crucial substance across the coleoptile could also block the phototropic curvature (Boysen-Jensen 1910, 1913). This was followed by experiments by Paal (1917–1919), who showed that the growth substance was asymmetrically distributed by decapitating a coleoptile and repositioning the cap such that a "phototropic" response could be obtained in the absence of light. Finally, the independent experiments of Cholodny (1927) and (Went 1926, 1928) determined that (1) auxin is moving laterally across the coleoptile tip, (2) the effects of asymmetric hormone distribution affect the longitudinal growth, and (3) in the case of shoot phototropism, growth on the side away from the light (shaded side) results in phototropic curvature. These proposals formed the basis of the Cholodny-Went hypothesis [152] that has guided much of the tropic research over the past 70 years.

Prior to summarizing the most recent findings on the perception and action of auxin and associated gene expression, it is noteworthy to highlight some of the developments with respect to the mechanism of auxin movement, which, of course, had been the basis for many of the experiments that culminated in the Cholodny-Went hypothesis. First and foremost has been the discovery of genes, by both forward and reverse genetic approaches, encoding facilitators of auxin transport such as the *PIN-FORMED* (*PIN*) family (Galweiler et al. 1998). Mutations in PIN1 and PIN3 have clear phenotypic affects on phototropism (Galweiler et al. 1998; Friml et al. 2002). Proper localization of PIN proteins appears to be essential for tropic responsiveness (Blakeslee et al. 2004). Moreover, the PIN proteins appear to influence tropic responses in combination with another class of auxin transport facilitators, the MULTIDRUG RESISTANCE (MDR) proteins (Noh et al. 2001). Mutants of *mdr* result in mislocalized PIN1 protein, resulting in enhanced tropisms (Noh et al. 2003). The continuing study of the MDR/P-glycoprotein proteins as auxin transporters will continue to shape our understanding of how hormone movement can directly affect tropisms. It is to the downstream effects of this auxin movement and the accompanying changes in gene expression that we turn next.

In the time since identification of the hormone auxin and the development of the Cholodny-Went hypothesis, it is only recently that the molecular evidence has begun to unravel regarding how this hormone goes about eliciting physiological changes. A number of genes have been described that are induced in the presence of auxin (Abel and Theologis 1996). However, identification of a mutant defective in both gravitropism and phototropism, called *non-phototropic hypocotyl 4 (nph4)* (Liscum and Briggs 1995), and subsequent cloning of the gene affected provided the first link between auxin-regulated gene expression and tropic responsiveness (Harper et al. 2000). NPH4 was found to be a member of a family an auxin responsive transcription factors, in this case AUXIN RESPONSE FACTOR 7 (ARF7) (Harper et al. 2000). Interestingly, *nph4/arf7* mutants exhibit conditional phenotypes (Harper et al. 2000), and these phenotypes have helped elucidate some of the cross-talk between phototropins and phytochrome, and also how light affects the flow of auxin (Stowe-Evans et al. 1998, 2001).

The movement of auxin was hypothesized to have numerous effects on plant physiological processes and considerable efforts to identify mutants that were resistant to its effects were undertaken to delineate what genes played a regulatory role for this hormone. One such screen identified two mutants, called *massugu* (Japanese for "straight"), *msg1* and *msg2*, that bore a phenotypic resemblance to *nph4* mutants in its tropic responses (Watahiki and Yamamoto 1997). In fact, *msg1* was found to be an allele of *NPH4*, though *msg2* represented a different complementation group (Stowe-Evans et al. 1998). It was observed that *msg2* demonstrated a number of auxin-related defects in its physiology though they were genetically distinct from those of other auxin-resistant mutants, suggesting a different role for these mutants in the pathway (Watahiki and Yamamoto 1997). When cloned, *MSG2* was found to represent the *IAA19* gene, with the *msg2* mutations representing a gain-of-function allele (Tatematsu et al. 2004). *IAA19* is a member of a family of genes, designated the *AUX/IAA* genes, whose transcription are rapidly induced in response to auxin (Guilfoyle 1998; Reed 2001; Liscum and Reed 2002). While an in-depth discussion of the biochemical nature of the proteins encoded by these genes is beyond the scope of this review, it is worthwhile to note that the Aux/IAA proteins function as dominant repressors of ARF protein function, and that all of the *msg2* alleles that were identified occurred within a highly conserved domain (domain II) among the Aux/IAA family. The QVVGWPPVRSYRK sequence within domain II of Aux/IAA proteins, and in particular the GWPPV sequence, appears to have a key role in the persistence of these proteins, and mutations in this domain result in the physiological aberrations associated with auxin resistance (Gray et al. 2001; Kepinski and Leyser 2002).

Our understanding of how Aux/IAA proteins regulate ARF function in a dynamic nature with respect to various auxin concentrations encountered within a plant has been dramatically enhanced by the recent discovery of a class of auxin receptor proteins. The first of these proteins identified is TRANSPORT INHIBITOR RESPONSE 1 (TIR1), which when mutated yields a weak auxin resistance phenotype

(Ruegger et al. 1998). Functionally TIR1 (and other members of its family) has been shown to be a component in an E3 ubiqutin-ligase complex consisting of an F-box protein (TIR1), cullin1, and *Arabidopsis* SKP1 (ASK1) that triggers protein destruction via the action of the 26S proteasome (Dharmasiri et al. 2005; Kepinski and Leyser 2005). These characterizations have shown the receptor to target domain II of the Aux/IAA family members - the so-called degron. While this type of ubiquitin-mediated destruction is typically triggered by a posttranslational modification, this appears not to be the case here. It is auxin itself that facilitates this interaction and acts as the "molecular glue" between the SCF^{TIR1} and domain II of the Aux/IAA protein (Tan et al. 2007). This provides a molecular paradigm for how auxin can actually induce transcription of a suite of genes, which act to repress auxin signaling while in high enough concentrations participating in a complex whose purpose it is to degrade these proteins.

In practice, data from a study comparing transcript levels between the shaded and lit sides of seedlings treated with blue light or the top and bottom sides of gravistimulated seedlings led to transcriptional changes that were accompanied by modest auxin movement (Esmon et al. 2006). Thus, a certain suite of genes at the onset of tropic stimulation, all of which possessed auxin responsive elements (AuxRE) in their promoters, are expressed to induce physiological changes. This expression, mediated by the interaction of *Aux/IAA* (in the case of gravitropism and phototropism, IAA19) and ARF (in this case *NPH4/ARF7*), is modulated through the destruction of the IAA protein by the 26S proteasome. Among the genes found to have these tropic-stimulus-induced (Wen et al. 1995) differences in expression are *SAUR* and *GH3*, which perhaps not surprisingly have been documented as auxin-stimulated (Hagen et al. 1984; McClure and Guilfoyle 1987) as well as expansins (*EXP1* and *EXP8*), which have long been implicated (McQueen-Mason et al. 1992) as enzymes involved in auxin-regulated cell elongation (Rayle and Purves 1967; Evans and Rayle 1970; Rayle and Cleland 1970; Rayle and Cleland 1992).

It is worth noting that although auxin clearly has a key role in the establishment of tropic responsiveness, there have been recent implications that suggest a link between auxin and brassinolide may also affect tropic responsiveness (Whippo and Hangarter 2005; Nakamoto et al. 2006). This may represent a coming phase of study - the consideration of additional hormones, their related transcriptional networks, and the associated cross-talk on the regulation of phototropism.

7 Summary

We have described some of the numerous aspects of phototropin signaling in plants, elaborating on the players discovered and characterized thus far. We have an understanding of the photoreceptor, key accessory proteins, and even the movement of associated hormones and their effects on gene expression. Phototropin signaling has come from general observations of plant physiology towards a clearer picture of the molecular aspects of phototropin activation. Still, there are many issues that require further investigation. A black box still hangs over the intermediate steps from phototropin activation to the events surrounding auxin movement and the associated transcriptional responses - there are still many aspects that have yet to be addressed, and many questions still need answering.

References

Abel S, Theologis A (1996) Early genes and auxin action. Plant Physiol 111:9–17

Ahmad M, Jarillo JA et al (1998) Cryptochrome blue-light photoreceptors of Arabidopsis implicated in phototropism. Nature 392:720–723

Albagli O, Dhordain P et al (1995) The BTB/POZ domain: a new protein-protein interaction motif common to DNA- and actin-binding proteins. Cell Growth Differ 6:1193–1198

Alexandre MT, Arents JC et al (2007) A base-catalyzed mechanism for dark state recovery in the *Avena sativa* phototropin-1 LOV2 domain. Biochemistry 46:3129–3137

Assmann SM (1993) Signal transduction in guard cells. Annu Rev Cell Biol 9:345–375

Babourina O, Newman I et al (2002) Blue light-induced kinetics of H+ and Ca2+ fluxes in etiolated wild-type and phototropin-mutant Arabidopsis seedlings. Proc Natl Acad Sci U S A 99:2433–2438

Ballario P, Talora C et al (1998) Roles in dimerization and blue light photoresponse of the PAS and LOV domains of *Neurospora crassa* white collar proteins. Mol Microbiol 29:719–729

Blaauw O, Blaauw-Jansen G (1964) The influence of red light on the phototropism of *Avena* coleoptiles. Acta Bot Neerl 13:541–552

Blaauw-Jansen G (1959) The influence of red and far red light on growth and phototropism of the *Avena* seedling. Acta Bot Neerl 8:1–39

Blakeslee JJ, Bandyopadhyay A et al (2004) Plant Physiol 134:28–31

Boccalandro H, De Simone S et al (2007) PHYTOCHROME KINASE SUBSTRATE1 regulates root phototropism and gravitropism. Plant Physiol 146:108–115

Boysen-Jensen P (1910) Ueber die Leitung der phototropischen Reizes in Avenakeimpflanzen. Ber Dtsch Bot Ges 28:118–120

Boysen-Jensen P (1913) Ueber die Leitung der phototropischen Reizes in der Avenakoleoptile. Ber Dtsch Bot Ges 31:559–566

Briggs WR, Beck CF et al (2001) The phototropin family of photoreceptors. Plant Cell 13:993–997

Chen X, Lin WH et al (2008) An inositol polyphosphate 5-phosphatase functions in PHOTOTROPIN1 signaling in Arabidopis by altering cytosolic Ca2+. Plant Cell 20:353–366

Cheng Y, Qin G et al (2007) NPY1, a BTB-NPH3-like protein, plays a critical role in auxin-regulated organogenesis in Arabidopsis. Proc Natl Acad Sci U S A 104:18825–18829

Cho H, Tseng T et al (2007) Physiological roles of the light, oxygen, or voltage domains of phototropin 1 and phototropin 2 in Arabidopsis. Plant Physiol 143:517–529

Cholodny N (1927) Wuchshormone und Tropismem bei den Planzen. Biol Zentralbl 47:604–626

Chon HP, Briggs W (1966) Effect of red light on the phototropic sensitivity of corn coleoptiles. Plant Physiol 41:1715–1724

Christie JM (2007) Phototropin blue-light receptors. Annu Rev Plant Biol 58:21–45

Christie JM, Corchnoy SB et al (2007) Steric interactions stabilize the signaling state of the LOV2 domain of phototropin 1. Biochemistry 46:9310–9319

Christie JM, Reymond P et al (1998) Arabidopsis NPH1: a flavoprotein with the properties of a photoreceptor for phototropism. Science 282:1698–1701

Christie JM, Salomon M et al (1999) LOV (light, oxygen, or voltage) domains of the blue-light photoreceptor phototropin (nph1): binding sites for the chromophore flavin mononucleotide. Proc Natl Acad Sci U S A 96:8779–8783
Christie JM, Swartz TE et al (2002) Phototropin LOV domains exhibit distinct roles in regulating photoreceptor function. Plant J 32:205–219
Corchnoy SB, Swartz TE et al (2003) Intramolecular proton transfers and structural changes during the photocycle of the LOV2 domain of phototropin 1. J Biol Chem 278:724–731
Correll MJ, Coveney KM et al (2003) Phytochromes play a role in phototropism and gravitropism in Arabidopsis roots. Adv Space Res 31:2203–2210
Cosgrove D (1981) Rapid suppression of growth by blue light. Plant Physiol 67:584–590
Crosson S, Moffat K (2001) Structure of a flavin-binding plant photoreceptor domain: insights into light-mediated signal transduction. Proc Natl Acad Sci U S A 98:2995–3000
Crosson S, Moffat K (2002) Photoexcited structure of a plant photoreceptor domain reveals a light-driven molecular switch. Plant Cell 14:1067–1075
Darwin C (1880) The power of movement in plants (London: John Murray)
DeBlasio SL, Luesse DL et al (2005) A plant-specific protein essential for blue-light-induced chloroplast movements. Plant Physiol 139:101–114
Dharmasiri N, Dharmasiri S et al (2005) The F-box protein TIR1 is an auxin receptor. Nature 435:441–445
Dittrich M, Freddolino PL et al (2005) When light falls in LOV: a quantum mechanical/molecular mechanical study of photoexcitation in Phot-LOV1 of *Chlamydomonas reinhardtii*. J Phys Chem B 109:13006–13013
Eitoku T, Nakasone Y et al (2005) Conformational dynamics of phototropin 2 LOV2 domain with the linker upon photoexcitation. J Am Chem Soc 127:13238–13244
Eitoku T, Nakasone Y et al (2007) Photochemical intermediates of Arabidopsis phototropin 2 LOV domains associated with conformational changes. J Mol Biol 371:1290–1303
Emi T, Kinoshita T et al (2005) Isolation of a protein interacting with Vfphot1a in guard cells of *Vicia faba*. Plant Physiol 138:1615–1626
Engelmann W, Simon K, Phen CJ (1992) Leaf movement rhythms in *Arabidopsis thaliana*. Z Naturforsch 47c:925–928
Esmon CA, Tinsley AG et al (2006) A gradient of auxin and auxin-dependent transcription precedes tropic growth responses. Proc Natl Acad Sci U S A 103:236–241
Evans ML, Rayle DL (1970) The timing of growth promotion and conversion to indole-3-acetic acid for auxin precursors. Plant Physiol 45:240–243
Faux MC, Scott JD (1996) Molecular glue: kinase anchoring and scaffold proteins. Cell 85:9–12
Fedorov R, Schlichting I et al (2003) Crystal structures and molecular mechanism of a light-induced signaling switch: the Phot-LOV1 domain from *Chlamydomonas reinhardtii*. Biophys J 84:2474–2482
Folta KM, Lieg EJ et al (2003) Primary inhibition of hypocotyl growth and phototropism depend differently on phototropin-mediated increases in cytoplasmic calcium induced by blue light. Plant Physiol 133:1464–1470
Friml J, Wisniewska J, et al. (2002) Lateral relocalization of auxin efflux regular PIN3 mediates tropism in Arabidopsis. Nature 445:806–809
Furukawa M, He YJ et al. (2003) Targeting of protein ubiquitination by BTB-Cullin 3-Roc1 ubiquitin ligases. Nat Cell Biol 5:1001–1007
Galen C, Huddle J et al. (2004) An experimental test of the adaptive evolution of phototropins: blue-light photoreceptors controlling phototropism in *Arabidopsis thaliana*. Evolution 58:515–523
Gallagher S, Short TW et al. (1988) Light-mediated changes in two proteins found associated with plasma membrane fractions from pea stem sections. Proc Natl Acad Sci U S A 85:8003–8007
Galweiler L, Guan C et al. (1998) Regulation of polar auxin transport by AtPIN1 in Arabidopsis vascular tissue. Science 282:2226–2230

Geldner N, Hyman D et al. (2007) Endosomal signaling of plant steroid receptor kinase BRI1. Genes Dev 21:1598–1602

Geyer R, Wee S et al. (2003) BTB/POZ domain proteins are putative substrate adaptors for cullin 3 ubiquitin ligases. Mol Cell 12:783–790

Gray WM, Kepinski S et al. (2001) Auxin regulates SCF(TIR1)-dependent degradation of AUX/IAA proteins. Nature 414:271–276

Guilfoyle T (1998) Aux/IAAproteins and auxin signal transduction. Trends Plant Sci 3:205–207

Guo H, Kottke T et al. (2005) The phot LOV2 domain and its interaction with LOV1. Biophys J 89:402–412

Haga K, Takano M et al. (2005) The Rice COLEOPTILE PHOTOTROPISM1 gene encoding an ortholog of Arabidopsis NPH3 is required for phototropism of coleoptiles and lateral translocation of auxin. Plant Cell 17:103–115

Hagen G, Kleinschmidt A, Guilfoyle T (1984) Auxin-regulated gene expression in intact soybean hypocotyl and excised hypoxotyl sections. Planta 162:147–153

Hanks SK, Hunter T (1995) Protein kinases 6. The eukaryotic protein kinase superfamily: kinase (catalytic) domain structure and classification. FASEB J 9:576–596

Harada A, Sakai T et al. (2003) Phot1 and phot2 mediate blue light-induced transient increases in cytosolic Ca2+ differently in Arabidopsis leaves. Proc Natl Acad Sci U S A 100:8583–8588

Harper RM, Stowe-Evans EL et al. (2000) The NPH4 locus encodes the auxin response factor ARF7, a conditional regulator of differential growth in aerial Arabidopsis tissue. Plant Cell 12:757–770

Harper SM, Neil LC et al. (2003) Structural basis of a phototropin light switch. Science 301:1541–1544

Harper SM, Neil LC et al. (2004) Conformational changes in a photosensory LOV domain monitored by time-resolved NMR spectroscopy. J Am Chem Soc 126:3390–3391

Iino M (1990) Phototropism: mechanisms and ecological implications. Plant Cell Environ 13:633–650

Iino M, Briggs WR, Shafer E (1984) Phytochrome-mediated phototropism in maize seedling shoots. Planta 160:41–51

Inada S, Ohgishi M et al. (2004) RPT2 is a signal transducer involved in phototropic response and stomatal opening by association with phototropin 1 in *Arabidopsis thaliana*. Plant Cell 16:887–896

Inoue S, Kinoshita T et al. (2008a) Blue light-induced autophosphorylation of phototropin is a primary step for signaling. Proc Natl Acad Sci U.S.A 105:5626–5631

Inoue S, Kinoshita T et al. (2008b) Leaf positioning of Arabidopsis in response to blue light. Mol Plant 1:15–26

Iwata T, Nozaki D et al. (2003) Light-induced structural changes in the LOV2 domain of Adiantum phytochrome3 studied by low-temperature FTIR and UV-visible spectroscopy. Biochemistry 42:8183–8191

Iwata T, Tokutomi S et al. (2002) Photoreaction of the cysteine S-H group in the LOV2 domain of Adiantum phytochrome3. J Am Chem Soc 124:11840–11841

Janoudi A, Poff KL (1990) A common fluence threshold for first positive and second positive phototropism in *Arabidopsis thaliana*. Plant Physiol 94:1605–1608

Janoudi AK, Gordon WR et al. (1997) Multiple phytochromes are involved in red-light-induced enhancement of first-positive phototropism in *Arabidopsis thaliana*. Plant Physiol 113:975–979

Jarillo JA, Ahmad M, Cashmore AR (1998) NPL1 (accession no. AF053941): a second member of the NPH serine/threonine kinase family of Arabidopsis (PGR98–100). Plant Physiol 117:719

Jarillo JA, Gabrys H et al. (2001) Phototropin-related NPL1 controls chloroplast relocation induced by blue light. Nature 410:952–954

Jones MA, Feeney KA et al. (2007) Mutational analysis of phototropin 1 provides insights into the mechanism underlying LOV2 signal transmission. J Biol Chem 282:6405–6414

Kagawa T, Sakai T et al. (2001) Arabidopsis NPL1: a phototropin homolog controlling the chloroplast high-light avoidance response. Science 291:2138–2141
Kang B, Grancher N et al. (2008) Multiple interactions between cryptochrome and phototropin blue-light signalling pathways in *Arabidopsis thaliana.* Planta 227:1091–1099
Kasahara M, Swartz TE et al. (2002) Photochemical properties of the flavin mononucleotide-binding domains of the phototropins from Arabidopsis, rice, and *Chlamydomonas reinhardtii.* Plant Physiol 129:762–773
Kepinski S, Leyser O (2002) Ubiquitination and auxin signaling: a degrading story. Plant Cell 14(Suppl): S81–S95
Kepinski S, Leyser O (2005) The Arabidopsis F-box protein TIR1 is an auxin receptor. Nature 435:446–451
Khurana JP, Poff KL (1989) Mutants of Arabidopsis thaliana with altered phototropism. Planta 178:400–406
Khurana JP, Ren Z et al. (1989) Mutants of *Arabidopsis thaliana* with decreased amplitude in their phototropic esponse. Plant Physiol 91:685–689
King SM (2000) The dynein microtubule motor. Biochim Biophys Acta 1496:60–75
Kinoshita T, Doi M et al. (2001) Phot1 and phot2 mediate blue light regulation of stomatal opening. Nature 414:656–660
Kinoshita T, Emi T et al. (2003) Blue-light- and phosphorylation-dependent binding of a 14-3-3 protein to phototropins in stomatal guard cells of broad bean. Plant Physiol 133:1453–1463
Kiss JZ, Ruppel NJ et al. (2001) Phototropism in Arabidopsis roots is mediated by two sensory systems. Adv Space Res 27:877–885
Knieb E, Salomon M et al. (2004) Tissue-specific and subcellular localization of phototropin determined by immuno-blotting. Planta 218:843–851
Knieb E, Salomon M et al. (2005) Autophosphorylation, electrophoretic mobility and immunoreaction of oat phototropin 1 under UV and blue Light. Photochem Photobiol 81:177–182
Kong SG, Suzuki T et al. (2006) Blue light-induced association of phototropin 2 with the Golgi apparatus. Plant J 45:994–1005
Kong SG, Kinoshita T et al. (2007) The C-terminal kinase fragment of Arabidopsis phototropin 2 triggers constitutive phototropin responses. Plant J 51:862–873
Konjevic R, Steinitz B et al. (1989) Dependence of the phototropic response of *Arabidopsis thaliana* on fluence rate and wavelength. Proc Nat Acad Sci U S A 86:9876–9880
Krek W (2003) BTB proteins as henchmen of Cul3-based ubiquitin ligases. Nat Cell Biol 5:950–951
Lariguet P, Schepens I et al. (2006) PHYTOCHROME KINASE SUBSTRATE 1 is a phototropin 1 binding protein required for phototropism. Proc Natl Acad Sci U S A 103:10134–10139
Lascève G, Leymarie J et al. (1999) Arabidopsis contains at least four independent blue-light-activated signal transduction pathways. Plant Physiol 120:605–614
Lawrence CJ, Morris NR et al. (2001) Dyneins have run their course in plant lineage. Traffic 2:362–363
Liscum E, Briggs WR (1995) Mutations in the NPH1 locus of Arabidopsis disrupt the perception of phototropic stimuli. Plant Cell 7:473–785
Liscum E, Briggs W (1996) Mutations of Arabidopsis in potential transduction and response components of the phototropic signaling pathway. Plant Physiol 112:291–296
Liscum E, Reed JW (2002) Genetics of Aux/IAA and ARF action in plant growth and development. Plant Mol Biol 49:387–400
Luesse DR, DeBlasio SL et al. (2006) Plastid movement impaired 2, a new gene involved in normal blue-light-induced chloroplast movements in Arabidopsis. Plant Physiol 141:1328–1337
Matsuoka D, Tokutomi S (2005) Blue light-regulated molecular switch of Ser/Thr kinase in phototropin. Proc Natl Acad Sci U S A 102:13337–13342
Matsuoka D, Iwata T et al. (2007) Primary processes during the light-signal transduction of phototropin. Photochem Photobiol 83:122–130
McClure B, Guilfoyle TJ (1987) Characterization of a class of small auxin-inducible soybean polyadenylated RNAs. Plant Mol Biol 9:611–623

McQueen-Mason S, Durachko DM et al. (1992) Two endogenous proteins that induce cell wall extension in plants. Plant Cell 4:1425–1433

Motchoulski A, Liscum E (1999) Arabidopsis NPH3: a NPH1 photoreceptor-interacting protein essential for phototropism. Science 286:961–964

Nakamoto D, Ikeura A et al. (2006) Inhibition of brassinosteroid biosynthesis by either a dwarf4 mutation or a brassinosteroid biosynthesis inhibitor rescues defects in tropic responses of hypocotyls in the Arabidopsis mutant nonphototropic hypocotyl 4. Plant Physiol 141:456–464

Nakasako M, Iwata T et al. (2004) Light-induced structural changes of LOV domain-containing polypeptides from Arabidopsis phototropin 1 and 2 studied by small-angle X-ray scattering. Biochemistry 43:14881–14890

Nakasako M, Zikihara K et al. (2008) Structural basis of LOV1-dimerization of Arabidopsis phototropins 1 and 2. J Mol Biol 381:718–733

Nakasone Y, Eitoku T et al. (2006) Kinetic measurement of transient dimerization and dissociation reactions of Arabidopsis phototropin 1 LOV2 domain. Biophys J 91:645–653

Nakasone Y, Eitoku T et al. (2007) Dynamics of conformational changes of Arabidopsis phototropin 1 LOV2 with the linker domain. J Mol Biol 367:432–442

Neiß C, Saalfrank P (2003) Ab Initio quantum chemical investigation of the first steps of the photocycle of phototropin: a model study. Photochem Photobiol 77:101–109

Noh B, Bandyopadhyay A et al. (2003) Enhanced gravi- and phototropism in plant *mdr* mutants mislocalizing the auxin efflux protein PIN1. Nature 423:999–1002

Noh B, Murphy AS et al. (2001) Multidrug resistance-like genes of Arabidopsis required for auxin transport and auxin-mediated development. Plant Cell 13:2441–2454

Nozaki D, Iwata T et al. (2004) Role of Gln1029 in the photoactivation processes of the LOV2 domain in adiantum phytochrome3. Biochemistry 43:8373–8379

Oikawa K, Kasahara M et al. (2003) Chloroplast unusual positioning1 is essential for proper chloroplast positioning. Plant Cell 15:2805–2815

Okada K, Shimura Y (1992) Mutational analysis of root gravitropism and phototropism of *Arabidopsis thaliana* seedlings. Aust J Plant Physiol 19:439–448

Onodera A, Kong SG et al. (2005) Phototropin from *Chlamydomonas reinhardtii* is functional in *Arabidopsis thaliana*. Plant Cell Physiol 46:367–374

Paal A (1917–1919) Ueber phototropische Reizung. Jahrb Wiss Bot 58:406–473

Palmer JM, Short TW et al. (1993a) Correlation of blue light-induced phosphorylation to phototropism in *Zea mays* L. Plant Physiol 102:1219–1225

Palmer JM, Short TW et al. (1993b) Blue light-induced phosphorylation of a plasma membrane-associated protein in *Zea mays* L. Plant Physiol 102:1211–1218

Parker K, Baskin TI et al. (1989) Evidence for a phytochrome-mediated phototropism in etiolated pea seedlings. Plant Physiol 89:493–497

Parks BM, Quail PH et al. (1996) Phytochrome A regulates red-light induction of phototropic enhancement in Arabidopsis. Plant Physiol 110:155–162

Pawson T, Scott JD (1997) Signaling through scaffold, anchoring, and adaptor proteins. Science 278:2075–2080

Pedmale UV, Liscum E (2007) Regulation of phototropic signaling in Arabidopsis via phosphorylation state changes in the phototropin 1-interacting protein NPH3. J Biol Chem 282:19992–20001

Perez-Torrado R, Yamada D et al. (2006) Born to bind: the BTB protein-protein interaction domain. Bioessays 28:1194–1202

Pintard L, Willis JH et al. (2003) The BTB protein MEL-26 is a substrate-specific adaptor of the CUL-3 ubiquitin-ligase. Nature 425:311–316

Pintard L, Willems A et al. (2004) Cullin-based ubiquitin ligases: Cul3-BTB complexes join the family. EMBO J 23:1681–1687

Raikhel N, Hicks G (2007) Signaling from plant endosomes: compartments with something to say! Genes Dev 21:1578–1580

Rayle DL, Cleland R (1970) Enhancement of wall loosening and elongation by acid solutions. Plant Physiol 46:250–253

Rayle DL, Cleland RE (1992) The acid growth theory of auxin-induced cell elongation is alive and well. Plant Physiol 99:1271–1274

Rayle DL, Purves WK (1967) Conversion of indole-3-ethanol to indole-3-acetic acid in cucumber seedling shoots. Plant Physiol 42:1091–1093

Reed JW (2001) Roles and activities of Aux/IAA proteins in Arabidopsis. Trends Plant Sci 6:420–425

Reymond P, Short TW et al. (1992a) Blue light activates a specific protein kinase in higher plants. Plant Physiol 100:655–661

Reymond P, Short TW et al. (1992b) Light-induced phosphorylation of a membrane protein plays an early role in signal transduction for phototropism in Arabidopsis thaliana. Proc Natl Acad Sci U S A 89:4718–4721

Ruegger M, Dewey E et al. (1998) The TIR1 protein of Arabidopsis functions in auxin response and is related to human SKP2 and yeast grr1p. Genes Dev 12:198–207

Ruppel NJ, Hangarter RP et al. (2001) Red-light-induced positive phototropism in Arabidopsis roots. Planta 212:424–430

Sachs J (1887) Lectures on the Physiology of Plants. Oxford, Clarenda Press

Sakai T, Kagawa T et al. (2001) Arabidopsis nph1 and npl1: blue light receptors that mediate both phototropism and chloroplast relocation. Proc Natl Acad Sci U S A 98:6969–6974

Sakai T, Wada T et al. (2000) RPT2. A signal transducer of the phototropic response in Arabidopsis. Plant Cell 12:225–236

Sakamoto K, Briggs WR (2002) Cellular and subcellular localization of phototropin 1. Plant Cell 14:1723–1735

Salomon M (2004) Dimerization of the plant photoreceptor phototropin is probably mediated by the LOV1 domain. FEBS Lett 572:8–10

Salomon M, Christie JM et al. (2000) Photochemical and mutational analysis of the FMN-binding domains of the plant blue light receptor, phototropin. Biochemistry 39:9401–9410

Salomon M, Knieb E et al. (2003) Mapping of low- and high-fluence autophosphorylation sites in phototropin 1. Biochemistry 42:4217–4225

Salomon M, Lempert U et al. (2004) Dimerization of the plant photoreceptor phototropin is probably mediated by the LOV1 domain. FEBS Lett 572:8–10

Sato Y, Iwata T et al. (2005) Reactive cysteine is protonated in the triplet excited state of the LOV2 domain in Adiantum phytochrome3. J Am Chem Soc 127:1088–1089

Schroeder JI, Allen GJ et al. (2001) Guard cell signal transduction. Annu Rev Plant Physiol Plant Mol Biol 52:627–658

Short TW, Reymond P, Briggs WR (1993) A pea plasma membrane protein exhibiting blue light-induced phosphorylation retains photosensitivity following triton solubilization. Plant Physiol 101:647–655.

Sorkin A, Von Zastrow M (2002) Signal transduction and endocytosis: close encounters of many kinds. Nat Rev Mol Cell Biol 3:600–614

Stoelzle S, Kagawa T et al. (2003) Blue light activates calcium-permeable channels in Arabidopsis mesophyll cells via the phototropin signaling pathway. Proc Natl Acad Sci U S A 100:1456–1461

Stowe-Evans EL, Harper RM et al. (1998) NPH4, a conditional modulator of auxin-dependent differential growth responses in Arabidopsis. Plant Physiol 118:1265–1275

Stowe-Evans EL, Luesse DR et al. (2001) The enhancement of phototropin-induced phototropic curvature in Arabidopsis occurs via a photoreversible phytochrome A-dependent modulation of auxin responsiveness. Plant Physiol 126:826–834

Suetsugu N, Kagawa T et al. (2005) An auxilin-like J-domain protein, JAC1, regulates phototropin-mediated chloroplast movement in Arabidopsis. Plant Physiol 139:151–162

Swartz TE, Corchnoy SB et al. (2001) The photocycle of a flavin-binding domain of the blue light photoreceptor phototropin. J Biol Chem 276:36493–36500

Swartz TE, Wenzel PJ et al. (2002) Vibration spectroscopy reveals light-induced chromophore and protein structural changes in the LOV2 domain of the plant blue-light receptor phototropin. Biochemistry 41:7183–7189

Tan X, Calderon-Villalobos LI et al. (2007) Mechanism of auxin perception by the TIR1 ubiquitin ligase. Nature 446:640–645

Tatematsu K, Kumagai S et al. (2004) MASSUGU2 encodes Aux/IAA19, an auxin-regulated protein that functions together with the transcriptional activator NPH4/ARF7 to regulate differential growth responses of hypocotyl and formation of lateral roots in *Arabidopsis thaliana*. Plant Cell 16:379–393

Thimann KV, Curry GM (1960) Phototropism and phototaxis. In: Comparative biochemistry, vol., Florkin M and Musan H, eds (Newyork: Academic Press), pp 243–309

Sullivan S, Thomson CE et al. (2008) *In vivo* phosphorylation site mapping and functional characterization of Arabidopsis phototropin 1. Mol Plant 1:178–194

Tokutomi S, Matsuoka D et al. (2008) Molecular structure and regulation of phototropin kinase by blue light. Biochim Biophys Acta 1784:133–142

Tsuchidamayama T, Nakano M et al. (2008) Mapping of the phosphorylation sites on the phototropic signal transducer, NPH3. Plant Sci 174:626–633

Tsunoda S, Sierralta J et al. (1998) Specificity in signaling pathways: assembly into multimolecular signaling complexes. Curr Opin Genet Dev 8:419–422

Ueno K, Kinoshita T et al. (2005) Biochemical characterization of plasma membrane H + -ATPase activation in guard cell protoplasts of *Arabidopsis thaliana* in response to blue light. Plant Cell Physiol 46:955–963

van den Heuvel S (2004) Protein degradation: CUL-3 and BTB--partners in proteolysis. Curr Biol 14:R59-R61

von Wiesner J (1878) Die Helio Erscheinungen im Pflanzenreiche: eine Physiologische Monographie. Kaiserlich-koniglichen Hof- und Staatsdruckerei, Vienna

Wada M, Kagawa T, Sato Y (2003) Chloroplast movement. Ann Rev Plant Biol l54:455–468

Wan YL, Eisinger W et al. (2007) The subcellular localization and blue light-induced movement of phototropin 1-GFP in etiolated seedlings of *Arabidopsis thaliana*. Mol Plant 1:103–117

Wang KL, Yoshida H et al. (2004) Regulation of ethylene gas biosynthesis by the Arabidopsis ETO1 protein. Nature 428:945–950

Watahiki MK, Yamamoto KT (1997) The massugu1 mutation of Arabidopsis identified with failure of auxin-induced growth curvature of hypocotyl confers auxin insensitivity to hypocotyl and leaf. Plant Physiol 115:419–426

Wen W, Meinkoth JL et al. (1995) Identification of a signal for rapid export of proteins from the nucleus. Cell 82:463–473

Went F (1926) On growth accelerating substances in the coleoptile of *Avena sativa*. Proc K Ned Akad Wet 30:10–19

Went F (1928) Wuchsstoff und Wachstum. Recl Trav Bot Neerl 25:1–116

Whippo CW, Hangarter RP (2003) Second positive phototropism results from coordinated coaction of the phototropins and cryptochromes. Plant Physiol 132:1499–14507

Whippo CW, Hangarter RP (2004) Phytochrome modulation of blue-light-induced phototropism. Plant Cell Environ 27:1223–1228

Whippo CW, Hangarter RP (2005) A brassinosteroid-hypersensitive mutant of BAK1 indicates that a convergence of photomorphogenic and hormonal signaling modulates phototropism. Plant Physiol 139:448–457

Whippo CW, Hangarter RP (2006) Phototropism: bending towards enlightenment. Plant Cell 18:1110–1119

Wilkins A, Ping Q et al. (2004) RhoBTB2 is a substrate of the mammalian Cul3 ubiquitin ligase complex. Genes Dev 18:856–861

Xu L, Wei Y et al. (2003) BTB proteins are substrate-specific adaptors in an SCF-like modular ubiquitin ligase containing CUL-3. Nature 425:316–321

Yamamoto A, Iwata T et al. (2008) Role of Phe1010 in light-induced structural changes of the neo1-LOV2 domain of Adiantum. Biochemistry 47:922–928

Zeiger E (1983) The biology of stomatal guard cells. Annu Rev Plant Physiol 34:441–474

Zimmerman BK, Briggs W (1963a) Phototropic dosage-response curves for oat coleoptiles. Plant Physiol 38:253–262

Zimmerman BK, Briggs WR (1963b) A kinetic model for phototropic responses of oat coleoptiles. Plant Physiol 38:253–261

Signaling in the Circadian Clock

Eleni Boikoglou and Seth J. Davis

Abstract The plant circadian clock in is an internal-timing mechanism that provides adaption to daily environmental changes. Successive changes in light and temperature are able to reset the oscillator. The most prominent signal studied is light; temperature is a signal that is less well characterized. A complex network consisting of interconnected loops was revealed in *Arabidopsis thaliana* on the basis of genetic data of a light-dark entrained oscillator. The basic loop generates the 24-h rhythmicity of the oscillating mechanism, whereas the morning and evening loops are fine-tuned rhythmicity in a temporal-specific manner. Here, we review the current understanding of the *A. thaliana* circadian clockwork with regards to the signaling of the input pathways to the oscillator, and from there, to downstream processes that converge on evolvability of the oscillator.

1 Introduction

Most organisms experience dramatic changes in light and temperature on a daily and seasonal timescale. Seasonal changes in photoperiod and temperature affect many physiological processes during the life cycle. However, the most regular changes occur under daily diurnal conditions, generating a linked light and dark cycle, and a warm and cool cycle (Barak et al. 2000). Specifically, during the light period, temperatures are generally warmer, whereas during the dark period, temperatures are generally cooler. These concerted changes take place essentially every day during the 24 h rotation of Earth around its axis. To anticipate these rhythmic changes, many organisms have evolved an internal-timing mechanism, called a "circadian clock," from the Latin *circa* and *dian*, meaning "about a day." The circadian clock generates rhythms used in anticipation of the ever-changing light and temperature environment.

E. Boikoglou
Department of Plant Developmental Biology, Max Planck Institute for Plant Breeding Research, Carl-von-Linné-Weg 10, 50829, Cologne, Germany

S.J. Davis (✉)
Carl-von-Linné-Weg 10, 50829 Cologne, Germany
e-mail: davis@mpiz-koeln.mpg.de

F. Baluška and S. Mancuso (eds.), *Signaling in Plants*, Signaling and Communication in Plants, 261
DOI: 10.1007/978-3-540-89228-1_13, © Springer-Verlag Berlin Heidelberg 2009

Circadian rhythms are manifest from cyanobacteria to mammals to plants. The first report of diurnal rhythmicity was made in the fourth century BC, when Androsthenes described that tamarind plants open and close their leaves periodically in course of 24 h (Bretzl 1903). However, the first experiments to directly test for circadian rhythms were recorded in 1729 by a French astronomer, de Mairan, who observed that the leaf movements of the plant *Mimosa pudica* were rhythmic in the absence of a changing environment. He thus conclusively demonstrated that an internal mechanism was involved (de Mairan 1729). In the early 1930s, Buenning and Stern studied leaf movements in bean plants and realized that exposure to brief dim red light was able to synchronize the circadian clock. When these experiments were performed in constant darkness, they observed that the period of leaf movement was not 24 h, but extended to a periodicity of 25.5 h. Therefore, the plants were not obliged to follow a 24-h period, and were free running (Buenning 1931). The persistence in the absence of environmental cues is thus considered one of the key defining hallmarks of circadian rhythms.

Genetic studies in the mid-twentieth century were performed in the fruit fly *Drosophila melanogaster* by one of the two intellectual founders of circadian rhythms, Colin Pittendrigh. The result of those studies led to the proposal that the circadian clock is a mechanism that measures the duration of light, which is termed "photoperiod" (Pittendrigh and Minis 1964). To explain the photoperiodic phenomena, two different models were introduced called the "external-coincidence model" and the "internal-coincidence model" (Fig. 1). There are two basic differences between the two models. The internal-coincidence model proposes that two or more oscillators are involved that may be coupled to inductive signals such as dawn or dusk. This suggests that while the photoperiod changes, the phase relationship also changes. In contrast, the external-coincidence model proposes that there is one oscillator and that the external inductive signal has a differential effect that varies with the steady state of the oscillator (Pittendrigh and Minis 1964). To date, these two models are still robustly debated in the scientific community.

Every circadian organism has its own endogenous period that may deviate from 24 h. As one example, the mean periodicity in beans was found to be 25.5 h (Buenning 1931). The only way for organisms to faithfully have rhythms of exactly 24 h is for them to be synchronized every day to the changes of the external environment. Studies in plants have shown that when the internal and the external period match, enhanced fitness is conferred (Michael et al. 2003b; Dodd et al. 2005). To match these two periods, a synchronization process that is also called "entrainment" is obviously an absolute requirement. The main cues for this are changes in light and temperature.

During the day-night cycle, dramatic changes in temperature occur over the 24-h day, with the light period thus coinciding with warmer temperatures and the dark period with cooler temperatures. The onset and offset of light is the main synchronization cue of the circadian clock. For some organisms, a dedicated photoreceptor mediates entrainment (Emery et al. 1998, 2000; Stanewsky et al. 1998). In contrast to a variety of organisms that detect only one wavelength of light, or do not differentiate between different light intensities, plants can detect variations in light

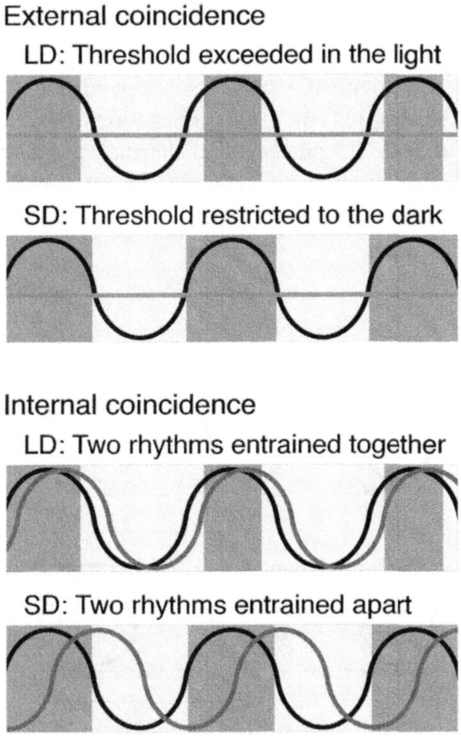

Fig. 1 Two models for the mechanism of photoperiodic timing. In the external-coincidence model, a physiological response is triggered when light perception coincides with a circadian-regulated gene that has exceeded a required threshold of expression. In the illustrations shown here, under long days (LD), light is perceived both at dawn and at dusk of the expression phase, whereas under short days (SD), threshold expression is restricted to the dark. The *bar* indicates the threshold, the *light* and *darker boxes* indicate the periods of light and darkness, respectively, and the *black curve* represents a circadian-regulated factor that is activated only in the presence of the threshold light phase. In the internal-coincidence model, the effect of light is simply to entrain two distinct circadian oscillators. It is illustrated that long days causes similar entrainment of two rhythms. This coincident expression could generate two regulatory molecules that require each other's activity for physiological function. Under short days, the entrainment of these two rhythms is phased apart. This could restrict the simultaneous expression of two factors, thus inhibiting their co-action. (Adapted from Davis 2002)

intensity, and wavelength, and set their physiology according to these changes (Fankhauser and Staiger 2002; Nagy and Schafer 2002). All these changes have a variable effect on oscillator activity (Somers et al. 1998; Yanovsky et al. 2001). The mechanistic effect of temperature on the circadian clock of several organisms has started to emerge (Diernfellner et al. 2005; Boothroyd et al. 2007; Busza et al. 2007); however, in what way temperature sets the plant circadian clock is not as well known as it is for light. However, changes in both signals, light and temperature, are able to synchronize the core oscillator and thereby set the proper timing of gene expression, and temporal regulation of many output rhythms (Millar 2004). These different outputs oscillate with the same 24-h period, but they occur at different phases (Fig. 2). For example, when two oscillations have a phase difference of 180°, they are in antiphase. Light-dark and thermal signals thus converge on the clock to generate intrinsic phases of gene expression in manners not fully understood.

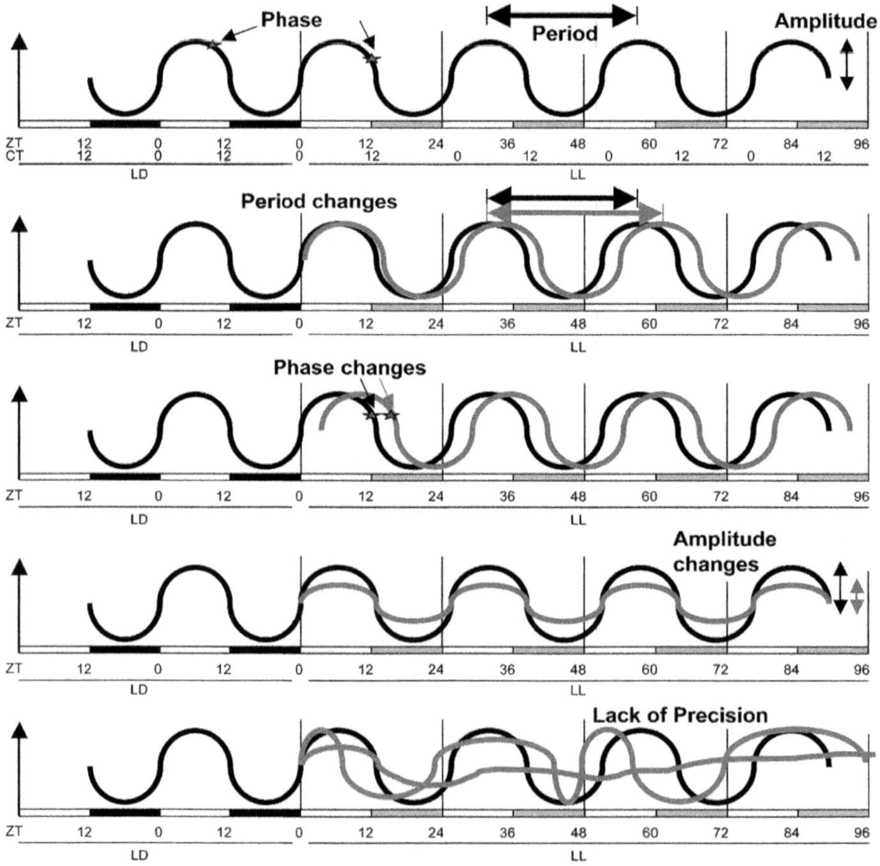

Fig. 2 Definition of circadian parameters. **a** A circadian rhythm with indicated parameters. Numbers indicate times (hours). **b** Example of period change. Note that the distance from peak to peak changes. **c** Example of phase change. Note that the distance from peak to peak does not change. **d** Example of amplitude change. Note the maxima to minima difference. **e** Two examples of lack of precision. A low-amplitude effect can generate precision changes. Wavering periodicity also results in a lack of precision. *CT* circadian time, *LD* light-dark cycles, *LL* constant light, *ZT Zeitgeber* time. (Adapted from Hanano et al. 2006)

The period of the entrained rhythms should remain the same for a range of mean temperatures. Unlike other processes which are affected by temperature increases or decreases, circadian rhythms are temperature-compensated. This temperature compensation, which is another defining hallmark of circadian rhythms, is a buffering mechanism of the circadian clock against long-term temperature variations (Edwards et al. 2005, 2006; Gould et al. 2006; Salathia et al. 2006). If the period of the clock is not temperature-compensated, then in two successive days, for example, a warm one followed by a cool one, the oscillator would be differentially and incorrectly entrained. The clock on the warm day would run faster and, therefore, would have a shorter period, whereas on the cool day it would run slower and, therefore, would have a longer period. Then the oscillator would be inappropriately at a different state between the two successive dawns. Therefore, the adaptive significance of the temperature-compensated period is that the clock has a preserved phase relationship to the environment, and thereby sets all processes in good time.

The generated circadian rhythms have a characteristic waveform, described by peaks and troughs. These rhythms are defined by three key parameters named in the chronobiology community as "period," "phase," and "amplitude" (Barak et al. 2000) (Fig. 2). Amplitude is the half magnitude of the circadian oscillation from peak to trough. Period is the time an oscillation needs to perform a complete cycle with the end point of one cycle as the beginning of the next cycle, and it is approximately 24 h. Phase is a time relative to a reference time point, usually a peak reference point taken after the last onset of light. Moreover, a recently added fourth parameter is precision robustness of the rhythms, considered as an atypical parameter; it is measured by relative amplitude error (RAE) (Doyle et al. 2002; McWatters et al. 2007). RAE is defined as the ratio of the measured, or observed, amplitude error in relation to the most probable, or expected, amplitude error. This means that RAE is a measure of how well the actual data fit to the cosine curve generated by the least-squares method. Of all four parameters, period is the most studied example from plants (Alabadi et al. 2001; Locke et al. 2006; Zeilinger et al. 2006; Ding et al. 2007a).

The basis of the circadian clock in most organisms studied consists of a negative transcriptional/translational feedback loop. This loop has a positive and a negative part. The positive part of the loop induces the expression of the negative, and in turn, the negative part suppresses expression of the positive. To be a component of the feedback loop, certain requirements should be fulfilled. First, morning- and evening-expressed components, and their products, should oscillate in a circadian manner. Additionally, constitutive expression or loss of expression of a core clock component might cause arrhythmic behaviors (Aronson et al. 1994; Wang and Tobin 1998; Barak et al. 2000), although this is not always the case (Green and Tobin 1999; Barak et al. 2000). Notably, in the plant-circadian system, mutations in any of the core-oscillator genes results in a short-period rhythmic phenotype, whereas constitutive expression of some of any of these genes can result in arrhythmic phenotypes (Wang and Tobin 1998; Strayer et al. 2000). Obviously, any changes in the level of a core clock component can, thereby, set the various outputs.

2 Genetic Framework for the *Arabidopsis thaliana* Oscillator

2.1 The Core Oscillator Machinery and Fine-Tuning of Rhythmicity

2.1.1 The Core Oscillator Machinery

A way to understand the circadian clock is by monitoring the several outputs that the oscillator controls. The most studied molecular output of the plant-circadian clock is the transcript accumulation of chlorophyll *a*/b-binding protein 2 (CAB2). CAB2 is an essential protein of the photosynthetic machinery whose transcript is expressed around mid-morning. The generation of transgenic *Arabidopsis thaliana* plants expressing the luciferase gene fused to the *CAB2* promoter contributed greatly to improving our knowledge of the circadian-clock function (Millar et al. 1995). The luciferase gene serves as a "reporter" in that it encodes a protein that produces bioluminescence as a result of the expression of the regulating promoter. When a circadian-regulated promoter is used, the resulting bioluminescence of transgenic plants is rhythmic (Millar et al. 1992). This technology is suited for precise measurements of the marker, and it is suited for high-throughput detection of clock outputs. Forward-genetic screens of mutagenized *CAB2::LUC* transgenic plants led to the identification of many circadian genes, and consequently to the current understanding of the circadian rhythms (Millar et al. 1995).

(Alabadi et al. 2001) proposed back in 2001 the basic framework of the *A. thaliana* clock. They presented a model that comprised three genes, two closely related Mybtranscription factors *Circadian Clock-Associated 1* (*CCA1*) and *Late elongated Hypocotyl* (*LHY*),and a pioneer protein called *Timing of CAB expression 1* (*TOC1*) work in a loop (Alabadi et al. 2001). An alternative hypothesis at that time was presented where the five pseudo-response regulator (PRR) genes drive each other's expression in a bucket-brigade time model (Makino et al. 2000, 2002). These models were not mutually exclusive, and a combined version of these hypotheses is shown in Fig. 3. Key to the development of the above-mentioned model of Alabadi et al. was the study of lines that had constitutive expression of *CCA1* or *LHY*, and here arrhythmicity of rhythms was detected (Schaffer et al. 1998; Wang and Tobin 1998). These overexpression lines also display a reduction in *TOC1* transcript accumulation supporting redundancy of *CCA1* and *LHY* in the repression of *TOC1*. Reduced function at *TOC1* resulted in short periodicity of rhythms and attenuation further also resulted in low *CCA1* and *LHY* transcript accumulation(Alabadi et al. 2001). This supports a positive genetic role for *TOC1* in induction of *CCA1* and*LHY*. From there, null mutants of any of these three genes exhibit a short-period phenotype (Millar et al. 1995; Alabadi et al. 2002; Mizoguchi et al. 2002). Single loss of function *cca1*and *lhy* mutants exhibit similar phenotypes under all conditions tested, and the *cca1 lhy* double mutant is more extreme in period phenotypes, supporting the notion of redundancy between these sequence-related factors (Alabadi et al. 2002; Mizoguchi et al. 2002).

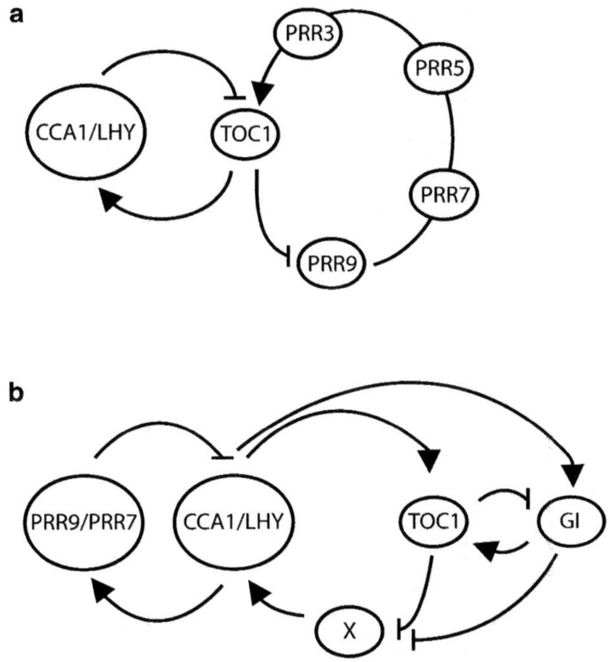

Fig. 3 Historical feedback models of the *Arabidopsis thaliana* oscillator. **a** A hybrid version of the CCA1/LHY-TOC1 feedback model, which includes the pseudo-response regulator feedback loop. **b** One possible multiple-feedback model. Each *arrow* is predicted from the gene expressions in respective mutants and overexpressers. The strengths of genetic interaction are not indicated. An *arrow* indicates that a given gene positively regulates its target. A *stopped arrow* indicates negative regulation of the target. (Adapted from Hanano and Davis 2005)

Under a 24-h diurnal cycle, *CCA1* and *LHY* are maximally expressed at early morning. Their encoding proteins repress *TOC1* transcription by direct binding to the evening element present in the *TOC1* promoter (Alabadi et al. 2001). This binding directs chromatin reorganization at the *TOC1* promoter and this is the causal basis for transcriptional repression (Perales and Mas 2007). Interestingly, in contrast to other circadian-model organisms, where the repressive elements form heterodimers to suppress the positive elements of the clock (Darlington et al. 1998; Gekakis et al. 1998), in *A. thaliana*, CCA1 and LHY can bind individually on the *TOC1* promoter (Yu et al. 2006). By suppressing *TOC1* expression, the morning Myb factors indirectly reduce their own expression over time, since TOC1 eventually promotes expression and low TOC1 levels lead to low *CCA1/LHY* levels (Alabadi et al. 2001). When CCA1/LHY protein levels reduce to release repression, *TOC1* messenger RNA (mRNA) levels progressively increase. Eventually, TOC1 protein levels peak levels in the evening. Since TOC1 is an activator of *CCA1/LHY*, an increase of TOC1 levels subsequently causes an increase in *CCA1/LHY* levels. When TOC1 protein levels reduce, with a mechanism that will be described later, during the night phase, TOC1 reaches its minimum level. In this way, the loop closes and a new cycle begins (Alabadi et al. 2001). This finalizes what has been referred to in the literature as the single-loop model (Locke et al. 2005; Zeilinger et al. 2006) (Fig. 3). However, there are several phenotypes that cannot be explained by the single-loop model (Hanano and Davis 2005). First, all three single mutants display a short-period phenotype, and not an arrhythmic phenotype, as is expected for a core-clock gene working in a close oscillator (Locke et al. 2005). Further, the triple mutant *cca1 lhy toc1*was found to be partially rhythmic for a cycle after both light-dark and temperature entrainment (Ding et al. 2007a; Niwa et al. 2007). Moreover, overexpressing *TOC1* is protein does not affect the mean levels of *CCA1/LHY* expression (Makino et al. 2002). Collectively, the CCA1/LHY-TOC1 loop readily explains most, but not all, aspects of the experimentally defined circadian behavior in *A. thaliana*.

Experimental validation with computational methods by two research groups suggested that the plant circadian clock comprises at least three interconnected feedback loops (Locke et al. 2005, 2006; Zeilinger et al. 2006). The main loop, described above, is responsible for generating rhythms during the 24-h day-night cycle, whereas the two additional loops are specialized for morning- or evening-specific fine-tuning, respectively. The morning loop comprises *CCA1/LHY*, *PRR9*, and *PRR7*. *PRR7* and *PRR9* are genes that encode polypeptides that are related in sequence to *TOC1*, and these are members of the five *PRR* genes (Makino et al. 2000; Matsushika et al. 2000). It was shown that CCA1 and LHY can also bind to promoter elements within *PRR9* and *PRR7* (Farre et al. 2005). Interestingly, the two Myb transcription factors have opposite effects on the different members of the *PRR* family. In contrast to the above-described repression on *TOC1*, CCA1/LHY genetically function to activate *PRR9* and *PRR7* (Farre et al. 2005). Conversely, *PRR9* and *PRR7* genetically function to repress *LHY/CCA1* (Farre et al. 2005; Salome and McClung 2005). In addition to the morning loop, a specialized evening loop also exists. This loop comprises the interaction of TOC1 and a "Y" factor whose phenotype is largely explained by GIGANTEA (GI) (Locke et al. 2005). Several studies showed that GI is involved in light input to the circadian clock (Martin-Tryon et al. 2007). Furthermore, GI shows an acute light induction, as predicted by simulation studies (Locke et al. 2005). This acute peak, and the circadian peak of GI, might explain the biphasic expression of *TOC1*, since it was shown mathematically that these two genes form a loop. Specifically, it was predicted that *GI* activates *TOC1*, and *TOC1* represses *GI* (Locke et al. 2005). Additionally, it was mathematically predicted that an "X" factor is required for *CCA1/LHY* induction at dawn (Locke et al. 2006; Zeilinger et al. 2006). This factor remains hypothetical to date. The above mathematical models have been termed by the Millar group the "three-loop model" and by the Doyle group the "four-loop model" (Locke et al. 2006; Zeilinger et al. 2006) (Fig. 3). The general concepts of these hypothetical models have largely been confirmed (Ding et al. 2007a; Ito et al. 2007; Niwa et al. 2007).

It was mentioned earlier that TOC1 is not the only positive element of the core oscillator. This suggests that another gene(s) might be responsible for the direct activation of CCA1/LHY (Fig. 2). In this respect, other evening-expressed genes, such as *LUX* and *ELF4*, have been implicated as part of the oscillator (Doyle et al. 2002; Hazen et al. 2005; Kikis et al. 2005; Onai and Ishiura 2005; Kolmos and Davis 2007a; McWatters et al. 2007; Kolmos et al. 2008). It was shown that in *elf4-1* mutant the expression of *CCA1/LHY* and *TOC1* was retained for a cycle and then turned to arrhythmic, in constant light conditions, suggesting that ELF4 is required to sustain rhythmicity in the CCA1/LHY-TOC1 loop (Doyle et al. 2002). ELF4 was shown to be required for red-light-mediated induction of *CCA1/LHY* (Kikis et al. 2005). Moreover, *CCA1/LHY* and *ELF4* genetically interact reciprocally, where *CCA1/LHY* suppress *ELF4* (Kikis et al. 2005). ELF4 overexpression studies have further supported the idea that ELF4 activates *CCA1/LHY* and represses *TOC1* (McWatters et al. 2007). A second gene that could fit as the X factor in the three-loop/four-loop model, although this has not been tested in the mathematical models, is *LUX*. Specifically, it was shown that *LUX* genetically suppresses *TOC1*, and activates *CCA1* and *LHY* (Hazen et al. 2005). Further, CCA1 and LHY bind to the evening element of the *LUX* promoter and this correlates with suppression of LUX transcript accumulation in a manner similar to *TOC1* repression (Hazen et al. 2005; Onai and Ishiura 2005). *LUX* can thus be considered as a core-clock component since it satisfies many of the properties of the circadian clock.

2.1.2 Tuning the Oscillator

Recently, a cytosolic signaling molecule, cyclic adenosine diphosphate ribose (cADPR), was incorporated within the circadian clock (Dodd et al. 2007). It was shown that inhibition of cADPR synthesis affects the circadian period. In turn, cADPR is an output of the clock, since its accumulation is circadian-regulated. Thus, small-molecule signaling is intrinsic to the oscillator. Moreover, microarray studies have shown that several core-clock and clock-controlled genes are regulated by the cADPR-derived signal. The only unaffected gene was *TOC1*. Furthermore, cADPR is a modulator of the oscillations of calcium (Dodd et al. 2007). Interestingly, modelingof the tertiary structure of TOC1 suggested that the C-terminal of TOC1 can serve as a metal binding site and particularly for Ca^{2+} (Kolmos et al. 2008). This would implicate cytosolic Ca^{2+} changes as an input to the oscillator and TOC1 as the decoder of the calcium signaling. That hypothesis remains to be tested.

Additional to the metabolic network control, different hormones modulate the circadian-clock parameters. Since hormones are trafficking molecules throughout the plant, an exciting hypothesis would be that synchronization of circadian clocks in different tissues is mediated by an array of hormones. Evidence was provided for the reciprocal regulation between the circadian clock and hormones (Hanano et al. 2006; Salome et al. 2006; Covington and Harmer 2007). In a survey report, it was shown that abscisic acid and brassinosteroids affect periodicity, whereas cytokinin severely affects the phase of the clock (Hanano et al. 2006). The effect of auxin on the clock was similar to that of the molecular phenotype of *elf3* and *elf4*, suggestingthat auxin could be involved in the gating mechanism of the light signaling

(Hanano et al. 2006). Interestingly, several steps of the auxin signaling are circadian-controlled as was found by genome-wide transcriptional profiling (Covington and Harmer 2007). The best studied hormonal input to the clock was cytokinin. This molecule was shown to activate a response regulator 4 (ARR4), which represses the dark conversion of inactive to active PHYB (Hanano et al. 2006). Furthermore, mutations in the cytokinin responders in the ARR family contribute to the clock periodicity (Salome et al. 2006). Cytokinin can thus be considered part of the mechanism that connects hormone- and light-signaling pathways to the circadian clock.

Posttranslational control of the circadian clocks appears to be a highly conserved feature among different species. *A. thaliana* is not an exception to this. Under diurnal conditions, the three core-clock proteins undergo cycles of posttranslational modifications. CCA1 was the first protein for which posttranslational control was identified. In yeast-two-hybrid experiments, Casein Kinase II subunit β (CKB3) was identified. This factor is the third regulatory subunit of Casein Kinase II (CKII) in *A. thaliana*, and thus works as a kinase that phosphorylates an interacting CCA1 (Sugano et al. 1998). In vitro studies showed that not only CKB3, but also all subunits of CKII interact with CCA1. This is intriguing as CCA1 has 21 putative CKII phosphorylation sites (Daniel et al. 2004). In addition, it was shown that overexpression of *CBK3* alters the period of *CCA1* and *LHY* transcriptaccumulation, as it also does for a number of output genes; no amplitude changes were observed (Sugano et al. 1999). Further, the CKB4 subunit is itself subjected to circadian regulation (Perales et al. 2006). Clearly, the consequence of CCA1 phosphorylation by CKII is to maintain the normal functioning of the clock (Daniel et al. 2004). Moreover, it was found that in plant extracts CKII-mediated phosphorylation of CCA1 is required for the formation of DNA-protein complexes, but also for the formation of CCA1 homodimer (Sugano et al. 1998; Daniel et al. 2004). Although phosphorylation of transcription factors is necessary to target their degradation, and various such examples exist in other model species, this hypothesis remain to be tested for CCA1.

Owing to the high degree of amino acid similarity between CCA1 and LHY, it was tested and shown that CKII also interacts with LHY in vitro (Sugano et al. 1999). However, it was shown in an in vitro protein-degradation assays that LHY protein turnover is controlled by proteasome in a light-independent manner, and specifically that De-Etiolated 1 (DET1) acts to inhibit proteolytic turnover of LHY protein (Song and Carre 2005). Furthermore, LHY translatability is also diurnally controlled (Kim et al. 2003a). Together it is clear that numerous posttranslation control elements exist for the morning Myb-like clock proteins.

The third core-clock protein, TOC1, was also found to be under posttranslational control. It was shown that it can interact with the clock-regulating protein Zeitlupe (ZTL). ZTL targets TOC1 for degradation through the proteasome pathway in darkness (Mas et al. 2003a). Later it was shown that ZTL assembles an Skp/Cullin/F-box (SCF) complex in vivo (Han et al. 2004; Kevei et al. 2006). SCF is a ubiquitin ligase complex that recruits specific substrates, and usually itself, for ubiquitination and subsequent proteolysis by the 26S proteasome (Vierstra 2003). In this way, ZTL has been proposed to dictate the protein levels of TOC1 dependent on the light-dark signals present around dusk (Kim et al. 2007). It has further been proposed that the evening protein GI works to stabilize ZTL in a complex activated by blue light, which leads to destruction of TOC1 (Kim et al. 2007). Recently, it was shown that in addition to TOC1, ZTL mediates the degradation of another member of the PRR protein, PRR5 (Kiba et al. 2007). Taken together, it appears that post-translational control of core-clock proteins sculpts the driven rhythms in response to altering environmental conditions.

2.2 Slave Oscillators

Slave-oscillator components posses two molecular properties (Barak et al. 2000). First, when they are constitutively expressed, they repress accumulation of their own endogenous transcript and thereby regulate their own expression through a negative-feedback loop. This relates to similar regulation of *trans*-target transcripts. Second, the slave oscillator is downstream of the core oscillator and its rhythm is highly dependent on the core oscillator. One example of a slave oscillator is the MYB factor early phytochrome-responsive 1 (EPR1). EPR1 regulates its own expression and other downstream genes, such as *CAB2*, but not the parameters of core-oscillator genes (Kuno et al. 2003). Therefore, slave oscillators work as intermediates between the core oscillator and the output pathway, perhaps contributing to proper phase maintenance.

A classic example of a slave-oscillator gene is *Cold and Circadian Regulated 2* (*CCR2*), also named *AtGRP7* for *A. thaliana Glycine-Rich Protein 7*. In *A. thaliana*, CCR2 is an RNA-binding protein whose transcriptional expression is induced by cold temperatures. Under cold conditions, CCR2 appears to involved in the regulation of abscisic acid signaling (Cao et al. 2006). It is abundant in the nucleus and the cytoplasm of guard cells, where it affects stomatal opening and closing in response to cold stress (Kim et al. 2008). Furthermore, it is considered as part of a slave oscillator since it regulates its own expression by binding to its own RNA (Staiger et al. 2003a). This binding might be responsible for producing an early stop codon in its endogenous transcript, preventing in this way its translation into a functional CCR2 protein (Staiger et al. 2003a). Additionally, it was found that CCR2 suppresses the rhythmic expression of the related RNA binding protein AtGRP8, also known as CCR1 (Heintzen et al. 1997). Importantly, when CCR2 is overexpressed, core-clock genes are normally expressed. So, CCR2 represents a slave that obtains timing information from the core oscillator, retains the rhythmicity through its negative feedback, and then transduces this rhythmicity, to thereby regulate a subset of clock-controlled transcripts. *CCR2* transcription is regulated by the clock in a circadian manner, with peak time 8–12 h after onset of illumination, whereas CCR2 protein oscillates with a delay of 4 h after the respective mRNA peak expression (Heintzen et al. 1997). *CCR2* transcription serves as an ideal tool to probe the oscillator function owing to the robust rhythmicity under a wide range of conditions, and its temperature regulation.

3 Entrainment

The circadian oscillator generates rhythmic outputs by using timing information given by the external environment, such as light and temperature onset. The process through which the oscillator synchronizes to the environment is called "entrainment" (Barak et al. 2000; Salome and McClung 2004). A characteristic of an entrained oscillator is that it must have a stable phase regarding to the entrainment cycle (Merrow et al. 2006). This can be tested in cycles with length different from 24 h for which the amount of light versus dark or warm versus cold is the same. This type of entrainment is called "symmetrical T cycles." When the phase of an output occurs at the same time regardless the cycle length indicates a driven rhythm (Merrow 2006). This strongly suggests that the light is able to mask entrained rhythms, meaning that light can affect rhythms without entraining the underlying oscillator. Therefore, masking could be due to strong stimulus intensity or a weak oscillator (Merrow et al. 2006).

One can appreciate the importance of entrainment when flying over many time zones. It takes several days in the new time zone for the human circadian clock to gradually reset itself to the local environmental signals (Waterhouse et al. 2007). Therefore, one can study entrainment by applying such experiments to plants grown in controlled conditions. For this, a phase shift between the old and the new entrainment protocol is required. Plants with a functionally entrained oscillator would gradually reset themselves to the new entrainment protocol, and thereby attain a phase angle, then the phase of the rhythms would show a stable shift equal to the difference of the onset of the two signals. This would indicate that these rhythms are entrained to the new environment. Moreover, by synchronizing plants to two synchronous, but phase-shifted, entrainment protocols, one could determine a preference for one of the two protocols.

Even pulses of entrainment cues could also trigger a phase shift to the oscillator. Depending on the state of the components of the oscillator, differential response at different time points throughout a circadian cycle can be observed. For example, in the plant circadian clock, a light pulse during subjective day would not trigger any phase shift to the clock. However, pulses of light during the dark period could either cause phase delays or advances depending on whether the pulse was given at early or late subjective night, respectively (Covington et al. 2001). This suggests that light input is gated by the oscillator, suggesting that the oscillator regulates its own resetting. The advantage of gating is that, in this way, resetting of the oscillator by inappropriate signals, such as clouds during daytime or flashes of moonlight during night, can be prevented. Plotting of phase delays/advances against circadian time give rise to a phase response curve (Michael et al. 2003a). This type of experiment reveals the entrainment-sensitive period of the circadian-gated cycle.

3.1 Light Regulation of the Circadian Clock

At the molecular level, light is perceived by photoreceptors (Somers et al. 1998; Franklin and Whitelam 2004). To be a photoreceptor, several criteria must be fulfilled. Mutants lacking such a protein should have aberrant phenotypes compared with wild-type plants in light-regulated processes, such as hypocotyl elongation, flowering time, and circadian rhythms. Additionally, a photoreceptor protein should bind to a chromophore, and that chromophore must undergo certain photochemistry under specific fluence and wavelengths of light. Plants can detect a broad range of wavelengths of the visible-light spectrum. For this, an array of photoreceptors are present within plants (Fankhauser and Staiger 2002; Chen et al. 2004).

To exploit the available light, a suite of photoreceptors in plants are used to monitor changes in light duration, quantity, and quality (Murtas and Millar 2000; Franklin and Whitelam 2004; Salome and McClung 2004; Kolmos and Davis 2007b). Depending on the quality of the light detected, photoreceptors are subdivided into structural and functional classes. Red/far-red light is perceived by phytochromes. Five phytochromes have been identified so far, and named PHYA–E (Devlin and Kay 2001). From these, PHYA is rapidly degraded upon light illumination, but all the other phytochromes are light-stable (Hennig et al. 1999). Phytochromes exist in two interconvertible forms, Pr and Pfr. Upon absorption of red light, Pr is converted to Pfr, the biologically active form, and moves to the nucleus, while upon absorption of far-red light and in darkness, the active form converts to the inactive form (Elich and Chory 1994; Quail et al. 1995). These form conversions are the result of conformational changes of the attached linear tetrapyrrole chromophore, the molecule where light is absorbed. The changes in conformation upon light absorption trigger the phytochrome-mediated light signalingin plants (Kim et al. 2002). The exact signal transduction path of phytochromes to the circadian clock has not been clarified yet. Still, it is clear that phytochromes are components of light input to the oscillator (Somers et al. 1998).

Blue light is mainly absorbed by cryptochromes (CRY1–2), and phototropins (PHOT1–2) (Kang et al. 2008). No data to date link phototropins to clock input. Cryptochromes show structural similarities to microbial photolyases, enzymes involved in blue-light-dependent DNA repair. Cryptochromes bind a flavine adenine dinucleotide chromophore, and this allows absorption of blue light, which initiates signaling (Lin and Todo 2005). Interestingly, cryptochromes not only mediate blue light, but they are also required for phytochrome-mediated signaling to the clock (Devlin and Kay 2000); thus, photoreceptor cross-talk exists. In plants, cryptochromes do not appear to be part of the core oscillator, a situation in contrast to animals (Siepka et al. 2007; Tamanini et al. 2007). Cryptochromes are components of light input to the clock, as mutations in cryptochrome genes lead to light-dependent-period defects (Yanovsky et al. 2001). Complex genetic interactions between the various photoreceptors exist, with either a synergistic or an antagonistic effect. Additionally, physical interactions between CRY1 and PHYA, and CRY2 and PHYB were shown (Ahmad et al. 1998; Mas et al. 2000). These interactions confirmed the role of PHYA in the blue-light entrainment pathway (Somers et al. 1998). However, the double *cry1 cry2* and the quadruple *phyA phyB cry1 cry2*mutant retain entrainable rhythmicity, suggesting that other photoreceptors mediate light input to the clock (Yanovsky et al. 2000). Inversely, the expression pattern of all phytochromes and cryptochromes was found to be circadian-regulated (Bognar et al. 1999; Hall et al. 2001; Toth et al. 2001; Hall et al. 2002). Collectively, photoreceptors affect the period length of the clock in a fluence-dependent manner, but they themselves are circadian-regulated. Photoreceptors are thus both inputs and outputs of the clock (Somers et al. 1998). Apparently, recruitment of the appropriate photoreceptor in different light environments enables advantageous resetting of the clock in uncertain spectral-quality environments.

Intermediate light signaling genes are involved in the input pathway. ELF3 plays such a key role in the phototransduction pathway to the clock. Under continuous light, *elf3* mutants become rapidly arrhythmic, but in constant darkness remain rhythmic (Hicks et al. 1996; Covington et al. 2001). Additionally, overexpression of ELF3 results in a long-period phenotype under constant light, but this line has no effect on the period in constant darkness (Covington et al. 2001). These phenotypes suggest that ELF3 functions in light input, by gating light during subjective night (McWatters et al. 2000). Moreover, yeast-two-hybrid and in vitro pull-down assays showed that ELF3 interacts with PHYB, therefore linking ELF3 to the phytochrome signal transduction (Liu et al. 2001). All these observations suggest a role of ELF3 to mediate light input to the circadian clock.

Another component of the light signaling pathway to the clock is Far-red elongated hypocotyl 3 (*fhy3*) (Allen et al. 2006). *fhy3* mutants displayed arrhythmic *CAB2*expression similar to that of *elf3*, but only in response to continuous red light. Furthermore, it disrupted the expression of all the circadian genes tested, including *CCR2*, *CCA1*, and *LHY*. During darkness all genes were robustly rhythmic. This phenotype was also described for the *elf3* mutant. Collectively, *fhy3* has a gating role in phytochrome-mediated signal to the circadian clock during the subjective day. Recently *fhy3* has been shown to function as a transcription factor (Lin et al. 2007). As such, this finding moves us towards a functional description of a gating mechanism and the role of transcriptional induction of target genes by the clock in a light-dependent fashion.

TIME FOR COFFEE (*TIC*) is another gene that functions as a potential photoentrainmentintermediate working close to the oscillator (Hall et al. 2003; Ding et al. 2007b). Interestingly, TIC and ELF3 function at different phases of the circadian cycle, with TIC being required at dawn and ELF3 at dusk (Hall et al. 2003). Therefore, it was foreseen that the *elf3 tic* double mutants would have a completely arrhythmic phenotype (Hall et al. 2003). Surprisingly, and in contrast to ELF3, neither *TIC*'s mRNA nor its protein is circadian-expressed. Ding et al. (2007b) have shown that TIC affects the clock by promoting transcriptional induction of *LHY*, rather than *CCA1*. This is the further evidence of an uncoupling function of *CCA1* and *LHY*, otherwise redundant transcription factors (Ding et al. 2007b). How *TIC* mediates light input has not yet been resolved.

Sensitivity to red light reduced (*SRR1*) is a gene of the PHYB signaling pathway that also acts independently of PHYB (Staiger et al. 2003b). Its effect on the circadian clock was confirmed by the short-period phenotype of the *srr1* mutant, and the dampened amplitude phenotype of several circadian-clock and clock-regulated genes. Although ELF3 and SRR1 are involved in the PHYB signaling, they have genetically separate roles. SRR1 is required for circadian oscillations, whereas ELF3 is required specifically for light input to the clock. The exact function of SRR1 in the PHYB-mediated light signaling is not yet known.

Other clock components are also involved in light transmission to the clock. Specifically, mutations in PRR7 and PRR9 lengthen the period of *CCR2* under constant light, whereas the *CCR2* period is unaffected in constant darkness (Farre et al. 2005). The evening gene *GI* is light-induced (Locke et al. 2005). Genes such as *CCA1*, *LHY*, and *PRR9* have been found to mediate red and also blue light to the clock. Furthermore, the GI protein stabilizes ZTL protein accumulation (Kim et al. 2007). ZTL is a putative blue-light photoreceptor that regulates its own protein expression in a circadian manner (Kim et al. 2003b). This ZTL-dependent degradation of TOC1 in darkness is probably indirectly mediated by the stabilization of ZTL by GI under blue light (Kim et al. 2007). Collectively, a number of genes and physiological roles in light input to the clock have been discovered. The next great task will be to uncover how the encoded proteins direct entrainment in response to altering light conditions.

3.2 Temperature Regulation of the Circadian Clock

Over 24 h, plants experience not only a light-dark cycle, but also a warm-cool cycle, where warm coincides in time with light and cool with darkness. Although light is the major factor responsible for the resetting of the circadian oscillator, temperature oscillations can also reset the oscillator (Barak et al. 2000; McClung 2006) Temperature can affect the oscillator differentially. Entrainment occurs after changes in ambient temperatures and not by extreme stress temperatures, such as chilling or heat shock. In experiments performed with ambient temperature cycles, the existence of two oscillators with differential response to the environmental entrainment cues was reported, since two genes showed preference for either light-dark or temperature entrainment (Michael et al. 2003a). However, knowledge of temperature perception and oscillator resetting is fairly poor.

Two members of the *PRR* family in *Arabidopsis*, *PRR7* and *PRR9*, have been implicated as targets for temperature entrainment. It was reported that the *prr7 prr9* double mutant when subjected to temperature cycles under constant light failed to be entrained (Salome and McClung 2005); however, when similar experiments were performed where the double mutant plants were entrained to temperature cycles, but in constant darkness, expression of *CCA1* and *LHY* was rhythmic. This suggests that additional gene(s) other than *PRR7* and *PRR9* are involved in maintaining residual rhythmicity in temperature cycles in the absence of light. To conclude, temperature cycles can entrain the oscillator, and *PRR7* and *PRR9* are two components that are required for the response to a temperature-entrained oscillator.

Temperature cycles can entrain the circadian oscillator, but the circadian oscillator is temperature-compensated, being buffered from mean temperature differences. Temperature-compensation studies in other model organisms - e.g., *Drosophila* and *Neurospora* - have shownthat clock components identified under light-dark entrainment also play a role in temperature compensation, and surprisingly, also in temperature entrainment (Diernfellner et al. 2005; Glaser and Stanewsky 2005). One could thus expect that some of the genetic components involved in temperature compensation of *A. thaliana* would also be involved in temperature entrainment of the circadian clock. An example in *A. thaliana* is the gene *GI*, which according to circadian modeling is part of a three-loop light-entrained oscillator (Locke et al. 2006; Zeilinger et al. 2006). Additionally, *GI* plays a role in temperature compensation in the circadian clock, especially, in higher and lower mean temperatures (Gould et al. 2006). Whether *GI* is part of thermal entrainment has not yet been established.

Another gene involved in the temperature compensation of the circadian clock is *FLC* (Edwards et al. 2006). Specifically, FLC is a transcription factor that maintains the same period over moderate temperature ranges, although the period is altered at 27°C (Edwards et al. 2005, 2006). Additionally, a dose-dependent effect of *FLC* on the circadian period after light-dark entrainment was shown (Salathia et al. 2006). The classic role of *FLC* is in repression of flowering time (Michaels and Amasino 1999). High levels of *FLC* suppress expression of downstream flowering-integrator genes (Searle et al. 2006). After extended low temperatures, *FLC* is downregulated, so the downstream positive regulators are activated and flowering is promoted (Bastow et al. 2004; Mylne et al. 2006). This pathway is also part of the periodicity effect of *FLC* on the clock (Salathia et al. 2006). Interestingly, as prolonged exposure to chilling conditions that occur during winter suppresses FLC, which leads to floral induction, this lowering of FLC also results in the speeding up of clock periodicity (Salathia et al. 2006). The relationship between *FLC* and the clock appears to involve the target gene *LUX* (Edwards et al. 2006). How FLC regulates LUX is unknown.

In addition to low temperatures, changes in mean ambient temperatures affect flowering. It was shown that at either 16 or 23°C photoreceptor genes such as PHYB, PHYD, PHYE, CRY1, and CRY2 display a differential flowering (Blazquez et al. 2003; Halliday et al. 2003). The differential control of flowering at various temperatures by these genes, and especially the exaggerated phenotype of CRY2, prompt us to think that photoreceptors could also be involved in temperature perception, as well as in light perception. It is intriguing that this light and temperature connection could extend to the entrainment of the clock. It is currently untested if photoreceptor mutants in *A. thaliana* have temperature-dependent clock phenotypes.

Temperature signaling and resetting of the oscillator could be rather complicated to resolve because many biochemical reactions are temperature sensitive. Normally, increases of 10°C will double the speed of metabolic reactions (McClung 2006). At very high temperatures, biochemical reactions are affected by heat shock proteins that are activated at these stressful temperatures (Rensing and Monnerjahn 1996). At low temperatures, plant growth and development is influenced, through direct inhibition of metabolism, and/or indirect induction of other stress responses (Strand et al. 1999; Stitt and Hurry 2002). The biophysical nature of a thermal reset of the oscillator can thus not be predicted.

4 Output

The findings of the first microarray experiments regarding the circadian clock were published in 2000, where it was shown that approximately 10% of the *A. thaliana* genome is under circadian control (Harmer et al. 2000; Davis and Millar 2001).

Systems establishment of the role of transcription factors within the oscillator has revealed that these factors are overrepresented as cycling components (Edwards et al. 2006; Hanano and Davis 2007; Hanano et al. 2008). Furthermore, Michael et al. (2008) have shown that the number of diurnally regulated genes is up to 90%, in at least one of the conditions tested; thus, the rhythmic habitat of plants can control essentially every biological process. As an overt developmental example, mutants within a core-clock gene display hypocotyl and flowering-time defects, directly implicating the circadian clock in the control of these two processes (Hicks et al. 1996; Schaffer et al. 1998; Wang and Tobin 1998; Mizoguchi et al. 2002). Hypocotyl elongation was studied extensively under different photoperiod, light-fluence, and quality regimes (Dowson-Day and Millar 1999; El-Din El-Assal et al. 2001; Maloof et al. 2001; Borevitz et al. 2002; Doyle et al. 2002; Balasubramanian et al. 2006; Nozue et al. 2007; Filiault et al. 2008). Regarding hypocotyl growth, several clock and clock-regulated genes have opposite effects when plants are grown under red light (Ito et al. 2007). Regarding the effect of *CCA1/LHY–TOC1* on flowering time and hypocotyl growth, it was shown that CCA1/LHY regulate hypocotyl growth, whereas flowering time is regulated by TOC1 in a CCA1/LHY-dependent way (Mas et al. 2003b; Ding et al. 2007a; Niwa et al. 2007).

The shade-avoidance syndrome is a clock-regulated process. The clock gates the rapid shade avoidance response, and this is mediated by TOC1 (Salter et al. 2003). It was previously shown that TOC1 interacts with the PIF3 like 1 (PIL1) gene (Makino et al. 2002), a gene that is rapidly induced under low red/far-red light (Salter et al. 2003). PIL1 upregulation and characterization of other early response transcripts of the light and hormonal pathways have also been described (Devlin et al. 2003). Interestingly, the *pil1* mutant is rhythmically altered in promoting and repressing the shade-avoidance response in a circadian-regulated fashion (Salter et al. 2003). Collectively, a number of genes that coordinate these physiological avoidance outputs serve as markers that can be used to answer various genetic and physiological questions.

5 Natural Variation

In nature, *A. thaliana* grows in a wide range of environments, distributed over the Northern Hemisphere at latitudes from the tropics to the arctic circle, and at altitudes from sea level up to 3,500 m (Loudet et al. 2002; Koornneef et al. 2004). To date, more than 600 accessions have been collected from various places around the world. Each accession is presumed to be adapted to the local environment rendering the best plant performance. This adaptation is reflected by selection upon genetic variation. Relevant to the clock, this genetic variation can be strongly selected by rhythmic environmental factors, such as light and temperature. This selection thus generates a useful genetic resource via natural resources. Often, genetic variation of multiple genes controls continuous traits. Continuous traits are of a quantitative nature, and this is why they also called "quantitative traits." Quantitative traits are under the control of several loci called "quantitative trait loci" (QTL). So, by scoring differences in quantitative traits, one can map allelic variation (Alonso-Blanco et al. 2005; Alonso-Blanco et al. 2006). Here, QTLs of light and temperature regulation of the clock will be explored.

In addition to natural accessions, for QTL analysis in *A. thaliana*, "immortal" mapping populations, such as recombinant inbred lines (RILs), can be used. RILs are generated by the cross of two accessions that originate from ecologically distinct environments (Koornneef et al. 2004; Alonso-Blanco et al. 2005). Several rounds of self-fertilization generations will produce lines that are essentially a homozygote mosaic of the two parents for 98% the genome. RILs can be used in replicate experiments under various environments to map QTL for several traits (Koornneef et al. 2004). In this case, the direct comparison of map positions (genetic location) of genes encoding for these different traits is allowed. To fine-map the detected QTL, near-isogenic lines are used. Near-isogenic lines are generated by repeated backcrosses, resulting in introgressions of part of the genome of one parent into the gene pool of the second parent. At the moment, various RILs have been generated and are publicly available for natural-variation studies (Alonso-Blanco et al. 2005, 2006).

Assessment of natural variation and subsequent QTL analysis facilitates the unraveling of the genetics of quantitative traits, such as hypocotyl length, clock parameters, and flowering time (Alonso-Blanco et al. 1998; Swarup et al. 1999; Maloof et al. 2001; Fankhauser and Staiger 2002; Edwards et al. 2005; Balasubramanian et al. 2006). The analysis of variation in flowering time in naturally late flowering accessions and RILs was used particularly to identifyrepressors of the floral transition. Most of the phenotypic variation in flowering time resulted from natural variation of various loci, such as *FLC*, *FRI*, *PHYC*, *CRY2*, *HUA*, and *FLM* (Johanson et al. 2000; El-Din El-Assal et al. 2001; Doyle et al. 2005; Shindo et al. 2005; Werner et al. 2005; Balasubramanian et al. 2006). Existing RILs can thus be exploited to uncover individual genes that coordinate particular processes being investigated.

Several QTL studies have been performed on a number of accessions and RILs for control of clock parameters. Swarup et al. (1999) were the first to map in *A. thaliana* allelic variation responsible for circadian-period variation, by measuring rhythmic leaf movement under constant light and temperature in a RIL set after light-dark entrainment. This was extended using various rhythmic outputs, including assaying leaf movement or luciferase expression as a reporter for clock-controlled genes (Swarup et al. 1999; Michael et al. 2003b; Edwards et al. 2005; Darrah et al. 2006). The most informative RIL population has been the one generated by the cross of Cvi to L*er* (CvL). CvL RILs transformed with *CAB2::LUC* have identified a first chromosome QTL for the circadian phase of plants entrained in short-day photoperiods (Darrah et al. 2006). Although *GI* is localized in that area, from this analysis it appears that *GI* was not the candidate for the QTL of this phase, suggesting another, yet unidentified, locus is responsible for this phenotype. However, period QTL of leaf movement of 48 CvL lines for temperature compensation at 12, 22, and 27°C also implicate *GI* as a temperature-responsive gene, and thus as a candidate gene for temperature compensation at 27°C (Edwards et al. 2005). From these two studies, it can be concluded that at least two different QTL exist in the first chromosome, from 15 to 40 cM (Edwards et al. 2005; Darrah et al. 2006). Additional natural-variation work has implicated *FLC* polymorphisms as a candidate QTL (Swarup et al. 1999; Edwards et al. 2005), and the allelic status at *FLC* does modulate periodicity. Still, most periodicity QTL in *A. thaliana* have not been molecularly identified, and such efforts will be a major focus for future natural-variation studies.

6 Final Thoughts

Future directions for the plant circadian community will lead towards an integrative model, where entrainment inputs of light, temperature, and humidity will be taken into consideration. Questions to be answered include whether the environmental information provided by changes in light and temperature converges to one oscillator or to many different oscillators that are coupled with phases. And if so, under which circumstances do such behaviors occur. Determination of whether there are times of the day when one input is stronger than the other will generate a physiological description of sensitivity phases, and mutant characterization to isolate the genetic influences of these sensitive moments can thus be conducted. Further characterization of temperature entrainment is required. Up to now, a thermoreceptor has not been identified. It is not known how temperature changes are signaled to the oscillator, or if there is any differential response of the oscillator to changes in cooler compared with changes in warmer temperatures. From these mechanistic studies one can move towards the analysis of the evolution of clock genes as the *A. thaliana* species migrates along lines of latitude. This genetic architecture will hopefully reveal the fitness benefit of the clock as it evolves.

References

Ahmad M, Jarillo JA, Smirnova O, Cashmore AR (1998) The CRY1 blue light photoreceptor of Arabidopsis interacts with phytochrome A in vitro. Mol Cell 1:939–948

Alabadi D, Oyama T, Yanovsky MJ, Harmon FG, Mas P et al (2001) Reciprocal regulation between TOC1 and LHY/CCA1 within the Arabidopsis circadian clock. Science 293:880–883

Alabadi D, Yanovsky MJ, Mas P, Harmer SL, Kay SA (2002) Critical role for CCA1 and LHY in maintaining circadian rhythmicity in Arabidopsis. Curr Biol 12:757–761

Allen T, Koustenis A, Theodorou G, Somers DE, Kay SA et al (2006) Arabidopsis FHY3 specifically gates phytochrome signaling to the circadian clock. Plant Cell 18:2506–2516

Alonso-Blanco C, El-Assal SE, Coupland G, Koornneef M (1998) Analysis of natural allelic variation at flowering time loci in the Landsberg erecta and Cape Verde Islands ecotypes of *Arabidopsis thaliana*. Genetics 149:749–764

Alonso-Blanco C, Mendez-Vigo B, Koornneef M (2005) From phenotypic to molecular polymorphisms involved in naturally occurring variation of plant development. Int J Dev Biol 49:717–732

Alonso-Blanco C, Koornneef M, van Ooijen JW (2006) QTL analysis. Methods Mol Biol 323:79–99

Aronson BD, Johnson KA, Loros JJ, Dunlap JC (1994) Negative feedback defining a circadian clock: autoregulation of the clock gene frequency. Science 263:1578–1584

Balasubramanian S, Sureshkumar S, Agrawal M, Michael TP, Wessinger C et al (2006) The PHYTOCHROME C photoreceptor gene mediates natural variation in flowering and growth responses of Arabidopsis thaliana. Nat Genet 38:711–715

Barak S, Tobin EM, Andronis C, Sugano S, Green RM (2000) All in good time: the Arabidopsis circadian clock. Trends Plant Sci 5:517–522

Bastow R, Mylne JS, Lister C, Lippman Z, Martienssen RA et al (2004) Vernalization requires epigenetic silencing of FLC by histone methylation. Nature 427:164–167

Blazquez MA, Ahn JH, Weigel D (2003) A thermosensory pathway controlling flowering time in *Arabidopsis thaliana*. Nat Genet 33:168–171

Bognar LK, Hall A, Adam E, Thain SC, Nagy F et al (1999) The circadian clock controls the expression pattern of the circadian input photoreceptor, phytochrome B. Proc Natl Acad Sci U S A 96:14652–14657

Boothroyd CE, Wijnen H, Naef F, Saez L, Young MW (2007) Integration of light and temperature in the regulation of circadian gene expression in Drosophila. PLoS Genet 3:e54

Borevitz JO, Maloof JN, Lutes J, Dabi T, Redfern JL et al (2002) Quantitative trait loci controlling light and hormone response in two accessions of *Arabidopsis thaliana*. Genetics 160:683–696

Bretzl H (1903) Botanische Forschungen des Alexanderzuges. Teubner, Leipzig

Buenning E (1931) Untersuchungen ueber die autonomen tagesperiodischen Bewungen der Primaerblaetter von *Phaseolus multiflorus*. Jahrb Wiss Bot 75:439–480

Busza A, Murad A, Emery P (2007) Interactions between circadian neurons control temperature synchronization of Drosophila behavior. J Neurosci 27:10722–10733

Cao S, Jiang L, Song S, Jing R, Xu G (2006) AtGRP7 is involved in the regulation of abscisic acid and stress responses in Arabidopsis. Cell Mol Biol Lett 11:526–535

Chen M, Chory J, Fankhauser C (2004) Light signal transduction in higher plants. Annu Rev Genet 38:87–117.

Covington MF, Harmer SL (2007) The circadian clock regulates auxin signaling and responses in Arabidopsis. PLoS Biol 5(8):e222

Covington MF, Panda S, Liu XL, Strayer CA, Wagner DR et al (2001) ELF3 modulates resetting of the circadian clock in Arabidopsis. Plant Cell 13:1305–1315

Daniel X, Sugano S, Tobin EM (2004) CK2 phosphorylation of CCA1 is necessary for its circadian oscillator function in Arabidopsis. Proc Natl Acad Sci U S A 101:3292–3297

Darlington TK, Wager-Smith K, Ceriani MF, Staknis D, Gekakis N et al (1998) Closing the circadian loop: CLOCK-induced transcription of its own inhibitors per and tim. Science 280:1599–1603

Darrah C, Taylor BL, Edwards KD, Brown PE, Hall A et al (2006) Analysis of phase of LUCIFERASE expression reveals novel circadian quantitative trait loci in Arabidopsis. Plant Physiol 140:1464–1474

Davis SJ (2002) Photoperiodism: the coincidental perception of the season. Curr Biol 12:R841–843

Davis SJ, Millar AJ (2001) Watching the hands of the Arabidopsis biological clock. Genome Biol 2:1008.1001–1008.1004

de Mairan J (1729) Observation botanique. Hist Acad R Sci 35–36

Devlin PF, Kay SA (2000) Cryptochromes are required for phytochrome signaling to the circadian clock but not for rhythmicity. Plant Cell 12:2499–2510

Devlin PF, Kay S (2001) Circadian photoreception. Annu Rev Physiol 63:677–694

Devlin PF, Yanovsky MJ, Kay SA (2003) A genomic analysis of the shade avoidance response in Arabidopsis. Plant Physiol 133:1617–1629

Diernfellner AC, Schafmeier T, Merrow MW, Brunner M (2005) Molecular mechanism of temperature sensing by the circadian clock of *Neurospora crassa*. Genes Dev 19:968–1973

Ding Z, Doyle MR, Amasino RM, Davis SJ (2007a) A complex genetic interaction between *Arabidopsis thaliana* TOC1 and CCA1/LHY in driving the circadian clock and in output regulation. Genetics 176:1501–1510

Ding Z, Millar AJ, Davis AM, Davis SJ (2007b) TIME FOR COFFEE encodes a nuclear regulator in the *Arabidopsis thaliana* circadian clock. Plant Cell 19:1522–1536
Dodd AN, Salathia N, Hall A, Kevei E, Toth R et al (2005) Plant circadian clocks increase photosynthesis, growth, survival, and competitive advantage. Science 309:630–633
Dodd AN, Gardner MJ, Hotta CT, Hubbard KE, Dalchau N et al (2007) The Arabidopsis circadian clock incorporates a cADPR-based feedback loop. Science 318:1789–1792
Dowson-Day MJ, Millar AJ (1999) Circadian dysfunction causes aberrant hypocotyl elongation patterns in Arabidopsis. Plant J 17:63–71
Doyle MR, Davis SJ, Bastow RM, McWatters HG, Kozma-Bognar L et al (2002) The ELF4 gene controls circadian rhythms and flowering time in *Arabidopsis thaliana*. Nature 419:74–77
Doyle MR, Bizzell CM, Keller MR, Michaels SD, Song J et al (2005) HUA2 is required for the expression of floral repressors in *Arabidopsis thaliana*. Plant J 41:376–385
Edwards KD, Lynn JR, Gyula P, Nagy F, Millar AJ (2005) Natural allelic variation in the temperature-compensation mechanisms of the *Arabidopsis thaliana* circadian clock. Genetics 170:387–400
Edwards KD, Anderson PE, Hall A, Salathia NS, Locke JC et al (2006) FLOWERING LOCUS C mediates natural variation in the high-temperature response of the Arabidopsis circadian clock. Plant Cell 18:639–650
El-Din El-Assal S, Alonso-Blanco C, Peeters AJ, Raz V, Koornneef M (2001) A QTL for flowering time in Arabidopsis reveals a novel allele of CRY2. Nat Genet 29:435–440
Elich TD, Chory J (1994) Initial events in phytochrome signaling: still in the dark. Plant Mol Biol 26:1315–1327
Emery P, So WV, Kaneko M, Hall JC, Rosbash M (1998) CRY, a Drosophila clock and light-regulated cryptochrome, is a major contributor to circadian rhythm resetting and photosensitivity. Cell 95:669–679
Emery P, Stanewsky R, Helfrich-Forster C, Emery-Le M, Hall JC et al (2000) Drosophila CRY is a deep brain circadian photoreceptor. Neuron 26:493–504
Fankhauser C, Staiger D (2002) Photoreceptors in Arabidopsis thaliana: light perception, signal transduction and entrainment of the endogenous clock. Planta 216:1–16
Farre EM, Harmer SL, Harmon FG, Yanovsky MJ, Kay SA (2005) Overlapping and distinct roles of PRR7 and PRR9 in the Arabidopsis circadian clock. Curr Biol 15:47–54
Filiault DL, Wessinger CA, Dinneny JR, Lutes J, Borevitz JO et al (2008) Amino acid polymorphisms in Arabidopsis phytochrome B cause differential responses to light. Proc Natl Acad Sci U S A 105:3157–3162
Franklin KA, Whitelam GC (2004) Light signals, phytochromes and cross-talk with other environmental cues. J Exp Bot 55:271–276
Gekakis N, Staknis D, Nguyen HB, Davis FC, Wilsbacher LD et al (1998) Role of the CLOCK protein in the mammalian circadian mechanism. Science 280:1564–1569
Glaser FT, Stanewsky R (2005) Temperature synchronization of the Drosophila circadian clock. Curr Biol 15:1352–1363
Gould PD, Locke JC, Larue C, Southern MM, Davis SJ et al (2006) The molecular basis of temperature compensation in the Arabidopsis circadian clock. Plant Cell 18:1177–1187
Green RM, Tobin EM (1999) Loss of the circadian clock-associated protein 1 in Arabidopsis results in altered clock-regulated gene expression. Proc Natl Acad Sci U S A 96:4176–4179
Hall A, Kozma-Bognar L, Toth R, Nagy F, Millar AJ (2001) Conditional circadian regulation of PHYTOCHROME A gene expression. Plant Physiol 127:1808–1818
Hall A, Kozma-Bognar L, Bastow RM, Nagy F, Millar AJ (2002) Distinct regulation of CAB and PHYB gene expression by similar circadian clocks. Plant J 32:529–537
Hall A, Bastow RM, Davis SJ, Hanano S, McWatters HG et al (2003) The TIME FOR COFFEE gene maintains the amplitude and timing of Arabidopsis circadian clocks. Plant Cell 15:2719–2729
Halliday KJ, Salter MG, Thingnaes E, Whitelam GC (2003) Phytochrome control of flowering is temperature sensitive and correlates with expression of the floral integrator FT. Plant J 33:875–885
Han L, Mason M, Risseeuw EP, Crosby WL, Somers DE (2004) Formation of an SCF(ZTL) complex is required for proper regulation of circadian timing. Plant J 40:291–301

Hanano S, Davis SJ (2005) Pseudo-response regulator genes "tell" the time of day: multiple feedbacks in the circadian system of higher plants. In: McWatters H, Hall A (eds) Endogenous plant rhythms. Blackwell, Sheffield

Hanano S, Davis SJ (2007) Mind the clock. Plant Signal Behav 2:1–3

Hanano S, Domagalska MA, Nagy F, Davis SJ (2006) Multiple phytohormones influence distinct parameters of the plant circadian clock. Genes Cells 11:1381–1392

Hanano S, Stracke R, Jakoby M, Merkle T, Domagalska MA et al (2008) A systematic survey in *Arabidopsis thaliana* of transcription factors that modulate circadian parameters. BMC Genomics 9:182

Harmer SL, Hogenesch JB, Straume M, Chang HS, Han B et al (2000) Orchestrated transcription of key pathways in Arabidopsis by the circadian clock. Science 290:2110–2113

Hazen SP, Schultz TF, Pruneda-Paz JL, Borevitz JO, Ecker JR et al (2005) LUX ARRHYTHMO encodes a Myb domain protein essential for circadian rhythms. Proc Natl Acad Sci U S A 102:10387–10392

Heintzen C, Nater M, Apel K, Staiger D (1997) AtGRP7, a nuclear RNA-binding protein as a component of a circadian-regulated negative feedback loop in *Arabidopsis thaliana*. Proc Natl Acad Sci U S A 94:8515–8520

Hennig L, Buche C, Eichenberg K, Schafer E (1999) Dynamic properties of endogenous phytochrome A in Arabidopsis seedlings. Plant Physiol 121:571–577

Hicks KA, Millar AJ, Carre IA, Somers DE, Straume M et al (1996) Conditional circadian dysfunction of the Arabidopsis early-flowering 3 mutant. Science 274:790–792

Ito S, Nakamichi N, Nakamura Y, Niwa Y, Kato T et al (2007) Genetic linkages between circadian clock-associated components and phytochrome-dependent red light signal transduction in *Arabidopsis thaliana*. Plant Cell Physiol 48:971–983

Johanson U, West J, Lister C, Michaels S, Amasino R et al (2000) Molecular analysis of FRIGIDA, a major determinant of natural variation in Arabidopsis flowering time. Science 290(5490): 344–347

Kang B, Grancher N, Koyffmann V, Lardemer D, Burney S et al (2008) Multiple interactions between cryptochrome and phototropin blue-light signaling pathways in *Arabidopsis thaliana*. Planta 227:1091–1099

Kevei E, Gyula P, Hall A, Kozma-Bognar L, Kim WY et al (2006) Forward genetic analysis of the circadian clock separates the multiple functions of ZEITLUPE. Plant Physiol 140:933–945

Kiba T, Henriques R, Sakakibara H, Chua NH (2007) Targeted degradation of PSEUDO-RESPONSE REGULATOR5 by an SCFZTL complex regulates clock function and photomorphogenesis in *Arabidopsis thaliana*. Plant Cell 19:2516–2530

Kikis EA, Khanna R, Quail PH (2005) ELF4 is a phytochrome-regulated component of a negative-feedback loop involving the central oscillator components CCA1 and LHY. Plant J 44:300–313

Kim JI, Kozhukh GV, Song PS (2002) Phytochrome-mediated signal transduction pathways in plants. Biochem Biophys Res Commun 298:457–463

Kim JS, Jung HJ, Lee HJ, Kim KA, Goh CH et al (2008) Glycine-rich RNA-binding protein7 affects abiotic stress responses by regulating stomata opening and closing in *Arabidopsis thaliana*. Plant J 55(3):455–466

Kim JY, Song HR, Taylor BL, Carre IA (2003a) Light-regulated translation mediates gated induction of the Arabidopsis clock protein LHY. EMBO J 22:935–944

Kim WY, Geng R, Somers DE (2003b) Circadian phase-specific degradation of the F-box protein ZTL is mediated by the proteasome. Proc Natl Acad Sci U S A 100:4933–4938

Kim WY, Fujiwara S, Suh SS, Kim J, Kim Y et al (2007) ZEITLUPE is a circadian photoreceptor stabilized by GIGANTEA in blue light. Nature 449:356–360

Kolmos E, Davis SJ (2007a) ELF4 as a central gene in the circadian clock. Plant Signal Behav 2:370–372

Kolmos E, Davis SJ (2007b) Circadian rhythms: rho-related signals in time-specific light perception. Curr Biol 17:R808–810

Kolmos E, Schoof H, P luemer M, Davis SJ (2008) Structural insights into the function of the core-oscillator factor TIMING OF CAB2 EXPRESSION I (TOC1). Circadian Rhythms 6:3

Koornneef M, Alonso-Blanco C, Vreugdenhil D (2004) Naturally occurring genetic variation in *Arabidopsis thaliana*. Annu Rev Plant Biol 55:141–172

Kuno N, Geir Moeller S, Shinomura T, Xu X, Chua N (2003) The novel MYB protein EPR1 is a component of a slave circadian oscillaotor in Arabidopsis. Plant Cell 15(10):2476–2488

Lin C, Todo T (2005) The cryptochromes. Genome Biol 6:220

Lin R, Ding L, Casola C, Ripoll DR, Feschotte C et al (2007) Transposase-derived transcription factors regulate light signaling in Arabidopsis. Science 318:1302–1305

Liu XL, Covington MF, Fankhauser C, Chory J, Wagner DR (2001) ELF3 encodes a circadian clock-regulated nuclear protein that functions in an Arabidopsis PHYB signal transduction pathway. Plant Cell 13:1293–1304

Locke JC, Southern MM, Kozma-Bognar L, Hibberd V, Brown PE et al (2005) Extension of a genetic network model by iterative experimentation and mathematical analysis. Mol Syst Biol 1:2005–2013

Locke JC, Kozma-Bognar L, Gould PD, Feher B, Kevei E et al (2006) Experimental validation of a predicted feedback loop in the multi-oscillator clock of *Arabidopsis thaliana*. Mol Syst Biol 2:59

Loudet O, Chaillou S, Camilleri C, Bouchez D, Daniel-Vedele F (2002) Bay-0 x Shahdara recombinant inbred line population: a powerful tool for the genetic dissection of complex traits in Arabidopsis. Theor Appl Genet 104:1173–1184

Makino S, Kiba T, Imamura A, Hanaki N, Nakamura A et al (2000) Genes encoding pseudo-response regulators: insight into His-to-Asp phosphorelay and circadian rhythm in *Arabidopsis thaliana*. Plant Cell Physiol 41:791–803

Makino S, Matsushika A, Kojima M, Yamashino T, Mizuno T (2002) The APRR1/TOC1 quintet implicated in circadian rhythms of *Arabidopsis thaliana*: I. Characterization with APRR1-overexpressing plants. Plant Cell Physiol 43:58–69

Maloof JN, Borevitz JO, Dabi T, Lutes J, Nehring RB et al (2001) Natural variation in light senvsitivity of Arabidopsis. Nat Genet 29:441–446

Martin-Tryon EL, Kreps JA, Harmer SL (2007) GIGANTEA acts in blue light signaling and has biochemically separable roles in circadian clock and flowering time regulation. Plant Physiol 143:473–486

Mas P, Devlin PF, Panda S, Kay SA (2000) Functional interaction of phytochrome B and cryptochrome 2. Nature 408:207–211

Mas P, Kim WY, Somers DE, Kay SA (2003a) Targeted degradation of TOC1 by ZTL modulates circadian function in *Arabidopsis thaliana*. Nature 426:567–570

Mas P, Alabadi D, Yanovsky MJ, Oyama T, Kay SA (2003b) Dual role of TOC1 in the control of circadian and photomorphogenic responses in Arabidopsis. Plant Cell 15:223–236

Matsushika A, Makino S, Kojima M, Mizuno T (2000) Circadian waves of expression of the APRR1/TOC1 family of pseudo-response regulators in *Arabidopsis thaliana*: insight into the plant circadian clock. Plant Cell Physiol 41:1002–1012

McClung CR (2006) Plant circadian rhythms. Plant Cell 18:792–803

McWatters HG, Bastow RM, Hall A, Millar AJ (2000) The ELF3 zeitnehmer regulates light signaling to the circadian clock. Nature 408:716–720

McWatters HG, Kolmos E, Hall A, Doyle MR, Amasino RM et al (2007) ELF4 is required for oscillatory properties of the circadian clock. Plant Physiol 144:391–401

Merrow M, Boesl C, RIcken J, Messerschmitt M, Goedel M, Roennenber T (2006) Entrainment of the *Neurospora* circadian clock. Chronobiol Int 23:71–80

Michael TP, Salome PA, McClung CR (2003a) Two Arabidopsis circadian oscillators can be distinguished by differential temperature sensitivity. Proc Natl Acad Sci U S A 100:6878–6883

Michael TP, Salome PA, Yu HJ, Spencer TR, Sharp EL et al (2003b) Enhanced fitness conferred by naturally occurring variation in the circadian clock. Science 302:1049–1053

Michaels SD, Amasino RM (1999) FLOWERING LOCUS C encodes a novel MADS domain protein that acts as a repressor of flowering. Plant Cell 11:949–956

Michael TP, Mockler TC, Breton G, McEntee C, Byer A et al (2008) Network discovery pipeline elucidates conserved Time-of-Day specific cis-regulatory modules. PLoS Genet 4:e14

Millar AJ (2004) Input signals to the plant circadian clock. J Exp Bot 55:277–283
Millar AJ, Short SR, Chua NH, Kay SA (1992) A novel circadian phenotype based on firefly luciferase expression in transgenic plants. Plant Cell 4:1075–1087
Millar AJ, Carre IA, Strayer CA, Chua NH, Kay SA (1995) Circadian clock mutants in Arabidopsis identified by luciferase imaging. Science 267:1161–1163
Mizoguchi T, Wheatley K, Hanzawa Y, Wright L, Mizoguchi M et al (2002) LHY and CCA1 are partially redundant genes required to maintain circadian rhythms in Arabidopsis. Dev Cell 2:629–641
Murtas G, Millar AJ (2000) How plants tell the time. Curr Opin Plant Biol 3:43–46
Mylne JS, Barrett L, Tessadori F, Mesnage S, Johnson L et al (2006) LHP1, the Arabidopsis homologue of HETEROCHROMATIN PROTEIN1, is required for epigenetic silencing of FLC. Proc Natl Acad Sci U S A 103:5012–5017
Nagy F, Schafer E (2002) Phytochromes control photomorphogenesis by differentially regulated, interacting signaling pathways in higher plants. Annu Rev Plant Biol 53:329–355
Niwa Y, Ito S, Nakamichi N, Mizoguchi T, Niinuma K et al (2007) Genetic linkages of the circadian clock-associated genes, TOC1, CCA1 and LHY, in the photoperiodic control of flowering time in *Arabidopsis thaliana*. Plant Cell Physiol 48:925–937
Nozue K, Covington MF, Duek PD, Lorrain S, Fankhauser C et al (2007) Rhythmic growth explained by coincidence between internal and external cues. Nature 448:358–361
Onai K, Ishiura M (2005) PHYTOCLOCK 1 encoding a novel GARP protein essential for the Arabidopsis circadian clock. Genes Cells 10:963–972
Perales M, Mas P (2007) A functional link between rhythmic changes in chromatin structure and the Arabidopsis biological clock. Plant Cell 19:2111–2123
Perales M, Portoles S, Mas P (2006) The proteasome-dependent degradation of CKB4 is regulated by the Arabidopsis biological clock. Plant J 46:849–860
Pittendrigh CS, Minis DH (1964) The entrainment of circadian oscillations by light and their role as photoperiodic clocks. Am Nat 98:261–322
Quail PH, Boylan MT, Parks BM, Short TW, Xu Y et al (1995) Phytochromes: photosensory perception and signal transduction. Science 268:675–680
Rensing L, Monnerjahn C (1996) Heat shock proteins and circadian rhythms. Chronobiol Int 13:239–250
Salathia N, Davis SJ, Lynn JR, Michaels SD, Amasino RM et al (2006) FLOWERING LOCUS C-dependent and -independent regulation of the circadian clock by the autonomous and vernalization pathways. BMC Plant Biol 6:10
Salome PA, McClung CR (2004) The Arabidopsis thaliana clock. J Biol Rhythms 19:425–435
Salome PA, McClung CR (2005) PSEUDO-RESPONSE REGULATOR 7 and 9 are partially redundant genes essential for the temperature responsiveness of the Arabidopsis circadian clock. Plant Cell 17:791–803
Salome PA, To JP, Kieber JJ, McClung CR (2006) Arabidopsis response regulators ARR3 and ARR4 play cytokinin-independent roles in the control of circadian period. Plant Cell 18:55–69
Salter MG, Franklin KA, Whitelam GC (2003) Gating of the rapid shade-avoidance response by the circadian clock in plants. Nature 426:680–683
Schaffer R, Ramsay N, Samach A, Corden S, Putterill J et al (1998) The late elongated hypocotyl mutation of Arabidopsis disrupts circadian rhythms and the photoperiodic control of flowering. Cell 93:1219–1229
Searle I, He Y, Turck F, Vincent C, Fornara F et al (2006) The transcription factor FLC confers a flowering response to vernalization by repressing meristem competence and systemic signaling in Arabidopsis. Genes Dev 20:898–912
Shindo C, Aranzana MJ, Lister C, Baxter C, Nicholls C et al (2005) Role of FRIGIDA and FLOWERING LOCUS C in determining variation in flowering time of Arabidopsis. Plant Physiol 138:1163–1173
Siepka SM, Yoo SH, Park J, Lee C, Takahashi JS (2007) Genetics and neurobiology of circadian clocks in mammals. Cold Spring Harb Symp Quant Biol 72:251–259

Somers DE, Devlin PF, Kay SA (1998) Phytochromes and cryptochromes in the entrainment of the Arabidopsis circadian clock. Science 282:1488–1490

Song HR, Carre IA (2005) DET1 regulates the proteasomal degradation of LHY, a component of the Arabidopsis circadian clock. Plant Mol Biol 57:761–771

Staiger D, Zecca L, Wieczorek Kirk DA, Apel K, Eckstein L (2003a) The circadian clock regulated RNA-binding protein AtGRP7 autoregulates its expression by influencing alternative splicing of its own pre-mRNA. Plant J 33:361–371

Staiger D, Allenbach L, Salathia N, Fiechter V, Davis SJ et al (2003b) The Arabidopsis SRR1 gene mediates phyB signaling and is required for normal circadian clock function. Genes Dev 17:256–268

Stanewsky R, Kaneko M, Emery P, Beretta B, Wager-Smith K et al (1998) The cryb mutation identifies cryptochrome as a circadian photoreceptor in Drosophila. Cell 95:681–692

Stitt M, Hurry V (2002) A plant for all seasons: alterations in photosynthetic carbon metabolism during cold acclimation in Arabidopsis. Curr Opin Plant Biol 5:199–206

Strand A, Hurry V, Henkes S, Huner N, Gustafsson P et al (1999) Acclimation of Arabidopsis leaves developing at low temperatures. Increasing cytoplasmic volume accompanies increased activities of enzymes in the Calvin cycle and in the sucrose-biosynthesis pathway. Plant Physiol 119:1387–1398

Strayer C, Oyama T, Schultz TF, Raman R, Somers DE et al (2000) Cloning of the Arabidopsis clock gene TOC1, an autoregulatory response regulator homolog. Science 289:768–771

Sugano S, Andronis C, Green RM, Wang ZY, Tobin EM (1998) Protein kinase CK2 interacts with and phosphorylates the Arabidopsis circadian clock-associated 1 protein. Proc Natl Acad Sci U S A 95:11020–11025

Sugano S, Andronis C, Ong MS, Green RM, Tobin EM (1999) The protein kinase CK2 is involved in regulation of circadian rhythms in Arabidopsis. Proc Natl Acad Sci U S A 96:12362–12366

Swarup K, Alonso-Blanco C, Lynn JR, Michaels SD, Amasino RM et al (1999) Natural allelic variation identifies new genes in the Arabidopsis circadian system. Plant J 20:67–77

Tamanini F, Chaves I, Bajek MI, van der Horst GT (2007) Structure function analysis of mammalian cryptochromes. Cold Spring Harb Symp Quant Biol 72:133–139

Toth R, Kevei E, Hall A, Millar AJ, Nagy F et al (2001) Circadian clock-regulated expression of phytochrome and cryptochrome genes in Arabidopsis. Plant Physiol 127:1607–1616

Vierstra RD (2003) The ubiquitin/26S proteasome pathway, the complex last chapter in the life of many plant proteins. Trends Plant Sci 8:135–142

Wang ZY, Tobin EM (1998) Constitutive expression of the CIRCADIAN CLOCK ASSOCIATED 1 (CCA1) gene disrupts circadian rhythms and suppresses its own expression. Cell 93:1207–1217

Waterhouse J, Reilly T, Atkinson G, Edwards B (2007) Jet lag: trends and coping strategies. Lancet 369:1117–1129

Werner JD, Borevitz JO, Warthmann N, Trainer GT, Ecker JR et al (2005) Quantitative trait locus mapping and DNA array hybridization identify an FLM deletion as a cause for natural flowering-time variation. Proc Natl Acad Sci U S A 102:2460–2465

Yanovsky MJ, Mazzella MA, Casal JJ (2000) A quadruple photoreceptor mutant still keeps track of time. Curr Biol 10:1013–1015

Yanovsky MJ, Mazzella MA, Whitelam GC, Casal JJ (2001) Resetting of the circadian clock by phytochromes and cryptochromes in Arabidopsis. J Biol Rhythms 16:523–530

Yu W, Zheng H, Houl JH, Dauwalder B, Hardin PE (2006) PER-dependent rhythms in CLK phosphorylation and E-box binding regulate circadian transcription. Genes Dev 20:723–733

Zeilinger MN, Farre EM, Taylor SR, Kay SA, Doyle FJ 3rd (2006) A novel computational model of the circadian clock in Arabidopsis that incorporates PRR7 and PRR9. Mol Syst Biol 2:58

Vesicle Trafficking in Plant Pathogen Defence

Hans Thordal-Christensen

Abstract Recent progress in our understanding of plant mechanisms has revealed important roles of vesicle trafficking in the cellular processes of pathogen defence. While most of the existing knowledge regarding the mechanistic details related to vesicle trafficking originates from non-plant systems, it is evident that most of these processes are also conserved in plants. Meanwhile, we are beginning to understand the general mechanisms of how plants and pathogens communicate and fight, and the present review summarises a collection of recent observations demonstrating how vesicle trafficking processes are linked to individual steps of this interaction. In this way, there is evidence showing that vesicle trafficking is involved both in regulating and in executing these defences. The vesicle-trafficking-related proteins documented to be associated with defence include syntaxins, soluble *N*-ethylmaleimide-sensitive factor attachment protein, vesicle-associated membrane protein dynamin, Rab GTPase, and a guanine nucleotide exchange factor of an ADP-ribosylation factor GTPase.

1 Introduction

Pathogens cause significant stress to plants and when untreated they can result in heavy crop losses; therefore, plant researchers have for decades had a major focus on understanding the mechanisms underlying the innate protection of plants against pathogens. We now understand that plants activate different types of defence when attacked by pathogens. These defences include secretion of antimicrobial proteins, many of them termed "pathogenesis-related (PR) proteins" (van Loon et al. 2006); synthesis of antimicrobial secondary metabolites, so-called

H. Thordal-Christensen (✉)
Plant and Soil Science Laboratory, Department of Agriculture and Ecology, The Royal Veterinary and Agricultural University, Thorvaldsensvej 40, 1871 Frederiksberg C, Copenhagen, Denmark
e-mail: htc@life.ku.dk

F. Baluška and S. Mancuso (eds.), *Signaling in Plants*, Signaling and Communication in Plants, 287
DOI: 10.1007/978-3-540-89228-1_14, © Springer-Verlag Berlin Heidelberg 2009

phytoalexins (Dixon 2001); cell wall fortification (Hardham et al. 2007) and the "hypersensitive response" (HR), i.e. host cells undergoing programmed cell death (Hofius et al. 2007).

Plants can sense the presence of pathogens, and thereby activate defences in a coordinated manner. A number of plant plasma membrane located receptor-like kinases are able to detect pathogen molecules that are strictly needed for the survival of the microbes. Detection of these "pathogen-associated molecular patterns" (PAMPs) results in activation of basal defence in the plant (Zipfel 2008). This leads to various levels of protection, depending on the plant/pathogen combination. Basal defence in plants resembles innate immunity in animals. Pathogens counteract basal defence by introducing effector molecules into the plant cell. These effectors can interfere with the signalling process leading to basal defence (Espinosa and Alfano 2004). For a few biotrophic pathogens that require living host cells for their growth, inventories with many diverging effector protein candidates have been collected (Catanzariti et al. 2007; Jiang et al. 2008; Wei et al. 2007) suggesting that effectors have a broad range of targets in the plant. Plants may subsequently be able to detect effectors, or the result of effector activities, via classic resistance genes. This recognition process activates HR, an efficient defence against biotrophic pathogens (Jones and Dangl 2006; de Wit 2007).

Many signalling genes involved in controlling how defences are regulated in response to the recognition processes have been identified. Of particular importance is the *NPR1* gene encoding a protein that plays a central role in salicylic acid signalling (Durrant and Dong 2004). Upon increase in the salicylic acid level in the cell, NPR1 enters the nucleus, where it facilitates induction of defence genes. Additional defence signalling components contribute to defence activation together with NPR1. For extensive reviews, see Hammond-Kosack and Parker (2003) and Glazebrook (2005).

Pathogen defence taken as a whole is highly crucial to the plant, and a large proportion of the plant mechanisms are dedicated to protection against pathogens. This is reflected in the observation that a remarkable proportion of the genes in the plant genome are regulated very rapidly when the plant recognises an invader. As much as 20% (approximately 4,500 genes) of the *Arabidopsis* genome has altered regulation in response to bacteria (Bartsch et al. 2006; Thilmony et al. 2006; Truman et al. 2006). About half of these genes are upregulated, presumably for defence purposes. The other half, which are downregulated long before symptoms appear, are dominated by primary metabolism genes. This rapid downregulation has received less attention, but may serve to alleviate cellular damage caused by the invading pathogen. Most of the upregulation of genes occurs as the end result of very complex signalling processes. However, a significant number of these genes are upregulated owing to their involvement in the signalling processes.

When 20% of the plant genes respond to pathogen attack, it is inevitable that cellular vesicle trafficking is involved. Vesicle traffic is highly complex, and the emerging picture is that each vesicle trafficking highway, from compartment A to compartment B, utilises a large suite of proteins. The different vesicle trafficking highways are characterised by different versions of the same proteins. This large suite of proteins includes numerous members of various GTPase classes. Initially, vesicles are formed by endocytosis regulated by compartment and destination-specific ADP-ribosylation factor (Arf) GTPases and their associated guanine nucleotide exchange factors (GEFs) and GTPase-activating proteins (D'Souza-Schorey and Chavrier 2006). A second class of GTPases, the dynamins, is responsible for pinching off the endocytotic vesicle from the membrane (Bonner and Skop 2008). Endocytotic vesicles initially carry coat proteins, such as clathrin, which play specific roles, e.g. in cargo recognition (Edeling et al. 2006). In electron micrographs, the presence of a vesicle coat is indicative of endocytosis. A third class of GTPases, the Rho proteins, is involved in actin reorganisation and the recruitment of vesicles to actin filaments, along which they are mobilised (Hehnly and Stamnes 2007). By the help of a fourth class of GTPases, the Rabs and their associated GEFs, GTPase-activating proteins and effectors, the vesicles find their way to the correct target compartment (Schwartz et al. 2007). By an unknown mechanism, Rab proteins are involved in mediating the formation of soluble *N*-ethylmaleimide-sensitive factor attachment protein receptor (SNARE) complexes (Graham et al. 2008). SNARE

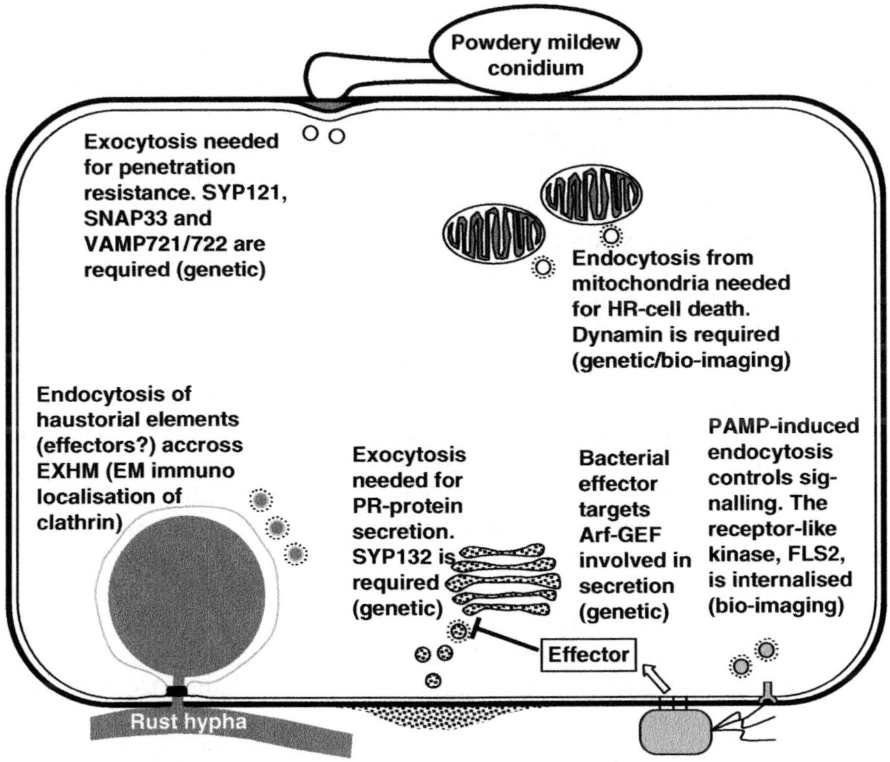

Fig. 1 Examples of vesicle trafficking in plant–pathogen interactions. *Rings surrounded by spots* signify coated endocytotic vesicles. Other rings signify exocytotic vesicles. Statements in *brackets* indicate the type of experimental evidence provided. See the text for detailed descriptions and references

proteins (syntaxin; soluble *N*-ethylmaleimide-sensitive factor attachment protein, SNAP; and vesicle-associated membrane protein,VAMP) are central players in the final step in vesicle trafficking (Lipka et al. 2007). Specific syntaxins, located on the target membrane, form complexes with specific VAMPs on the vesicles. These complexes involve four SNARE domain helices originating from three proteins. In addition to the target membrane syntaxin and the vesicle-bound VAMP, each providing one SNARE domain, a soluble SNAP provides two SNARE domains. There are multiple members of each of the SNARE proteins (Sanderfoot et al. 2000) which correspond to each other, and this specificity is involved in guiding the vesicles to the correct target membrane. The formation of the SNARE complex subsequently establishes a close contact between the two membranes, eventually leading them to merge (Fig. 1).

2 Endocytosis in PAMP Signalling

An early step in plant defence signalling is PAMP recognition by receptor-like kinases. Studies of metazoan plasma membrane receptor kinases have demonstrated important roles of endocytosis in controlling downstream signalling. Receptors can be endocytosed to transmit the signal after ligand binding, and they can be endocytosed to lysosomal degradation to arrest the signalling (Le Borgne 2006; Kirisits et al. 2007). For instance, the mammalian innate immunity receptor, Toll-like receptor 4, that recognises bacterial lipopolysaccharides needs to be endocytosed to stimulate a certain type of response (Kagan et al. 2008). Endocytotic vesicles carrying internalised receptors are taken up by late endosomes that thereby form multivesicular bodies. This emerging compartment subsequently fuses with lysosomes in which the activated receptors are degraded (Kirisits et al. 2007).

In plants, the best described PAMP-mediated basal defence, analogous to metazoan innate immunity, is activated by flagellin (Zipfel 2008). The flagellin receptor FLS2 is a leucine-rich repeat receptor kinase, sharing characteristics with metazoan Toll-like receptors (Zipfel 2008). Flagellin treatment of plants expressing a green fluorescent protein-FLS2 fusion protein induces removal of FLS2 from the plasma membrane to an intracellular compartment. Sensitivity of this removal to wortmannin confirms that this takes place via an endocytotic process (Robatzek et al. 2006). An indication that ligand-induced endocytosis is required for FLS2 signalling comes from studies of a mutant version of green fluorescent protein–FLS2. This mutant protein is plasma-membrane-located and binds flagellin. However, it does not support downstream signalling and it is not internalised after flagellin binding (Robatzek et al. 2006).

Another recent example of defence signalling that depends on endocytosis is found in tobacco BY-2 cells treated with the pathogen elicitor cryptogein (Leborgne-Castel et al. 2008). This protein stimulates uptake of the plasma membrane dye FM4-64 and formation of clathrin-coated pits on the tobacco plasma membrane. As a downstream response, an NADPH oxidase rbohD-mediated oxidative burst was recorded. Verification of these results was obtained by applying the endocytosis inhibitor tyrphostin A23. This inhibitor selectively targets clathrin-dependent endocytosis, as it inhibits tyrosine phosphorylation required for cargo loading of clathrin-coated vesicles.

3 SNAREs in Penetration Resistance

Basal defence is manifested in different ways and has distinct appearances. "Penetration resistance" is a type of basal defence mounted against fungal pathogens that penetrate directly through plant cell walls. This has primarily been studied using powdery mildew fungi. Once inside the plant cell, these fungi develop a specialised structure, the haustorium, with the purpose of taking up nutrients from the host. The penetration resistance involves formation of a cell wall fortification at the site of attack. There is early evidence in the literature, obtained by electron microscopy, suggesting that exocytosis plays a role in this form of basal defence. At the local area of the secondary cell wall, called the "papilla", Zeyen and Bushnell (1979) showed images of vesicles merging with the plasma membrane, indicative of exocytosis. This evidence generated in barley stood alone for 24 years until Paul Schultze-Lefert's, Shauna Somerville's and my laboratories characterised the barley *ror2* and *Arabidopsis pen1* mutants. The *PEN1* gene from *Arabidopsis* and its orthologue *ROR2* in barley were shown to be important for penetration resistance (Collins et al. 2003). Plants with mutant alleles of these genes are more easily penetrated by the barley powdery mildew fungus. *PEN1* and *ROR2* encode the plasma membrane syntaxin SYP121, supporting the idea that exocytosis plays a role in penetration resistance.

It remains unknown what the exact role of SYP121 in penetration resistance is. It has not been possible to observe qualitative differences between the papillae (Assaad et al. 2004, unpublished results), the cytoskeletal response or the endoplasmic reticulum response (Takemoto et al. 2006) of the wild type and *pen1–1*. However, Assaad et al. (2004) demonstrated that a mutation in *PEN1* causes a 2-h delay in the formation of papillae in response to the powdery mildew fungal attack. Such a delay will be important in the timing of the papilla formation relative to the progress of the fungal penetration hypha, and should be detrimental for the defence of the plant. This observation, confirmed by Kwon et al. (2008), suggests that SYP121 is involved in the regulation of the papilla formation, rather than controlling exocytosis of one or more papilla components. Both studies observed the delay by assaying the callose component of the papilla. The particular *Arabidopsis* callose synthase responsible for making the papilla callose has been identified (Jacobs et al. 2003; Nishimura et al. 2003). SYP121 potentially could serve a SNARE function to carry this callose synthase to the plasma membrane and allow it to focally accumulate at the site of attack, but no localisation studies are available for this enzyme. Interestingly, callose only plays a marginal role in penetration resistance (Jacobs et al. 2003). Therefore, whether or not SYP121 is directly involved in focal accumulation of the callose synthase cannot alone explain the severe loss of penetration resistance in the *pen1* mutants. This suggests that SYP121 plays a regulatory role through which it can control simultaneous build-in of several components into the papillae.

The SNARE partners of SYP121 have been identified. After a number of barley SNAP candidates had been tested, silencing barley SNAP34 was found to result in breakdown of penetration resistance, suggesting this protein forms SNARE complex with ROR2 (Collins et al. 2003). SNAP34–ROR2 interaction was confirmed by yeast two-hybrid analysis. The significance of HvSNAP34 in penetration was rediscovered in a shotgun approach where barley genes, upregulated after attack by the powdery mildew fungus, were silenced using RNA interference (Douchkov et al. 2005). Recently, the orthologous SNAP33 in *Arabidopsis* was also identified on the basis of its upregulation after pathogen attack (Kwon et al. 2008). The corresponding vesicle-associated SNAREs (VAMP721 and VAMP722) were identified via a systematic screen for ternary complex formation between SYP121, SNAP33 and a collection of VAMPs (Kwon et al. 2008).

A recent paper addressed the interaction between Rab-GTPase and the syntaxins SYP121 and SYP122. On the basis of functional analyses in tobacco protoplasts expressing dominant negative mutant proteins of the two syntaxins and of Rab11, Rehman et al. (2008) were able to show that inhibiting both SYP121 and Rab11 had an additive effect on reducing secretion of a marker protein. Meanwhile, inhibiting both SYP122 and Rab11 had the same effect on reducing secretion as inhibiting SYP122 and Rab11 individually. This suggests that Rab11 functions together with SYP122 and in a pathway different from that of SYP121.

Interestingly, during the manifestation of penetration resistance, multivesicular bodies have been observed. They form in the vicinity of the site at which penetration is attempted. Subsequently, they can either merge with the central vacuole or merge with the plasma membrane beneath the papilla and deliver the internal vesicle inside the papilla structure as exosomes (An et al. 2006). This observation may suggest that the papillae conferring penetration resistance are in part cellular waste fields where redundant material is thrown at the invading enemy.

4 Cellular Polarisation and Penetration Resistance

When a pathogen attempts to penetrate through a plant cell wall, cell polarity is triggered. The cytoskeleton is reoriented and the actin filaments become directed towards the site of attack (Kobayashi et al. 1997; Opalski et al. 2005; Shimada et al. 2006; Takemoto et al. 2006). This reorientation of the actin filament is required for penetration resistance. Inhibition of actin polymerisation by the use of cytochalasins increased the penetration rate by several fungi (Kobayashi et al. 1997; Mellersh et al. 2002; Miklis et al. 2007; Shimada et al. 2006; Yun et al. 2003).

This requirement for actin polymerisation in penetration resistance was confirmed molecularly by transiently overexpressing an actin-depolymerising factor in single barley epidermal cells attacked by powdery mildew spores (Miklis et al. 2007). Labelling of the actin filaments demonstrated that the cytochalasin indeed interfered with the rearrangement of the actin filaments (Kobayashi et al. 1997).

The Rho–GTPases are involved in the control of actin reorganisation (Hehnly and Stamnes 2007). This is reflected in the effect barley Rho proteins have on penetration resistance. The group of Ralph Hückelhoven has made single cell transient and stable transgenic alterations in gene expression of several Rac proteins that belong to the Rho-GTPases. They demonstrated that Rac activity suppresses actin reorganisation, and as a consequence also suppresses penetration resistance (Eichmann and Huckelhoven 2008; Opalski et al. 2005; Schultheiss et al. 2002, 2003, 2005, 2008).

5 SNAREs in Exocytosis of Defence Components

While the role of SYP121 in penetration resistance has not been fully established, the syntaxin required for PR-protein exocytosis has been identified. Production of extracellular proteins, which can be washed out of the intercellular space of leaves, has for many years been accepted as a major defence response (van Loon et al. 2006). The most commonly used defence response marker, PR-1, is an example of such a secreted protein. The molecular players of the secretory pathway, which brings these defence response proteins out of the cell, are now emerging. Expression of many of the genes encoding defence-related extracellular proteins is under control of NPR1, a central defence signalling component (Durrant and Dong 2004). Furthermore, Wang et al. (2005) found that expression of a number of secretory pathway genes is also dependent on NPR1. Knocking out some of these genes hampered the secretion of PR-1 and disease resistance (Wang et al. 2005). Kalde et al. (2007) identified the syntaxin AtSYP132 to be phosphorylated in response to pathogen elicitor treatment, and subsequently demonstrated it to be required for PR-1 secretion. These authors also studied tobacco and observed that suppressing *NbSYP132*, using RNA interference, caused the PR-1 protein to accumulate intracellularly rather than extracellularly. Notably, suppressing the penetration-resistance-related *SYP121* did not affect secretion of this protein.

6 Is Vesicle Trafficking Involved in Effector Secretion?

An emerging and fascinating area in plant-pathogen interaction studies concerns the pathogen's secretion of effectors to control processes in the plant cell. After successful penetration by powdery mildew and rust fungi and by oomycetes, these pathogens establish haustoria inside the plant cell. As haustoria grow, they stimulate the plant cell to generate a membrane that will surround this pathogen structure. Notably, these biotrophic pathogens avoid killing the plant cell during this drastic invasion. The nature of the extrahaustorial membrane is unknown. It is possible that the haustorium establishes itself in an existing plant cell compartment. However, it is inevitable that the formation of this membrane involves vesicle traffic. Similar "symbiotic" situations exist in animals and humans, where intracellular pathogens live in cells of the host. A fascinating example is the *Legionella* bacteria that live inside human cells. These bacteria secrete a protein that mimics human Rab1-interacting proteins and thereby stimulate the growth of a compartment that becomes a niche for the bacteria (Ingmundson et al. 2007; Machner and Isberg 2007). Interestingly, Rab homologues have been found associated with the plant-derived peribacteroid membrane in *Legume* root nodules (Cheon et al. 1993; Schiene et al. 2004; Wienkoop and Saalbach 2003).

The haustoria secrete large numbers of effector proteins destined for the plant cytoplasm. N-terminal signal peptides ensure that these proteins cross the pathogen plasma membrane, and are brought to the extrahaustorial matrix (Catanzariti et al. 2007; Jiang et al. 2008; Kemen et al. 2005; Whisson et al. 2007). However, most of the effector proteins will end up in the plant cell, and therefore also need to cross the plant-derived extrahaustorial membrane. The process for how this takes place has not been fully established. A large class of oomycetes effector proteins have an RxLR and a dEER motif in the N-terminal of the mature sequence (Jiang et al. 2008). It has been documented that these motifs are required for the transfer of one of these proteins across the extrahaustorial membrane (Whisson et al. 2007). The mechanism by which this takes place remains unknown; however, there is ultrastructural evidence in the literature suggesting that endocytosis might be a mechanism by which effector proteins are transferred across the extrahaustorial membrane. Studies of rust haustoria have revealed signs of coated vesicles on the plant cytoplasm side of the extrahaustorial membrane, indicative of endocytosis. Furthermore, immunolabelling confirmed the presence of clathrin, a vesicle coat protein, at these sites (Stark-Urnau and Mendgen 1995).

7 Vesicle Trafficking Targeted by Pathogen Virulence Effector

The significance of vesicle trafficking in defence is underpinned by the discovery that this process is targeted by a bacterial effector. In an attempt to find host targets for the *Pseudomonas syringae* pv. *tomato* type III secretion system effector, HopM1, a yeast two-hybrid screen was made on *Arabidopsis* (Nomura et al. 2006). The target protein AtMIN7 was identified. This protein, which is required for full defence in the wild type, is a GEF of an Arf-GTPase. AtMIN7–Arf GEF is destabilised in the presence of HopM1, and it is apparently localised to the Golgi-apparatus. The effect of HopM1 can be mimicked by brefeldin A, an inhibitor of exocytosis, suggesting that AtMIN7 plays a role in secretion of defence components.

8 Vesicle Trafficking in the HR

Evidence for involvement of vesicle traffic in the execution of HR comes from the discovery of a GTPase involved in this process. The *Arabidopsis* mutant *edr3* is more predisposed to develop HR lesions in response to fungal pathogens owing to a mutation in a dynamin-GTPase gene. While a knockout mutation in this dynamin gene causes no obvious defence-associated phenotype, the recessive *edr3* mutation causes an amino acid substitution in the GTP-hydrolysing active site (Tang et al. 2006). This is predicted to arrest the release of endocytotic vesicles. EDR3 has been localised to mitochondria and is known to play a central function in programmed cell death. Even though the EDR3-mediated process is not fully understood, the result points to an HR-related role of vesicle traffic from mitochondria.

9 SNAREs Are Negative Regulators of Defence

As described already, SYP121 is essential for penetration resistance; however, the closely related *Arabidopsis* SYP122 cannot be found to play a role in penetration resistance, neither when studied as single mutant nor when studied as a double mutant *syp121 syp122* (Zhang et al. 2007). On the other hand, these two syntaxins, as well as the SNARE proteins that complex with them, appear to be required to avoid overstimulation of a number of defences, different from penetration resistance. This is best documented in the syntaxin double mutant *syp121 syp122*. This line has a severe lesion-mimic phenotype, reflecting constitutive expression of a broad set of defences, including those controlled by the salicylic acid signalling pathway (Assaad et al. 2004; Zhang et al. 2007, 2008). Three hundred and sixty five genes are upregulated at least twofold in *syp121 syp122* relative to the wild type, and most of these genes are also upregulated during expression of pathogen defence. Furthermore, the lesion-mimic phenotype can be rescued by knocking out ten different genes encoding positive defence signalling proteins, as discussed below (Zhang et al. 2007, 2008). The severe phenotype of *syp121 syp122* indicates a shared function of these two syntaxins as negative regulators of defence. The single mutant *syp121*, unlike *syp122*, has a significant, albeit mild expression of defence relative to *syp121 syp122* (Zhang et al. 2007). This shows that SYP121 is the principal negative regulator, as its role in this phenotype can only be partially replaced by that of SYP122. Heese et al. (2001) described a similar lesion-mimic phenotype of a *snap33* knockout line. To document that SNAP33 is also a negative regulator of defence, we have introduced mutant alleles of genes in the salicylic acid signalling pathway. We found that these double mutants of positive defence signalling mutants and *snap33* are rescued, confirming that defence expression is essential for the lesion-mimic phenotype (Z. Zhang and H. Thordal-Christensen, unpublished results). Furthermore, Kwon et al. (2008) were unable to obtain a *vamp721 vamp722* double mutant in *Arabidopsis*, and they observed that RNA interference lines targeting both *VAMP721* and *VAMP722* had a lesion-mimic phenotype similar to that of *syp121 syp122*. This suggests that VAMP721 and VAMP722 also are negative regulators of defence.

It remains unsolved how this set of SNARE proteins can be required for penetration resistance and at the same time be negative regulators of other defences. However, the fact that SYP122 only has the negative defence regulator role may indicate that these two phenotypes result from separate functions of these SNARE proteins. The negative defence regulator function may be related to observed syntaxin phosphorylation. In *Arabidopsis*, SYP121 and SYP122 are phosphorylated on an N-terminal serine residue in response to the PAMP flg22 (Benschop et al. 2007; Nühse et al. 2003). The same is true for the closely related tobacco syntaxin NtSYP121 during resistance-gene-mediated resistance (Heese et al. 2005). It is uncertain what role phosphorylation of syntaxin plays in PAMP-induced defence. However, one model could be that phosphorylation serves to alleviate negative regulatory functions of unphosphorylated versions of these proteins; therefore, phosphorylation would activate defence. From studies of the *syp121 syp122* double mutant we have indications that these syntaxins are involved in PAMP signalling. There is significant overlap between genes upregulated in response to PAMP treatment and the 365 genes constitutively upregulated in *syp121 syp122* (Zhang et al. 2008). Furthermore, introducing mutant alleles of the genes *EDS1*, *PAD4*, *NDR1*, *SGT1b* and *RAR1* partially rescues the lesion-mimic phenotype of *syp121 syp122* (Zhang et al. 2008). These five genes are known to be required for resistance-gene-mediated defence, which is stimulated after activation of PAMP signalling (Glazebrook 2005; Jones and Dangl 2006).

10 SNARE Mutants Unravel Pathogen Defence Signalling Mechanisms

Studies of the broad activation of defences in *syp121 syp122* have contributed to our understanding of the complex defence signalling processes. Traditionally, defence signalling has been studied in the most obvious manner – in response to pathogens. This direct activation of defence is restricted by the defence-eliciting compounds presented by the pathogen in use, limiting, e.g., studies of interaction between separate signalling processes. Finding mutants that are hampered in their defence responses also suffers from such limitations, and at the same time requires that disease development and/or defence response can be quantified. Even though these methods have been successful, and led to discovery of a significant number of defence signalling components during the last 15 years, they certainly have not saturated the discovery of defence signalling components.

The use of lesion-mimic lines offers an alternative to pathogen inoculations in defence signalling studies (Lorrain et al. 2003). The rescuing effects of mutations in positive defence signalling genes are visible to the naked eye and can be documented using a regular digital camera. This has tremendous advantages over tedious analyses of transcript and protein expression patterns, pathogen growth, etc. Therefore, we have used the lesion-mimic phenotype of *syp121 syp122* to isolate 240 partially rescued mutants. In addition to the two syntaxin gene mutations, these triple mutants carry mutations in *SUPPRESSOR OF SYNTAXIN-RELATED DEATH* (*SSD*) genes, which are defence signalling genes (Zhang et al. 2008). Many of the *SSD* genes are well-known defence signalling genes, e.g. *SID2* and *PAD4*, whereas others have not previously been associated with defence signalling. So far we have characterised 12 *SSD* genes, two of which are novel.

11 SNAREs Dedicated to Environmental Stress

Arabidopsis has 24 syntaxins, three SNAPs and 14 VAMPs (Sanderfoot et al. 2000), and an interesting question relates to what extent these are dedicated to specific roles or clusters of roles. We have reason to believe that SYP121 and SYP122 define a set of syntaxins that is restricted to environmental stress responses. This is implied by the observed roles of these proteins as required for penetration resistance, their roles as negative regulators of other defences and their phosphorylation in response to pathogen signals (see earlier). Furthermore, SYP121 has been implicated in stomatal closure in response to drought stress (Leyman et al. 1999). This function is suggested to be due to this syntaxin influencing a calcium channel responding to abscisic acid (Sokolovski et al. 2008). Interestingly, we have discovered that the *syp121 syp122* plant is also oversensitive to salt stress. This is seen as stunted growth and poor root hair development (A. Feechan, D. Godfrey and H. Thordal-Christensen, unpublished results).

What allows us to suggest that SYP121 and SYP122 are dedicated to environmental stress responses is the fact that we can almost completely rescue the lesion-mimic phenotype of *syp121 syp122* by mutating a collection of genes required for the downstream defence signalling (Zhang et al. 2008). This strongly suggests that these two syntaxins are not involved in growth and development of unstressed plants.

12 Conclusion

There was a period in the history of plant science when unravelling the function of pathogen defence associated genes and proteins was attempted without a great deal of reference to cellular compartments and the communication between them via vesicle trafficking. However, the recent discovery of the roles of these processes in defence has opened an additional dimension in our understanding of how plants combat pathogens. Eukaryotic organisms have an overwhelmingly complex machinery dedicated to the many cellular highways of vesicle traffic, and there is evidence to suggest that parts of this protein machinery are set aside for managing processes related to external stress handling. It will be interesting to see whether in fact entire highways are specialised in stress-related mechanisms. If this is the case, it can be envisaged that modifying these, aiming to obtain improved pathogen defence, will have less negative pleiotropic effects on plant development and growth in general.

Acknowledgements I am grateful to Mads Eggert Nielsen, Carsten Pedersen, Dale Godfrey and Andrea Lenk for their critical reading of this manuscript.

References

An QL, Huckelhoven R, Kogel KH, van Bel AJE (2006) Multivesicular bodies participate in a cell wall-associated defence response in barley leaves attacked by the pathogenic powdery mildew fungus. Cell Microbiol 8:1009–1019

Assaad FF, Qiu JL, Youngs H, Ehrhardt D, Zimmerli L, Kalde M, Wanner G, Peck SC, Edwards H, Ramonell K, Somerville CR, Thordal-Christensen H (2004) The PEN1 syntaxin defines a novel cellular compartment upon fungal attack and is required for the timely assembly of papillae. Mol Biol Cell 15:5118–5129

Bartsch M, Gobbato E, Bednarek P, Debey S, Schultze JL, Bautor J, Parker JE (2006) Salicylic acid-independent ENHANCED DISEASE SUSCEPTIBILITY1 signaling in Arabidopsis immunity and cell death is regulated by the monooxygenase FMO1 and the Nudix hydrolase NUDT7. Plant Cell 18:1038–1051

Benschop JJ, Mohammed S, O'Flaherty M, Heck AJR, Slijper M, Menke FLH (2007) Quantitative phosphoproteomics of early elicitor signaling in Arabidopsis. Mol Cell Proteomics 6:1198–1214

Bonner MK, Skop AR (2008) Cell division screens and dynamin. Biochem Soc Trans 36:431–435

Catanzariti AM, Dodds PN, Ellis JG (2007) Avirulence proteins from haustoria-forming pathogens. FEMS Microbiol Lett 269:181–188

Cheon CI, Lee NG, Siddique ABM, Bal AK, Verma DPS (1993) Roles of plant homologs of Rab1P and Rab7P in the biogenesis of the peribacteroid membrane, a subcellular compartment formed *de novo* during root-nodule symbiosis. EMBO J 12:4125–4135

Collins NC, Thordal-Christensen H, Lipka V, Bau S, Kombrink E, Qiu JL, Huckelhoven R, Stein M, Freialdenhoven A, Somerville SC, Schulze-Lefert P (2003) SNARE-protein-mediated disease resistance at the plant cell wall. Nature 425:973–977

de Wit PJGM (2007) How plants recognize pathogens and defend themselves. Cell Mol Life Sci 64:2726–2732

Dixon RA (2001) Natural products and plant disease resistance. Nature 411:843–847

Douchkov D, Nowara D, Zierold U, Schweizer P (2005) A high-throughput gene-silencing system for the functional assessment of defense-related genes in barley epidermal cells. Mol Plant Microbe Interact 18:755–761

Durrant WE, Dong X (2004) Systemic acquired resistance. Annu Rev Phytopathol 42:185–209

Edeling MA, Smith C, Owen D (2006) Life of a clathrin coat: insights from clathrin and AP structures. Nat Rev Mol Cell Biol 7:32–44

Eichmann R, Huckelhoven R (2008) Accommodation of powdery mildew fungi in intact plant cells. J Plant Physiol 165:5–18

Espinosa A, Alfano JR (2004) Disabling surveillance: bacterial type III secretion system effectors that suppress innate immunity. Cell Microbiol 6:1027–1040

Glazebrook J (2005) Contrasting mechanisms of defense against biotrophic and necrotrophic pathogens. Annu Rev Phytopathol 43:205–227

Graham ME, Handley MTW, Barclay JW, Ciufo LF, Barrow SL, Morgan A, Burgoyne RD (2008) A gain-of-function mutant of Munc18–1 stimulates secretory granule recruitment and exocytosis and reveals a direct interaction of Munc18–1 with Rab3. Biochem J 409:407–416

Hammond-Kosack KE, Parker JE (2003) Deciphering plant-pathogen communication: fresh perspectives for molecular resistance breeding. Curr Opin Biotechnol 14:177–193

Hardham AR, Jones DA, Takemoto D (2007) Cytoskeleton and cell wall function in penetration resistance. Curr Opin Plant Biol 10:342–348

Heese A, Ludwig AA, Jones JD (2005) Rapid phosphorylation of a syntaxin during the Avr9/Cf-9-race-specific signaling pathway. Plant Physiol 138:2406–2416

Heese M, Gansel X, Sticher L, Wick P, Grebe M, Granier F, Jurgens G (2001) Functional characterization of the KNOLLE-interacting t-SNARE AtSNAP33 and its role in plant cytokinesis. J Cell Biol 155:239–249

Hehnly H, Stamnes M (2007) Regulating cytoskeleton-based vesicle motility. FEBS Lett 581:2112–2118

Hofius D, Tsitsigiannis DI, Jones JDG, Mundy J (2007) Inducible cell death in plant immunity. Semin Cancer Biol 17:166–187

Ingmundson A, Delprato A, Lambright DG, Roy CR (2007) *Legionella pneumophila* proteins that regulate Rab1 membrane cycling. Nature 450:365–3U1

Jacobs AK, Lipka V, Burton RA, Panstruga R, Strizhov N, Schulze-Lefert P, Fincher GB (2003) An Arabidopsis callose synthase, GSL5, is required for wound and papillary callose formation. Plant Cell 15:2503–2513

Jiang RHY, Tripathy S, Govers F, Tyler BM (2008) RXLR effector reservoir in two Phytophthora species is dominated by a single rapidly evolving superfamily with more than 700 members. Proc Natl Acad Sci U S A 105:4874–4879

Jones JD, Dangl JL (2006) The plant immune system. Nature 444:323–329

Kagan JC, Su T, Horng T, Chow A, Akira S, Medzhitov R (2008) TRAM couples endocytosis of Toll-like receptor 4 to the induction of interferon-beta. Nat Immunol 9:361–368

Kalde M, Nuhse TS, Findlay K, Peck SC (2007) The syntaxin SYP132 contributes to plant resistance against bacteria and secretion of pathogenesis-related protein 1. Proc Natl Acad Sci U S A 104:11850–11855

Kemen E, Kemen AC, Rafiqi M, Hempel U, Mendgen K, Hahn M, Voegele RT (2005) Identification of a protein from rust fungi transferred from haustoria into infected plant cells. Mol Plant Microbe Interact 18:1130–1139

Kirisits A, Pils D, Krainer M (2007) Epidermal growth factor receptor degradation: an alternative view of oncogenic pathways. Int J Biochem Cell Biol 39:2173–2182

Kobayashi Y, Kobayashi I, Funaki Y, Fujimoto S, Takemoto T, Kunoh H (1997) Dynamic reorganization of microfilaments and microtubules is necessary for the expression of non-host resistance in barley coleoptile cells. Plant J 11:525–537

Kwon C, Neu C, Pajonk S, Yun HS, Lipka U, Humphry M, Bau S, Straus M, Kwaaitaal M, Rampelt H, El Kasmi F, Jurgens G, Parker J, Panstruga R, Lipka V, Schulze-Lefert P (2008) Co-option of a default secretory pathway for plant immune responses. Nature 451:835–840

Le Borgne R (2006) Regulation of Notch signalling by endocytosis and endosomal sorting. Curr Opin Cell Biol 18:213–222

Leborgne-Castel N, Lherminier J, Der C, Fromentin J, Houot V, Simon-Plas F (2008) The plant defense elicitor cryptogein stimulates clathrin-mediated endocytosis correlated with reactive oxygen species production in bright yellow-2 tobacco cells. Plant Physiol 146:1255–1266

Leyman B, Geelen D, Quintero FJ, Blatt MR (1999) A tobacco syntaxin with a role in hormonal control of guard cell ion channels. Science 283:537–540

Lipka V, Kwon C, Panstruga R (2007) SNARE-ware: the role of SNARE-domain proteins in plant biology. Annu Rev Cell Dev Biol 23:147–174

Lorrain S, Vailleau F, Balague C, Roby D (2003) Lesion mimic mutants: keys for deciphering cell death and defense pathways in plants? Trends Plant Sci 8:263–271

Machner MP, Isberg RR (2007) A bifunctional bacterial protein links GDI displacement to Rab1 activation. Science 318:974–977

Mellersh DG, Foulds IV, Higgins VJ, Heath MC (2002) H_2O_2 plays different roles in determining penetration failure in three diverse plant-fungal interactions. Plant J 29:257–268

Miklis M, Consonni C, Bhat RA, Lipka V, Schulze-Lefert P, Panstruga R (2007) Barley MLO modulates actin-dependent and actin-independent antifungal defense pathways at the cell periphery. Plant Physiol 144:1132–1143

Nishimura MT, Stein M, Hou BH, Vogel JP, Edwards H, Somerville SC (2003) Loss of a callose synthase results in salicylic acid-dependent disease resistance. Science 301:969–972

Nomura K, DebRoy S, Lee YH, Pumplin N, Jones J, He SY (2006) A bacterial virulence protein suppresses host innate immunity to cause plant disease. Science 313:220–223

Nühse TS, Boller T, Peck SC (2003) A plasma membrane syntaxin is phosphorylated in response to the bacterial elicitor flagellin. J Biol Chem 278:45248–45254

Opalski KS, Schultheiss H, Kogel KH, Huckelhoven R (2005) The receptor-like MLO protein and the RAC/ROP family G-protein RACB modulate actin reorganization in barley attacked by the biotrophic powdery mildew fungus *Blumeria graminis* f.sp hordei. Plant J 41:291–303

Rehman RU, Stigliano E, Lycett GW, Sticher L, Sbano F, Faraco M, Dalessandro G, Di Sansebastiano GP (2008) Tomato Rab11a characterization evidenced a difference between SYP121-dependent and SYP122-dependent exocytosis. Plant Cell Physiol 49:751–766

Robatzek S, Chinchilla D, Boller T (2006) Ligand-induced endocytosis of the pattern recognition receptor FLS2 in Arabidopsis. Genes Dev 20:537–542

Sanderfoot AA, Assaad FF, Raikhel NV (2000) The Arabidopsis genome. An abundance of soluble N-ethylmaleimide-sensitive factor adaptor protein receptors. Plant Physiol 124:1558–1569

Schiene K, Donath S, Brecht M, Puhler A, Niehaus K (2004) A Rab-related small GTP binding protein is predominantly expressed in root nodules of *Medicago sativa*. Mol Gen Genet 272:57–66

Schultheiss H, Dechert C, Kogel KH, Huckelhoven R (2002) A small GTP-binding host protein is required for entry of powdery mildew fungus into epidermal cells of barley. Plant Physiol 128:1447–1454

Schultheiss H, Dechert C, Kogel KH, Huckelhoven R (2003) Functional analysis of barley RAC/ROP G-protein family members in susceptibility to the powdery mildew fungus. Plant J 36:589–601

Schultheiss H, Hensel G, Imani J, Broeders S, Sonnewald U, Kogel KH, Kumlehn J, Huckelhoven R (2005) Ectopic expression of constitutively activated RACB in barley enhances susceptibility to powdery mildew and abiotic stress. Plant Physiol 139:353–362

Schultheiss H, Preuss J, Percher T, Eichmann R, Hückelhoven R (2008) Barley RIC171 interacts with RACB in planta and supports entry of the powdery mildew fungus. Cell Microbiol 10:1815–1826

Schwartz SL, Cao C, Pylypenko O, Rak A, Wandinger-Ness A (2007) Rab GTPases at a glance. J Cell Sci 120:3905–3910

Shimada C, Lipka V, O'Connell R, Okuno T, Schulze-Lefert P, Takano Y (2006) Nonhost resistance in Arabidopsis-Colletotrichum interactions acts at the cell periphery and requires actin filament function. Mol Plant Microbe Interact 19:270–279

Sokolovski S, Hills A, Gay RA, Blatt MR (2008) Functional interaction of the SNARE protein NtSyp121 in Ca^{2+} channel gating, Ca^{2+} transients and ABA signalling of stomatal guard cells. Mol Plant 1:347–358

D'Souza-Schorey C, Chavrier P (2006) ARF proteins: roles in membrane traffic and beyond. Nat Rev Mol Cell Biol 7:347–358

Stark-Urnau M, Mendgen K (1995) Sequential deposition of plant glycoproteins and polysaccharides at the host-parasite interface of *Uromyces vignae* and *Vigna sinensis* - evidence for endocytosis and secretion. Protoplasma 186:1–11

Takemoto D, Jones DA, Hardham AR (2006) Re-organization of the cytoskeleton and endoplasmic reticulum in the Arabidopsis *pen1–1* mutant inoculated with the non-adapted powdery mildew pathogen, *Blumeria graminis* f. sp *hordei*. Mol Plant Pathol 7:553–563

Tang D, Ade J, Frye CA, Innes RW (2006) A mutation in the GTP hydrolysis site of *Arabidopsis* dynamin-related protein 1E confers enhanced cell death in response to powdery mildew infection. Plant J 47:75–84

Thilmony R, Underwood W, He SY (2006) Genome-wide transcriptional analysis of the *Arabidopsis thaliana* interaction with the plant pathogen *Pseudomonas syringae* pv. tomato DC3000 and the human pathogen Escherichia coli O157:H7. Plant J 46:34–53

Truman W, de Zabala MT, Grant M (2006) Type III effectors orchestrate a complex interplay between transcriptional networks to modify basal defence responses during pathogenesis and resistance. Plant J 46:14–33

van Loon LC, Rep M, Pieterse CMJ (2006) Significance of inducible defense-related proteins in infected plants. Annu Rev Phytopathol 44:135–162

Wang D, Weaver ND, Kesarwani M, Dong X (2005) Induction of protein secretory pathway is required for systemic acquired resistance. Science 308:1036–1040

Wei CF, Kvitko BH, Shimizu R, Crabill E, Alfano JR, Lin NC, Martin GB, Huang HC, Collmer A (2007) A *Pseudomonas syringae* pv. tomato DC3000 mutant lacking the type III effector HopQ1-1 is able to cause disease in the model plant *Nicotiana benthamiana*. Plant J 51:32–46

Whisson SC, Boevink PC, Moleleki L, Avrova AO, Morales JG, Gilroy EM, Armstrong MR, Grouffaud S, van West P, Chapman S, Hein I, Toth IK, Pritchard L, Birch PRJ (2007) A translocation signal for delivery of oomycete effector proteins into host plant cells. Nature 450:115–118

Wienkoop S, Saalbach G (2003) Proteome analysis. Novel proteins identified at the peribacteroid membrane from *Lotus japonicus* root nodules. Plant Physiol 131:1080–1090

Yun BW, Atkinson HA, Gaborit C, Greenland A, Read ND, Pallas JA, Loake GJ (2003) Loss of actin cytoskeletal function and EDS1 activity, in combination, severely compromises non-host resistance in Arabidopsis against wheat powdery mildew. Plant J 34:768–777

Zeyen RJ, Bushnell WR (1979) Papilla response of barley epidermal-cells caused by *Erysiphe graminis* – rate and method of deposition determined by microcinematography and transmission electron-microscopy. Can J Bot 57:898–913

Zhang Z, Feechan A, Pedersen C, Newman MA, Qiu JL, Olesen KL, Thordal-Christensen H (2007) A SNARE-protein has opposing functions in penetration resistance and defence signalling pathways. Plant J 49:302–312

Zhang Z, Lenk A, Andersson MX, Gjetting T, Pedersen C, Nielsen ME, Newman M-A, Hou B-H, Somerville SC, Thordal-Christensen H (2008) A lesion-mimic syntaxin double mutant in *Arabidopsis* reveals novel complexity of pathogen defense signaling. Mol Plant 1:510–527

Zipfel C (2008) Pattern-recognition receptors in plant innate immunity. Curr Opin Immunol 20:10–16

Index

D

E

F

GPSR Compliance

The European Union's (EU) General Product Safety Regulation (GPSR) is a set of rules that requires consumer products to be safe and our obligations to ensure this.

If you have any concerns about our products, you can contact us on ProductSafety@springernature.com

In case Publisher is established outside the EU, the EU authorized representative is:

Springer Nature Customer Service Center GmbH
Europaplatz 3
69115 Heidelberg, Germany

Batch number: 10755053

Printed by Printforce, the Netherlands